"Filotas has exposed the f_____
that saturated high levels of the CASB [Canadian Aviation Safety Board]. He has shown that the Gander investigation was either incredible incompetence or a cover-up."

> — L. Leighton Decore
> Noted Aviation Lawyer

". . . a profound and interesting effort to untangle the knot about the Arrow Gander crash."

> — Jerome Lederer
> President Emeritus, Flight Safety Foundation
> and Former Director of Safety, NASA

"In a most readable style, Les Filotas has done a painstaking, fully-documented job of telling the inside story of the shameful and inept coverup of this major U.S. military tragedy caught up in the Iran-Contra affair.

> — Frank G. McGuire
> Editor, *Counter-Terrorism and Security Intelligence Report*

"Filotas has produced a textbook case of what happens when the citizenry fails to pay attention to its government agencies. It is well-researched, well-written, and shocking in its findings. Highly recommended."

> — Dr. Richard Gabriel
> Professor of Political Science
> U.S. Army War College

IMPROBABLE CAUSE

*Dissent and Deceit in the Investigation of
Canada's Worst Air Disaster*

Les Filotas

McClelland-Bantam, Inc.
Toronto

IMPROBABLE CAUSE

IMPROBABLE CAUSE

A Seal Book / October 1991

ISBN 0-7704-2488-0

Canadian Cataloguing in Publication Data
Filotas, Les
Improbable cause : dissent and deceit in the
investigation of Canada's worst air disaster

Includes index.
ISBN 0-7704-2488-0

1. Aeronautics — Newfoundland — Gander — Accidents —
1985. 2. Aeronautics — Newfoundland — Gander —
Accident investigation. I. Title.

TL553.5.F55 1991   363.12'465'09718   C91-095367-8

Seal Books are published by McClelland-Bantam, Inc. Its trademark,
consisting of the words "Seal Books" and the portrayal of a seal, is
the property of McClelland-Bantam, Inc., 105 Bond Street, Toronto,
Ontario M5B 1Y3, Canada. This trademark has been duly registered in
the Trademark Office of Canada. The trademark consisting of the
words "Bantam Books" and the portrayal of a rooster is the property
of and is used with the consent of Bantam Books, 666 Fifth Avenue,
New York, New York 10103. This trademark has been duly registered in
the Trademark Office of Canada and elsewhere.

PRINTED IN CANADA
COVER PRINTED IN CANADA

0 9 8 7 6 5 4 3 2 1

Dedication

This book is dedicated to the victims of the Gander crash and to all those who grieve for them and seek the truth.

Return of 101st Airborne from Peacekeeping Mission in the Sinai

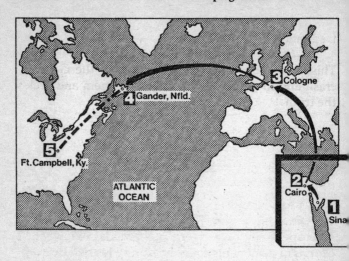

1 *Sinai—MFO South Camp at Sharm El Sheik:* The journey began with customs inspection on December 10, 1985. A short flight from Ras Nasrani Airport to Cairo International aboard two Egyptair 737's. Baggage was transported overland by truck.

2 *Cairo:* Troops arrive from Sinai 11:00–14:00Z (Greenwich Mean Time) December 11, 1985. Arrow Air 950JW arrives with Captain A Schoppaul and crew, 17:34Z; departs for Cologne 22:35Z. Egyptian provided baggage handlers and a single guard. A loss of electric power left the plane in darkness for half an hour.

3 *Cologne, West Germany:* Flight arrives 1:21Z December 12 for first refueling stop. No one guarded the aircraft as the baggage holds were opened and troops changed to uniforms. Captain John Griffin and crew took off for Gander at 3:20Z.

4 *Gander, Newfoundland:* Flight arrives for second refueling stop at 9:04Z (5:34 local time) and takes off for Hopkinsville, Kentucky at 10:15Z. Crash at 10:16.

5 *Ft. Campbell, Kentucky:* Home base of the 101st Airborne. Home coming celebrations replaced by memorial service.

ACKNOWLEDGEMENTS

Ross Stevenson's courage and uncompromising search for the truth has been a source of inspiration ever since I met him. Ross, an inveterate note taker and paper collector, has generously made available the cartons of documents collected during his six years on the Canadian Aviation Safety Board. I deeply appreciate his friendship and encouragement.

I am also deeply grateful for the continuing kind encouragement and unstinting help of Mrs. Zona Phillips who sent me every scrap of publicly or privately gathered information about the Gander crash available in the United States. Zona's extraordinary energy and dedication to seeking out the "truth about Gander" served as an inspiration to all of us.

This book could not have happened without the steadfast support of Norman Bobbitt and Dave Mussallem. I greatly appreciate their friendship through trying times. The friendship and support of Roger Lacroix was also invaluable. Without Roger's help, our dissenting opinion would never have seen the light of day.

I would like to thank Dean Cooke, publisher of Seal Books, and Gena Gorrell, my editor, for their help in shaping my manuscript into a book under a very tight schedule.

My wife Judy and son Zoli saw less of me during the months I was at home writing this book then during the years leading to it. I could not have stayed the course without their loving support.

L.F.
Sept. 1991

CONTENTS

PRELUDE

On May 6, 1937 while landing at Lakehurst
N.J., on the first trip of its 1937 schedule of
North Atlantic crossings, the hydrogen-inflated
"Hindenburg" burst into flames and was
completely destroyed, with a loss of 36 lives—
the first passenger casualties in the history of
commercial airship operations. The fire was
generally attributed to a discharge of
atmospheric electricity in the vicinity of a
hydrogen leak; however, the possibility of
sabotage cannot be overlooked.

Encyclopedia Britannica, 1970

In June 1987, six months after accepting government
appointment to the Canadian Aviation Safety Board
(CASB), I took the opportunity to attend a course
given to aircraft accident investigators at the University
of Southern California. During the day my mind
was filled with aerodynamics, metal fatigue and the
limits of human performance. After classes I rushed
to soak up the latest episode of the wildest soap opera
in recent memory—the joint committee hearings into

the Iran-Contra affair. Like millions of other viewers, I was entranced by the bizarre revelations of Lieutenant-Colonel Ollie North, his beautiful secretary and the improbable cast of supporting characters. But I never imagined that the drama of deceit, intrigue and shredded documents would be linked to my professional concern with a mysterious air crash at a remote Canadian airport.

Before returning to Ottawa, I took a few days of vacation to visit my parents in Léon, Mexico. I told my father of the controversy brewing on the CASB board over the investigation of the crash of an American military charter flight at Gander, Newfoundland, in the early morning of December 12, 1985. The rickety sixteen-year-old DC-8, operated by an obscure Florida airline, Arrow Air, was crammed with American troops heading home for Christmas after a tour of duty in the Middle East. The airplane refueled at Gander in preparation for the last leg of a flight from Cairo to Fort Campbell, Kentucky, but seconds after lift-off it crashed and burned. All on board were killed. The official toll of 256 victims made the crash the worst-ever aviation disaster in Canada and one of the ten worst anywhere; was also the worst-ever air disaster for the American military.

Although the CASB had not yet issued its report, the investigators had determined that the Arrow Air DC-8 had crashed because ice had not been removed from the wings. The ice theory was scientifically untenable, but — although there were strong indications of an in-flight explosion — the board seemed strangely reluctant to acknowledge the possibility of sabotage. Why, I asked my father, would everyone be so reluctant to even discuss the matter?

My father walked over to his bookshelf and handed me a volume. It was *The Hindenburg*, by Michael M. Mooney[1].

* * *

In 1937 the *Hindenburg*, last and most advanced of the German Zeppelins, was as much the pride of Germany as the *Queen Mary* was the pride of Britain. Commissioned in March 1936, the great airship had entered service just in time to drop leaflets praising Hitler in the Rhineland election campaign. Swastikas adorned the *Hindenburg*'s tail, on the personal orders of Dr. Joseph Goebbels, and as the airship travelled around the globe these symbols of the emerging Nazi superstate would draw the world's attention to the technological prowess of the Third Reich.

In 1936 the *Hindenburg* had completed ten successful round trips between Frankfurt and Lakehurst, New Jersey. But as it was docking in Lakehurst on May 6, 1937, after the first trip of its second season — about an hour and a half late due to a local thunderstorm — there was an explosion. The ship's seven million cubic feet of hydrogen ignited and destroyed the *Hindenburg* as thoroughly as the ignition of a hundred thousand pounds of jet fuel would destroy the Arrow Air DC-8 forty-eight years later. Photos of the wreckage show a scorched swastika on a burned tail surface — a chilling reminder of the scorched stars and stripes on a burned section of fuselage at Gander.

According to the *New York Times'* main story on May 7, 1937, "U.S. experts" attributed the explosion to the valving of hydrogen gas. But a small item on page 21 hints at another cause. Count C. G. von Zeppelin, nephew of the dirigible's inventor, interpreted eye-witness descriptions of a forward-surging internal blast as evidence of deliberate sabotage.

The next day, the *Times* noted that "Widespread reports that the destruction of the dirigible *Hindenburg* was due to sabotage were discounted . . . by the German embassy". The story doesn't elaborate or disclose the origins of these "widespread reports", but when the Assistant Secretary of Commerce appointed an investigation board, he felt it prudent to emphasize that "no evidence of sabotage had been discovered." By May 9 the coroner of Ocean County, New Jersey, had given a verdict of accidental death in all cases — and announced that an inquest was not necessary.

Twenty-five years after the *Hindenburg* tragedy, historian A. A. Hoehling presented persuasive circumstantial evidence that the airship had been sabotaged[2]. Extensive interviews with survivors and observers had convinced Hoehling that an anti-Nazi member of the *Hindenburg*'s own crew had planted a time bomb in one of the gas cells. In *The Hindenburg* — the book my father handed me when I told him of the controversy over the Gander crash — journalist Michael Mooney elaborated on Hoehling's sabotage hypothesis and claimed that Eric Spehl, a rigger with access to the gas cells, had boobytrapped the airship because the Gestapo had tortured his girlfriend's ex-husband. Spehl, an amateur photographer, had supposedly constructed his device from photographic equipment and magnesium. He had planned to blow up the ship after the passengers disembarked but, according to Mooney, the delay in arrival and a fault in his timing device made him a victim of his own hand.

In researching his book, Michael Mooney reviewed a mountain of documentation that had accumulated in the U.S. National Archives over thirty five years.

He interviewed survivors and others with personal knowledge of the circumstances. He identified Eric Spehl as the saboteur largely on the basis of interviews with Spehl's girlfriend, a pseudonymous "Beatrice Friedrich". European journalists soon discovered that Friedrich was Carla Jaekel, who still lived near Frankfurt. Jaekel had indeed been Spehl's lover, and the Gestapo had investigated her anti-Nazi connections, but she refused to talk to other reporters, or even admit that she had spoken to Mooney.

What interested me about Mooney's research was not so much his success in proving the case for sabotage as his evidence that authorities who publicly discounted the possibility had conspired behind the scenes to mislead the public. The leaders of the Third Reich had good reason, after all, to fear that their enemies would seek to destroy this grand symbol of their technology. Two years earlier, a time bomb had been discovered on the dirigible *Graf Zeppelin*. Since then, there had been numerous threats against the *Hindenburg*, and a team of undercover security officers was on board to ferret out potential saboteurs. Yet the board of inquiry didn't even raise the question of sabotage. Instead, the commissioners heard weeks of testimony on how ball lightning, static electricity or St. Elmo's fire could have ignited a hydrogen leak at the exterior surface. No matter that hundreds of witnesses had seen the fire start *inside* the hull, making the *Hindenburg* momentarily appear — as Hoehling reported — like a "thick silvery fish with a rosy glow in the abdomen". According to Hoehling, the German witnesses had been ordered not to cooperate; "Entirely unwittingly, [the commissioners] had been deceived, hoodwinked and blandished with a partial volunteering of information. The august government agency was culpable of the charge of naïveté as the greenest farm boy who is taken at the three-card monte game at the country fair''.

The commission's report, released on July 21, 1937, surprised no one. It concluded that ''a relatively small amount of explosive mixture could have been ignited by St. Elmo's fire or a similar electric phenomenon . . . .'' The report did caution that ''This explanation is not to be taken as a completely certain one'', but the warning didn't deter the authorities from enthusiastically embracing the finding — especially the authorities in Berlin.

Mooney cites diaries of off-the-record discussions of the evidence of sabotage among the commissioners and their advisers. He also cites letters from the secretaries of Commerce and Interior advising Commissioner South Trimble Jr. that ''a finding of sabotage might be the cause for an international incident.'' It was bad enough to lose the most celebrated symbol of the Nazis' technological invincibility through an act of God; sabotage by a Jew or Bolshevik would have been an intolerable affront.

* * *

December 16, 1985; Fort Campbell, Kentucky. A grim-faced President Ronald Reagan stood at a lectern in front of a hangar, looking pale and anguished and close to tears. He was giving a eulogy at the memorial service for the hundreds of young soldiers struck down at Gander. ''We wonder how this could be,'' he said. ''How it could have happened, and why.''

In accordance with international convention, the investigation to answer the President's questions was being conducted under the jurisdiction of Canada, where the airplane had crashed. While no one expected a definitive answer in the immediate future, early news reports emphasized the bleak winter weather and that the ageing airliner had a history of mechanical problems and had not been de-iced before takeoff.

Six weeks later, President Reagan had to cancel a State of the Union address for another day of mourning; one of the space shuttle *Challenger*'s solid-fuel boosters had exploded shortly after blastoff from Cape Kennedy. The spectacular end of the seven astronauts before the eyes of millions of television viewers overshadowed the deaths of 250 soldiers on a remote Canadian hillside. The investigation of the Gander crash would proceed with scant publicity.

* * *

The Canadian Aviation Safety Board was set up in October 1984 to end conflicts of interest in the investigation of air crashes. The new agency, stripped from the Department of Transport, consisted of headquarters and a laboratory in Ottawa and five regional offices. Most of its field investigators were former air force pilots trained to investigate air crashes through specialist courses and on-the-job experience. They were assisted by a dozen or so engineers and scientists in the laboratory, and the usual administrators and clerical staff. The agency, set up to be independent of all government departments, was headed by a government-appointed board of directors. The Gander investigation was the CASB's first probe of a major crash—and, as it turned out, its last.

When I was appointed as a director—a member of the board—in December 1986, all I knew of the Gander tragedy was what I had gleaned from news reports —the initial stories in the days following the crash and the reports of the CASB's seven-day public inquiry. I knew that the airplane had been old and that there were dark mutterings of slipping safety standards in the wake of the recent deregulation of American aviation. Media reports left the impression that the experts had pinpointed ice on the wings as the likely cause.

As it turned out, the board members who had been there all along knew very little more. The investigators had made progress reports, but these had revealed little more than what was in the media; no details of the investigation or tentative conclusions had been disclosed. In April 1987, the board finally received the investigators' draft report, concluding that the Arrow Air DC-8 had crashed because the crew had tried to take off with a small amount of ice on the wings. The report was presented in final form, requiring nothing but our approval. I was troubled by it.

As an aeronautical engineer and former university professor, I had specialized in the study of wings and the dynamics of flight. Although the doomed jet's antiquated flight data recorder yielded little useful information, even these meagre data sufficed to show that the DC-8 had accelerated normally during its takeoff run and the first seconds of its brief flight. The abrupt deceleration before the crash suggested a sudden catastrophe—not reduced performance due to ice on the wings.

The theory made just as little sense to the other two aeronautical engineers on the board, Norman Bobbitt and David Mussallem. Bobbitt had been a test pilot with Canada's National Research Council, and his work in cloud-seeding experiments gave him useful insights into the effects of icing. Mussallem was not only a graduate aeronautical engineer, but an airline captain with extensive experience on Boeing 707 and 747 aircraft; his doubts about the ice-on-the-wing theory, based on his training as an engineer, were reinforced by doubts about the presumed behaviour of the Arrow Air crew.

While Mussallem's experience with heavy jets surpassed that of any of the investigators, no one could talk about DC-8 operations with as much authority as board member Ross Stevenson. Stevenson, a trim, erect racketball fanatic with piercing grey eyes and just a touch of white to his hair, looked every inch a professional airline captain and much younger than his sixty-six years. He began flying with the Royal Canadian Air Force in 1942 — in Gander, as it happens — and flew Canso flying boats and Liberators on sub patrol and escort duty until April 1944. At that time he joined Canada's fledgling national airline, Trans-Canada Airlines, later renamed Air Canada. During his 38-year career with Air Canada, Stevenson logged some 30,000 hours as pilot in command, on airplanes ranging from the 10-passenger Lockheed 1408 to the 400-passenger Boeing 747. His flying time included over 9,000 hours as pilot in command on DC-8s.

Thus Ross Stevenson was viscerally familiar with both the DC-8's flying characteristics and the weather conditions at Gander. While I was pondering the data from the flight recorder, he studied weather records and the crew's schedule and professional record. If there had been ice on the aircraft, he felt, the crew would have detected it and had it removed. More important, his examination of weather data and witness statements convinced him that there could not have been ice on it. Yet the investigators' report seemed to rule out mechanical failure. So what kind of catastrophe could have caused the crash?

The months preceding the Gander tragedy had been marked by an unprecedented tempo of terrorism. The investigators' report noted "considerable speculation that the accident occurred as a result of the detonation, either accidental or deliberate, of some explosive device." But it dismissed such speculation on the grounds that "Detailed examination of the wreckage . . . revealed no evidence of an explosion or pre-impact fire." Initially I accepted this assessment on the assumption that the search for evidence of a pre-impact fire had been thorough. But clearly such evidence might have been obliterated by the fire and the plane's disintegration.

Those of us who attempted to discuss our objections to the investigators' report met inexplicable resistance from the CASB's chairman, Bernard Deschênes. Deschênes seemed determined to adopt the report without seriously considering the evidence against it.

While I continued to ponder the report and the chairman's strange attitude, I read Michael Mooney's book on the *Hindenburg,* and I started to detect eerie parallels between the official reaction to the destruction of the DC-8 at Gander and that of the *Hindenburg* almost fifty years earlier:

- motives and means for sabotage were readily apparent to anyone aware of current events;
- there was a risk that a finding of sabotage would create an international incident;
- long before the facts could be known, there were official disclaimers of sabotage and simultaneous promotion of an implausible theory absolving the authorities of all responsibility;
- a hastily convened public inquiry was preoccupied with unconvincing theories based on natural phenomena while ignoring striking circumstantial

evidence and withholding pertinent physical
evidence;
  –eye-witness evidence and lines of inquiry leading
   away from the officially sanctioned theory were
   treated with contempt.

As these similarities gnawed their way into my con-
scious mind, I had to confront the thought that the
parallel might extend further. Could there be more
than errors and arrogance behind the botched investigation of the Gander crash? Were the members of
the Canadian Aviation Safety Board, no less than the
members of the board of inquiry into the *Hindenburg*
crash, "deceived, hoodwinked and blandished with
a partial volunteering of information"?

The search for an answer lasted more than three
years and grew to preoccupy hundreds of people,
many seeking to expand the search, others determined to curtail it. It led down many twisted paths
and blind alleys. The story of the search includes
charges and countercharges of cover-up and incompetence; concealment and fabrication of evidence; resignation of senior officials; a police investigation of
my personal conduct; intervention of a retired justice
of the Canadian Supreme Court; and the ignominious
obliteration of the whole Canadian Aviation Safety
Board.

INTRODUCTION

Part 1

Gander — Fire and Ice

> It is an error to argue in front of your data. You
> find yourself insensibly twisting them around
> to fit your theories.
>
> *Sherlock Holmes*[1]

David Owen, a taciturn former bush pilot, was the senior investigator at the CASB's regional office in Moncton, New Brunswick, the board's closest office to Gander. Owen, who had also trained as an air traffic controller, was one of the CASB's top field investigators. On hearing of the Arrow Air crash on the morning of December 12, he immediately took off for Gander with Department of Transport pilot George Dewar, in a government aircraft. They arrived about four hours after the crash and circled the site before landing. Dewar later recalled their amazement at not being able to identify a wrecked airplane in the field of burning rubble[2].

Owen's arrival opened the CASB's on-site investigation of the crash. Investigators arrived from headquarters about an hour later and soon displaced him from the investigation, but it was Dave Owen who sent out the first official description of Canada's worst aviation disaster:

> JW950 [*sic*] crashed shortly after departure. No
> apparent survivors. Aircraft departed Rwy 22.

3

> Disappeared below the line between TCH [Trans-
> Canada Highway] and Gander Lake. Aircraft exploded
> on contact near the edge of Gander Lake.

Other early reports also emphasized the explosion,
and its magnitude. The fireball could be seen miles
away. Waitress Anne Hurley saw it about seven miles
in front of her while she was driving to work on the
Trans-Canada Highway[3]. Hedley Gill, who saw it
while parking his car in front of the airport, noted that
the airport's other runway passed right over Gander;
had the airplane tried to take off that way, "I would
say half the town would have been blown up".[4]

Seconds before the explosion, the DC-8 passed
directly over two trucks on the Trans-Canada Highway, about a thousand feet beyond the end of the
runway. A few hours later, one of the drivers, Cecil
Mackie, told reporters that "there was a flame on the
bottom" of the airplane, which brightly illuminated
the interior of his truck. Mackie, a soft-spoken resident
of Cape Breton, lived near Sydney airport, and airplanes flew over him every day. He was positive he
saw flames, not just the aircraft's lights. The second
trucker, Leonard Loughren, had a similar story. "I
think the right-hand side of the aircraft was afire," he
told Doreen Harty of the RCMP. "It was pitch-dark
and I could see the tail and parts of the plane from the
glow."

The DC-8 then vanished from the truckers' sight,
and moments later it exploded. "The explosion was
unbelievable," Loughren recalled. "It lit the sky up
like daylight."

Another trucker getting an early start on the morning, Gerald MacWhirter, was driving westbound on
the Trans-Canada Highway, in the opposite direction

to Mackie and Loughren. He noticed an aircraft crossing low over the highway about a quarter-mile ahead of him. "I couldn't see the right-hand side of the airplane," he recalled, "but I could tell it was very bright on that side of the plane, like something was on fire. I couldn't see the flame, but that side was brighter than it should be. . . . The plane went off to my left and I lost sight of it because of the trees. . . . "Later I heard a bang and saw a big puff of smoke rise up like a mushroom cloud."

Car rental agent Judith Parsons watched the DC-8 takeoff from the Tilden parking lot at the airport. A few seconds after the plane disappeared down the runway, she saw a flash and what appeared as a "large orange oval object" moving through the sky. The object then "blew up, it just went in a million pieces . . . . It was definitely not on the ground." Seconds later she saw the fireball. Parson's observations — which clearly pointed to an in-flight fire or explosion — were widely reported on the day of the accident.

On the day of the crash an anonymous caller told the Reuters agency in Beirut that the Islamic Jihad had sabotaged the Arrow Air jet in collaboration with an exiled Egyptian terrorist organization. The caller said the bomb was set to blow up as the plane landed in the United States, but exploded prematurely in Canada because the flight had been delayed. The Jihad's claim was repeated in a telephone call to the American consul-general in Oran, Algeria.[5] Yet another anonymous caller told an Italian news agency that the DC-8 had been sabotaged by a different Egyptian group.[6]

Crank calls follow air disasters as surely as insurance claims and legal suits, but the anonymous claim by the Islamic Jihad followed a pattern established after bombings of the American embassy and Marine barracks in Beirut. The call also resembled one received after bombings wounded twenty-seven people in Copenhagen five months earlier. That call warned that the Islamic Jihad's attacks should now be expected in any part of the world.

Other claims were not publicized. A few hours after the accident Rick Gibbons, representative of Canadian Press in London, took a call from a man with a thick Middle Eastern accent. Identifying himself as an official in an embassy he wouldn't name, the caller claimed that intelligence traffic from American and British sources indicated that an American jetliner had been sabotaged in Gander, Newfoundland.[7]

A representative of Reuters promptly relayed the Islamic Jihad's claim to the Canadian Aviation Safety Board and duty officer John Issenman took the overseas call. Issenman left the CASB in 1986. When I interviewed him in August 1990 he couldn't remember when the call had arrived during the hectic hours after the crash, but he was certain he had noted the call in his log. He couldn't explain why his log entries had been recopied with the record of the call omitted.

Despite the anonymous calls and eye-witness statements, representatives of the Canadian and American governments discounted the possibility of sabotage. Chief Pentagon spokesman Robert Sims told reporters, "We have no indications of explosions prior to the crash or of hostile action." "Initial reports indicate no evidence of sabotage or an explosion in flight," explained White house spokesman Larry Speakes. "There is no evidence that the plane blew up before it crashed," said the Canadian Minister of Transport, Don Mazankowski. Official sources closer to the scene were singing from the same choirbook. RCMP Staff Sergeant Hank Johnston told reporters that "Obviously you are going to have an explosion on impact with fuel on board" but "there is nothing to indicate an act of terrorism." Transport Canada spokesman Bruce Reid was of the same opinion: "Where it came down, it obviously exploded on impact. . . . We have no indication it exploded in the air."

On the record, the investigators themselves were non-committal. Peter Boag, designated as investigator in charge by the CASB, "would not comment on the possibility of sabotage, saying only that it will take time to determine the cause".[8] But by the day after the accident it was clear to reporters that the investigators had "all but ruled out early rumors that a bomb may have caused the explosion".[9] Though the investigators wouldn't speculate on record, the media did pick up a clear sense of where they were headed. "The question of icing is potentially significant," the *New York Times* reported in its first story from Gander, and another *Times* article datelined December 13 explained that "there was speculation that the plane was unable to climb because de-icing fluid had not been applied to its wing and fuselage." The *Edmonton Journal* headlined its December 13 feature article "Aircraft's wings not de-iced." The duty officer in the Pentagon had formed the same idea even earlier; the hand-written log from the morning of December 12 reads, "Suspect icing of the wings as the cause of the crash."*

---

*Some five weeks after the crash another claim of sabotage was sent to the U.S. embassy in Mauritius, a small island off Madagascar that is a centre for American intelligence operations because of its proximity to vital oil routes. A mysterious letter claimed that the crash was, in fact, a cold-blooded premeditated act which involved an expert sabotage a few hours before takeoff with the complicity of several Egyptian and Libyan mechanics . . . ." The letter was signed "the Sons of Zion", reported to be a front for the Israeli secret service, the Mossad. *St. Petersburg Times*, March 18, 1989.

In contrast to the non-committal position of the investigators, other officials speculated openly about ice on the wings. According to the December 13 *Globe and Mail*, Sergeant Roger Tinkham of the RCMP pointed out that "despite the freezing rain, the plane had not been de-iced", while William Slaughter, the Transport department's director of aviation safety programs, explained that there might have been "a thin coating of ice that the pilot did not realize could cause the plane to stall in the air after takeoff." Slaughter noted that "the matter of ice detection on the aircraft is left entirely to flight crews" and, alluding to Arrow Air's headquarters in sunny Florida, regretted that "pilots in some parts of the world still aren't familiar with the problem."

Slaughter was referring to the fact that ice or frost on a wing's curved upper surface will drag along a layer of stagnant air that, in effect, reduces the curvature, thus reducing the upward pull of the air streaming past this stagnant layer. To maintain the wing's lift the pilot must then pull back the control column to increase the angle between the plane of the wing and the line of flight. If this "angle of attack" becomes too great, further increase does not produce additional lift, and the wing is said to have stalled. But pilots are trained to recognize the onset of stall, which is marked by shuddering of the airplane, a sloppy feel to the controls and a variety of warnings built in by the designers. Before student pilots can solo, they must recognize an incipient stall and react at once by lowering the nose to reduce the angle of attack. With experience, this reaction becomes automatic — and airline captains practise it repeatedly in simulators throughout their careers.

Professional flight crews—even those based in Florida—are also trained to beware of ice on the wings. They know that airliners have heating systems to remove ice accumulated in flight, and that de-icing facilities are available at major airports to remove ice accumulated on the ground. It's true that crews have been known to try to take off without ensuring that wings were free of ice. And freezing rain had, in fact, fallen in Gander on December 12. Still, none of the ground crew saw ice on the aircraft—and the Arrow Air cockpit crew was highly experienced. So the lack of evidence pointing to ice could have been asserted as convincingly as the lack of evidence of terrorist attack. Yet on the second day after the crash, chief investigator Peter Boag was already suggesting that the cause of the accident had been narrowed to two factors: "the weight of the aircraft and its cargo, and the fact that it was not sprayed with de-icing solution before it took off in a freezing drizzle".[10]

While the airplane could have weighed a few per cent over the 330,000 pounds declared on the flight manifest, it had made the previous takeoff at Cologne without difficulty, under less favorable conditions and with 17,000 pounds more fuel; thus weight was soon ruled out as a significant factor. But the notion that the Arrow Air DC-8 took off from Gander "in a freezing drizzle" — groundless as it proved to be — was never abandoned. In contrast, the giant explosion highlighted in David Owen's initial dispatch and early eyewitness reports faded from subsequent accounts.

Part 2

The World—A Year of Terrorism

> . . . it is impossible for me to say whether the
> present case is an instance of crime or not, but
> the course of events is certainly among the
> most singular that I have ever listened to.
>
> *Sherlock Holmes*

The explosion of the Arrow Air DC-8 at Gander in late 1985 added 256 victims to what was already the worst year ever for aviation disasters. Some forty commercial aircraft disasters, including three of the world's four worst, had claimed over 2,000 lives and—not all had been accidents.

On January 23 a bomb had exploded in the forward lavatory of a Bolivian 727; one passenger had been killed but the aircraft had landed safely. On March 9 a bomb had exploded in the baggage compartment of a Royal Jordanian L-1011 while it was on the ground in Dubai; there were no injuries On October 30 an explosion in the baggage compartment of an American Airlines 727 on the ground at Dallas caused no injuries. But on June 23, 329 people had died when an explosion destroyed an Air India Boeing 747 over the Atlantic near Ireland. The source was a powerful bomb, probably disguised as a radio, smuggled on in the baggage of a passenger who didn't board.

The pattern continued after December 1985, as the CASB investigated the Gander crash, and for each bloody incident there were probably a number of near-misses thanks to security measures or sheer chance. Indeed, the January 24 issue of the trade journal *Flight International* noted that terrorist bombings had become as much a threat to civil aviation as human error or equipment failure. Henceforth, *Flight International* would include terrorist acts in its safety surveys.

The CASB's investigators, who were still working on their draft report on the crash of the Arrow Air DC-8, should have been all the more alert to the terrorist angle because of the board's work with the commission of inquiry into the Air India disaster. The government of India had made the CASB an accredited representative on the investigation "based on the large number of Canadians who lost their lives in this accident".[11] The Canadian government allocated the CASB an additional $7.4 million — some 50 per cent over the regular budget — for independent work on that crash.

The CASB produced a "detailed submission to the Indian Commission of Inquiry", tantamount to an independent report, which "suggested that the commission consider" a number of "cause-related findings" including "considerable circumstantial and other evidence to indicate that the initial event was an explosion occurring in the forward cargo compartment".[12] Technical experts from the American National Transportation Safety Board who assisted the Indian investigation also speculated publicly on the cause of the crash.

One might have expected the interest of the American government in the accident at Gander — which involved an American-manufactured and -registered

11

aircraft chartered by the American government and carrying only American citizens—to be as great as the interest of the Canadian government in the Air India disaster. Yet, rather than follow the precedent of the Canadian submission on the Air India bombing, the U.S. government seemed content to sit back and accept the theory of ice on the wing.

Washington's refusal to concede even the possibility of a terrorist strike seems especially puzzling in view of the Reagan administration's strident warnings about the growing menace of international terrorism —a menace said to have been neglected by the outgoing Carter administration and sorely needing tough new policies; a menace so grave that on January 26, 1981, the first Monday after the inauguration of his new administration, the President called a special Cabinet discussion on terrorism. Both he and Secretary of State George Shultz described "the evil scourge of terrorism" as a plague spread by "depraved opponents of civilization itself".[13]

Six months before the Gander crash this hard line against terrorists had been put to a practical test in the first major airline terrorist incident since the Reagan Administration had taken office. On June 14, 1985, hijackers had commandeered a TWA jetliner en route from Athens to Rome with 135 people on board. The gun-wielding terrorists had diverted the flight in the name of the "Islamic Jihad".

The Islamic Jihad

The Islamic Jihad or Islamic Holy War, is the best-recognized signature of the fighters of the Hezbollah (Party of God), Shiite fundamentalists seeking to export the Ayatollah Khomeini's Islamic revolution. The Hezbollah sprang up in the Bekaa valley of eastern

Lebanon in response to the Israeli invasion of June 1982; some 1,500 Iranian volunteers penetrated Lebanon and began to work with local Shiite groups and fanatic volunteers from other Islamic countries to drive Israel out of Lebanon and end the influence of the Unites States in the Middle East, and the Hezbollah was a loosely knit umbrella organization for them. The Hezbollah had no formal hierarchy — it supposedly derived its authority from the Khomeini and from Allah — and both Iran and Syria provided financial support.

The Iranian presence in Lebanon became evident in early 1983, just as the U.S. Marines in the multinational peacekeeping forces began to lose the appearance of neutrality. The American government's efforts to unite the Lebanese army accentuated rivalry between various militias, and Moslem and Shiite factions increasingly saw the Marines as supporting the Christian forces of Amin Gemayel. On March 16, 1983, a hand-grenade attack wounded five Marines patrolling a Shiite neighbourhood on the perimeter of Beirut airport. On April 18 a bomb in a delivery van, exploded under the front portico of the U.S. embassy in Beirut, killing seventeen Americans and forty other nationals. The suicide attack occurred while a CIA regional meeting was taking place. The planning was masterly, the bomb expertly assembled and placed at the exact spot where the force of the blast would be channelled upward for maximum effect. The American victims included the CIA station chief, Robert Ames, and three other CIA officials. An anonymous phone call claimed credit in the name of the Jihad.

Attacks against U.S. peace-keeping forces escalated, and climaxed on October 23 with the (then) worst-ever peacetime loss of American troops — a suicide attack against the Marine barracks at Beirut airport. A yellow Mercedes flatbed truck was loaded with 12,000 pounds of high explosives wrapped around butane-gas canisters, with a slab of marble on the truckbed to deflect the blast upward. The destructive effect from what FBI explosives experts later claimed was "the largest non-nuclear explosion that we had ever seen" ripped the building from its foundation and reduced it to rubble, killing 241 Marines.[14] American intelligence sources determined that the attack had been led by Imad Mughniyeh, leader of the Jihad's action group, who had also been behind the bombing of the American embassy and would later lead the attack on TWA Flight 847.*

Simultaneous attack on the French military headquarters in Beirut killed fifty-eight French paratroopers; once again, an anonymous call to a news agency claimed credit for the Jihad. In a widely quoted remark, President Reagan identified Iran as the leader of "a confederation of terrorist states" forming a "new international version of Murder Incorporated".[15]

The Islamic Jihad also opened a new phase in Lebanon's history of terrorism — political kidnapping. On March 16, 1984, members of the Jihad seized 57-year-old William Buckley, the CIA station chief, who had been sent to Lebanon to rebuild the organization that had been destroyed by the embassy bombing. Although he was publicly described as a political officer, Buckley's cover had been blown and the Jihad knew exactly whom they had; Buckley was tortured

---

*Mughniyeh was also implicated in the kidnappings of Terry Anderson and David Sutherland, Segev p.141; see also Emerson p.191.

to gain top secret information about CIA operations in the Middle East.[16] Concern about Buckley, a personal friend of CIA director William Casey, and about the information extracted from him became a main factor leading to the secret arms sales to Iran in contravention of the Reagan administration's declared policy.*

Seven other Americans and a number of citizens of other countries were kidnapped and held hostage in Lebanon between March 1984 and June 1985.** During this same period another truck bomb hit the American embassy, which had been moved to east Beirut after the 1983 bombing, but thanks to the quick action of the visiting British ambassador's bodyguard, only fourteen were killed.

During the seventeen-day ordeal that followed the TWA hijacking of June 14, 1985, the B-727 was forced

---

*Oliver North testified to the Iran-Contra Committee, "there was not only a professional relationship between Mr. Buckley and Director Casey but a personal one . . . to the very end, Director Casey was anxious to get the body of Bill Buckley home, and certainly the tortured confession." (I-C, p.529.)

**The other American hostages were: Jeremy Levin, Beirut bureau chief of Cable News Network: seized March 7, 1984; freed Feb. 13, 1985, through the intervention of Jesse Jackson and H. Ross Perot; Benjamin Weir, Presbyterian minister and long-time resident of Beirut: seized May 8, 1984; released Sept. 14, 1985; Peter Kilburn, librarian at the American University of Beirut: seized Dec. 3, 1984; murdered in retaliation for the U.S. raid on Libya in April 1986; Father Lawrence Jenko, head of the Catholic Aid Service in Labanon: kidnapped Jan. 8, 1985; released July 26, 1986; Terry Anderson, Associated Press correspondent in Beirut: seized March 16, 1985; still held as of this writing; David Jacobsen, director of the American University Hospital in Beirut: seized May 28, 1985; released Nov. 2, 1986; Thomas Sutherland, director of the American University's School of Agriculture: seized June 9, 1985.

to fly from Athens to Beirut to Algiers. The hijackers shot 23-year-old American navy diver Robert Stethem point-blank in the left temple and dumped his body on the tarmac at Beirut. The thirty-two remaining American captives sent a letter to President Reagan pleading that he mediate their release by prevailing on Israel to release jailed supporters of the PLO. But Reagan affirmed the hard-line policy, stating that the U.S. would "never make concessions to terrorists" because to do so "could only invite more terrorism".[17]

The TWA hijacking received widespread publicity and provoked bitter criticism — even from the administration's staunchest supporters. The Deputy Assistant Secretary of Defense, Noel Koch, later explained that the incident "demonstrated that our preparation for dealing with terrorists was egregiously inadequate. . . . The terrorists handled the hostages in a masterful way from the media standpoint . . . making the United States look like an absolutely feckless player on the world stage. . . . We became the gang that couldn't shoot straight."[18]

As terrorist attacks continued unabated, President Reagan declared that "limits had been reached" with terrorists and urged Western nations to draconian anti-terrorist measures.[19] "The United States gives terrorists no rewards," he stated. "We make no concessions. We make no deals"[20] But his tough talk didn't deter the Islamic Jihad. After the two terrorist bombs exploded in Copenhagen in July 1985, an anonymous caller claimed that the Jihad had carried out the attack in retaliation for Israeli raids in southern Lebanon, and warned that reprisals would no longer be confined to the Middle East but would henceforth "be aimed at every Zionist, American or reactionary establishment in various parts of the world".[21] The caller bragged that future attacks would be "immediate, surprising and lightning."

President Reagan had no problem identifying the targets of these promised attacks. "This is a war," he said, "in which innocent civilians are intentional victims and our servicemen have become specific targets".[22] The servicemen most vulnerable would be those stationed near the epicentre of terrorist activity in the Middle East—servicemen such as the members of the 101st Airborne Division on board the ill-fated Arrow Air DC-8 on December 12, 1985.

Business Opportunities in Egypt

The Islamic fundamentalists who swore to rid the Middle East of the Western powers did not distinguish between the Multinational Force and Observers (MFO) in Egypt and the more celebrated multinational peacekeeping force in Lebanon. The MFO, a contingent of about 2,600 soldiers and civilian observers from ten nations, was deployed along the eastern edge of the Sinai to patrol the border and ensure freedom of passage through the Strait of Tiran. The force had been set up in August 1981—almost, but not quite, in accordance with the Israeli-Egyptian peace treaty deriving from the accord between presidents Jimmy Carter and Anwar Sadat at Camp David in 1978. The accord had called for a United Nations force similar to those in Cyprus and Lebanon, but the threat of a veto from the Soviet Union had blocked the proposal at the U.N. The United States had then spearheaded the force, which included some 800 American troops, and had established the MFO's headquarters in Alexandria, Virginia.

While most of the world rejoiced at the prospect of peace brought on by the Israeli-Egyptian peace treaty, the Arab world was not as happy. On October 6, 1981, assassins claiming to represent the "Egyptian Islamic Jihad" shot and killed Anwar Sadat. Nevertheless, the success of the Camp David accord was assured, at least in part, by President Carter's subsidiary commitment of $4 billion worth of military aid to Egypt. Arms merchants and other entrepreneurs quickly spotted the profit potential in the coming flow of people and goods.

One such entrepreneur was Edwin Wilson, a renegade CIA agent later sentenced to fifty-two years in prison for illegally supplying arms to Libyan dictator Muammar Gadaffi and for conspiring to murder federal prosecutors and witnesses. Wilson advanced $500,000 to his former CIA case officer, Thomas Clines, to exploit the opportunity by setting up Egyptian-American Transport Services, Inc. (EATSCO). Using insider contacts, EATSCO obtained exclusive rights for shipping certain military supplies to Egypt, and by 1982 the fledgling enterprise had rung up $8 million in illegal overcharges.[23]

Another American entrepreneur quick to recognize the business opportunities presented by the MFO was George C. Batchelor, 61-year-old patriarch of what the *New York Times* termed "a large family-held aviation empire that has been involved in several government inquiries and lawsuits over safety and management issues in the last dozen years".[24]

Sensing an expanding potential for military and civil contracts, Batchelor established Arrow Air Inc. in Miami in 1981.* Arrow immediately succeeded in obtaining passenger and charter service contracts from the U.S. Department of Defense. Arrow's air-

---

*Arrow Air is sometimes incorrectly referred to as Arrow Airways, a defunct company previously owned by George Batchelor.

craft were leased from International Air Leases and maintained by Batch Air — neighboring Batchelor-owned enterprises.

Notwithstanding the military contracts, Arrow Air got off to a rocky start. The airline lost money every year from 1981 to 1984, during which time the company was assessed some $40,000 in fines for fifty-six breaches of air regulations. To keep the Federal Aviation Administration (FAA) from suspending Arrow's operating certificate, Batchelor agreed to reorganize its corporate structure. The reorganization involved hiring Richard P. Skully as senior vice president for Operations and Maintenance.

Dick Skully had spent thirty years with the FAA, rising from assistant administrator of the FAA Academy in Oklahoma City to number-two man in Washington. In 1978 he suddenly left the FAA for an operational position at a little upstart carrier in Miami; he had been lured away by C. Edward Acker, former president of Braniff, who had left two years earlier in the face of a South American ticket scandal and was making a comeback in the airline industry as head of Miami-based Air Florida. In the words of former Braniff captain John Nance, "it became clear why a former senior official of the FAA could be such an effective vice-president of operations of a small carrier growing frantically under the freedom of deregulation: The man knew the ropes. Who would know better how to handle the FAA than the FAA, which was what Skully personified".[25]

How George Batchelor lured Dick Skully from Air Florida is yet to be reported, but Arrow Air's fortunes took a turn for the better after Skully's arrival. During 1985, prior to the crash at Gander, Arrow made a profit of $12 million on a gross income that included $33.7 million from military charters.[26]

Meanwhile, to minimize the appearance of American domination, the MFO headquarters were moved to Rome. On February 15, 1984, a sub-machine gun burst fired through the window of his car killed the MFO's director, Leamon "Ray" Hunt. Hunt had previously served as chargé d'affaires in Beirut. At the time of his murder, the United States was "redeploying" its troops out of Lebanon—the final consequence of the Islamic Jihad's attack on the Marine barracks the previous October. Italian police linked Hunt's murder to Lebanese terrorist groups.

The retreat of American forces from Lebanon and the kidnapping of William Buckley provide the backdrop to Arrow Air's charter operations in the Sinai. Buoyed by success, the Islamic Jihad and other terrorist organizations were encouraged to continue the attacks and further reduce the American presence. American intelligence sources in the Middle East were left in disarray and, by default, the MFO became the only large concentration of American military assets in the Middle East. So when the Arrow Air flight bringing U.S. troops home from the Sinai for Christmas disintegrated in a gigantic fireball on December 12, 1985, the well-publicized attacks, claims of responsibility, and the Islamic Jihad's demonstrated expertise with explosives would have suggested sabotage to anyone who had followed the headline news — even without eye-witness reports that the DC-8 was on fire before it crashed.

And there were other factors lending credence, for those in the know, to the idea that the Arrow Air DC-8 might have been targeted by saboteurs. In December 1985 it was not generally known, for example, that U.S. troops might have been targeted in retaliation for a bungled arms delivery to Iran engineered from the

White House basement by Lieutenant-Colonel Oliver North. Few would have supposed that the "peace-keeping" duties of the Multinational Force and Observers included plans for rescuing the hostages in Lebanon, stockpiling weapons and other provocations to the Islamic Jihad.

Yet the Reagan administration not only chose to downplay the possibility of terrorist involvement, but dismissed it categorically — a reaction at inexplicable odds with its own strident anti-terrorist rhetoric. Indeed, the shock of an attack on peacekeeping forces coming home for Christmas might well have helped secure support for the administration's war against the "scourge of terrorism".

If, after all the warning signs, the Arrow Air DC-8 had been brought down by terrorists, the United States Army would stand guilty of another deplorable lapse in security, a lapse as traumatic as the débâcle at the Beirut airport; congressional inquiries and courts martial would surely follow. Even if the crash was due to mechanical causes or human error, the investigation could still expose the army's inadequate security arrangements in the Sinai — with equally unattractive political fallout and consequences to military careers. And there were other, less obvious reasons pointing to other, less obvious motives for restraining the impulse to cry, "Terrorist!"

As it turned out, neither the U.S. Army nor anyone else had to fear that the investigation would expose embarrassing circumstances. The Canadian Aviation Safety Board, barely a year old, anaemic from bureaucratic battles attending its creation, had neither the means nor the inclination to delve into the United States' embarrassing secrets.

Part 3
The CASB — A Rude Awakening

Although the Canadian Aviation Safety Board had an Ottawa mailing address, the head office was actually located in Hull, Quebec — across the Ottawa River, about a mile north of the Parliament Buildings. The CASB occupied the fifth and sixth floors of "Phase II" of Place Portage, a forbidding grey fortress of offices and shops.

"Take the escalator by the liquor store," the chairman, Bernard Deschênes, had instructed me during our brief telephone conversation, "up one flight, then the elevator." I killed a few minutes in the lobby and arrived exactly on time. The receptionist directed me past a maze of cubicles to Deschênes' corner office, where I thumbed through old issues of *Aviation Week* until the chairman could see me.

Deschênes was small, trim, and elegantly dressed in a brown suit with subdued pinstripes; he had a neat triangular moustache. As we shook hands, he examined me carefully through brown tortoiseshell glasses.

After a few pleasantries, he said we had to hurry to the meeting of the board, and explained that it was convenient to hold meetings in a building across the street, near the offices of ordinary board members. He conducted me back past the cubicles to the lobby, across the street through a covered overpass, up some stairs, through a number of unmarked corridors to a large empty space about the size of a gymnasium.

Dark lines on the stained grey carpet marked the positions of former partitions. The Canadian Aviation Safety Board was meeting in a corner set off by an assortment of shabby cloth screens.

Some fifteen participants were already seated on three sides of a rectangular assemblage of Formica-topped tables. Frank Thurston, the only board member I already knew, was in animated conversation with executive director Ken Johnson. Members of the board's staff huddled at the foot of the makeshift table; the place at the head was reserved for Chairman Deschênes.

The chairman waved me to an empty spot, took his place and introduced me by reading extracts from my résumé before continuing the agenda. He used an impromptu process that went right along with the ratty surroundings; agenda items came and went after desultory discussion, without a vote. Ken Johnson's interventions were indistinguishable from those of the board members, and staff members interrupted each other and board members, including the chairman, seemingly at will. It was not an encouraging beginning.

* * *

The more I saw of the workings of the CASB, the more I felt that the segregation of the board members in a separate building and the anarchy of the meetings were part of a strategy to restrict the flow of information.

Chairman Deschênes was an aloof and somewhat enigmatic figure. Prior to his appointment as the CASB's first chairman, he had been a Montreal lawyer principally noted for his backroom activities in support of the then-governing Liberal Party.[27] Liberal Transport minister Lloyd Axworthy had announced Deschênes' appointment in February 1984, though the CASB would not begin official operation for another eight months. Don Mazankowski, transportation critic in the opposition Conservative Party, had waxed furious that "a Co-Chairman of the Constitution and Legal Affairs Committee of the Liberal Party of Canada" had been parachuted into the chairmanship of the newly created board. To add insult to injury, the CASB had been created in consequence of an initiative from Mazankowski himself, one of the few during his brief tenure as Transport minister in the short-lived government of Joe Clark.

As it turned out, the Liberals lost the general election in September 1984 and Don Mazankowski once again became Minister of Transport. Thus, a few weeks before the CASB was formally inaugurated, Bernard Deschênes lost his status as a political insider. He quietly shelved plans for an elaborate public relations campaign to launch his new agency.

But Deschênes did not lose all his friends. Unlike the United States, where a new administration brings wholesale changes to the senior bureaucracy, Canada has a permanent cadre of civil servants to provide continuity and impartial service to whatever stripe of government the people may elect. The result is a powerful permanent bureaucracy run by senior "mandarins", which, according to Lloyd Axworthy, the outgoing Transport minister, exercises a virtual stranglehold on the machinery of government through its "near monopoly on information".[28] With his party swept out of office, Deschênes had little choice but to rely on this powerful network. He was forced into an alliance with the permanent bureaucracy, mainly through his chief lieutenant, Ken Johnson, a senior official inherited from the Transport department along with the rest of the new agency's employees. The alliance left Deschênes as a sort of governor-general of the CASB — retaining the outward trappings and perks of an agency head at the price of ceding bureaucratic control to the permanent civil service. Johnson would have absolute control over administrative matters, particularly hiring and firing of staff. Deschênes would keep members of the board on a tight leash.

This entente allowed Johnson and his staff to attain the nirvana of every good bureaucrat — an exclusive franchise on what he would be doing. Since nobody else investigated air crashes, nobody else would be able to measure how well accidents were being investigated. Who else could say how long it took to do an investigation? Or what kind of equipment the lab needed? Or how many conventions an investigator should attend? Success would be measured by the size of staff and budget, and would be assured by staying busy, generating paper, cultivating media friends and not offending political masters.

* * *

As for the members of the board, they would be kept in their place. Physical and procedural arrangements would isolate them from the administration, deny them significant information about investigations and prevent informed contact with the board's staff. They would be told to restrict their contacts with politicians, officials of other departments and even the board's own staff, in the name of ''independence''. Deschênes could always remind them that he had served as a consultant to the government in drafting the legislation that had set up the CASB, which he insisted qualified him uniquely to interpret the legislators' intent. As for inconvenient details—who knew better than he the "spirit" behind the words? "The touchstones of every good accident investigation process," he expounded in public speeches, "can be summarized under the headings of competence, equity or fairness, openness and integrity."[29]

I challenged Deschênes more than once on his autocratic dictates, but he justified his position by citing Section 9 of the Canadian Aviation Safety Board Act: "A Director of Investigation shall be appointed to have the exclusive authority to direct the conduct of investigations by the Board under this Act and to report to the Board with respect thereto." His interpretation of this scant clause was simple: The chairman would do all the appointing, and "exclusive authority" meant responsible exclusively to the chairman and exclusive of oversight or supervision by other members of the board. It followed that neither the details of investigations under way nor the investigators' tentative conclusions could be disclosed until the director of investigations was prepared to "report to the Board with respect thereto". Reports would be submitted complete with conclusions and recommendations, requiring nothing but *pro forma* approval.

Some board members liked the process just fine.

* * *

The director of investigations, who, by the doctrine of Bernard Deschênes, had exclusive authority to conduct the CASB's investigations, had in fact been selected by Ken Johnson before Deschênes arrived on the scene. He was a former fighter pilot, Tom Hinton.

In the Canadian air force Hinton had distinguished himself as a member of an elite aerobatics team, but no one would mistake him for Chuck Yeager. A man of medium build and reddish complexion, Hinton was prematurely bald, with a reddish moustache matching the ring of reddish hair remaining above his ears. I originally thought him forthright and cooperative, but as I got to know him better, that ring of hair began to look more and more like a slipped halo.

Hinton was a born bureaucrat. After leaving the air force he joined the Department of Transport as a civil aviation inspector, became head of the union of government pilots, obtained a Master's degree in Public Administration and worked his way up the management ladder to head the division updating aeronautical charts. He was an erudite speaker with an excellent memory for details, and his reports, sprinkled with names, dates, serial numbers and other minutiae, projected an impression of quiet competence. But under the air of reason lay a quick temper, easily aroused by those who lacked power over him —and this definitely included the appointed members of the CASB.

The choice of Tom Hinton as director of investigations exemplified Ken Johnson's approach to the selection of staff. Johnson wanted to surround himself with "good managers" who didn't allow technical quibbling to impede the flow of paper. To him, a stint in the air force and two weeks of classroom instruction would turn those with management potential into expert air crash investigators, true professionals. Those with extensive field experience or specialized technical qualifications could stay on tap in the laboratory or in a regional office — never on top at headquarters.

Peter Boag, a rising young star on Hinton's staff, was typecast as Johnson's ideal investigator. Boag had joined the air force in 1975, immediately after graduating from the University of Waterloo with a degree in Urban and Regional Planning. During his two tours of duty, he had flown and eventually instructed on light twin-engine turboprops. In 1982 Boag joined the Transport department and took a three-week orientation course for pilots who wanted to become accident investigators. Two years later he transferred, along with most of his group, to the newly created CASB.

Between 1982 and December 1985 Peter Boag assisted in several field investigations; on one occasion he was named investigator in charge for an accident of a Twin Otter, a model he had flown in the air force. During this period he also attended a two-week accident investigation course in Stockholm. On the whole, his work consisted of reviewing and editing reports written by others—to ensure consistency with standards set out in various manuals.

Like his boss, Tom Hinton, Peter Boag was an articulate, efficient bureaucrat. Unlike Hinton, he was not prone to flashes of anger. His style was that of long-suffering condescension, rocking back in his chair and casting his eyes heavenward when his patience was tried by a foolish question from a member of the board.

The selection of a neophyte like Peter Boag to head the investigation of Canada's worst air crash provoked considerable behind-the-scenes controversy. But Boag was not at all modest about his lack of experience. ''I have attended several internationally recognized accident investigation courses,'' he told the public inquiry into the Gander crash, ''and I have participated in the conduct, analysis or supervision of over one hundred aircraft accident investigations, some as the investigator in charge.''[30] More seasoned investigators described him as inexperienced and arrogant.

Despite the carping of older investigators, Peter Boag's appointment to head the probe into the Gander disaster proved a great success with the media. Boag's neat black hair, carefully trimmed aviator mustache and intense dark eyes complemented his cautious, deliberate answers to questions, and those who saw him on television invariably formed a good impression. His non-committal generalities did not seem like the evasions of a public relations flack, but the methodical thoroughness of a professional not about to make rash judgements.

The impression in the boardroom was somewhat different. Hinton, Boag and their staff made little attempt to disguise their opinion that reporting to the board was a somewhat distasteful formality, like a press conference where one had to put on a good face and suffer the rude or misinformed questions of ignorant interlocutors.

It's certainly true that in many organizations the board of directors serves only a ceremonial purpose. It's also true that many of the board's technical specialists in the lab and its field investigators in the regions produced solid, professional work. But Johnson, Hinton, Boag and the *apparatchik*s surrounding them at the Hull headquarters were bureaucrats specializing in coordination, liaison, the propagation of organizational charts and other skills serving to propel budgets and reorganization plans through the Ottawa labyrinth.

Those with inside knowledge were well aware that the CASB was not living up to the high hopes that had attended its creation. In September 1986, Walter Gadzos, a consultant hired by the Minister of Trans-

29

port, noted after ''extensive discussions with DOT [Department of Transport] staff, airline operators, airline safety officers, aviation consultants and lawyers'' that the performance of the CASB was ''disappointing''. ''The conduct of the investigations and the production of reports,'' Gadzos wrote in a confidential report to Transport minister John Crosbie, ''is almost universally judged to be below standards established prior to the formation of the Board''.[31]

Gadzos' appraisal of the CASB was uncovered after controversy over the Gander crash had already erupted. But I started to pick up the bad vibes at the very first meeting I attended. The board, I learned, had recently received a report from C.H. Rolf, Assistant Chief Judge of the Province of Alberta, on his public inquiry into the crash of a twin-engine Piper near the town of High Prairie on October 19, 1984. The passengers had included prominent provincial politicians, and Grant Notley, the leader of the opposition New Democratic Party in Alberta had been killed. Rolf was harshly critical of the CASB. The High Prairie crash had occasioned the first public inquiry of the CASB according to Judge Rolf the inquiry had ''left many questions unanswered'' and made it necessary to hold a further public inquiry under provincial legislation. ''Had the Canadian Aviation Safety Board held an 'open and public' inquiry,'' Rolf wrote, ''the convening of this Fatal Inquiry would not have been necessary.''[32] Rolf's blast hit at Deschênes' most tender ''touchstone'' — independence from the Transport department as required by law. The CASB had, moreover, allowed ''the local authority to be summarily preempted by headquarters staff,'' Judge Rolf complained.

Rolf's report forewarned of the charge that was to come in the wake of the Gander investigation. The investigators, he wrote, "appear to make an independent decision and then look to justify their decision." He was also dismayed at the CASB's refusal to recognize the next of kin as "interested parties"; "I was also shocked to learn that persons who have specific interest . . . are not allowed representation or participation before such inquiry." The judge's censure was an unheeded omen of what to expect from the families of the Gander victims, whose frustration and anger would play a dominant role in the future controversy.

Bernard Deschênes dismissed Judge Rolf's opinions as unfounded nonsense.

* * *

Yet another warning of problems in the CASB's operations could be found in the investigation—or rather the non-investigation—of an incident involving one of the Transport department's aircraft being flown by two of the department's most senior officials. On Saturday August 30, 1986, the two officials and two passengers flew the Beech King Air to the Belcher Islands, on the east coast of Hudson Bay. While the airplane was taxiing, the wheels got mired on the unpaved runway. A spray of gravel churned up by the propeller damaged the prop blades and cracked the windshield. The pilots swallowed hard and flew the damaged aircraft to Montreal. I first heard the story from an engineer in the Transport department who had examined the aircraft and told me, "They were damned lucky to make it back."

Opposition Members of Parliament got wind of the Belcher Islands incident and raised questions in the House of Commons. As a result, the licences of the pilots were lifted for two weeks for flying an aircraft in an unsafe condition. Many people, myself among them, wondered how the CASB was going to handle the investigation.

I was in a good position to find out, I thought, when I was appointed to the CASB a few months later. To my surprise, however, the board members knew nothing of the investigation. I asked Tom Hinton how the investigation of the Belcher Islands incident was coming along; he said in a casual tone that, since there had been no significant damage or injuries, no investigation was warranted.

At first I accepted this, but later, when I came to take all such explanations with a large grain of salt, I checked for myself. I learned that the CASB's own laboratory had examined the King Air's propeller. A memo in the files noted that "Severe foreign object damage was observed in the form of deep nicks and gouges along the length of the leading edges of all four blades, from the edge of the de-icing boot to the blade tip. There were also leading edge tears in the de-icing boots of the four blades from foreign object damage."[33] The damage clearly met the criteria for an investigation and was more severe than that in many minor accidents that were routinely investigated.

The seemingly casual attitude towards an incident to a Transport department King Air in the custody of senior officials was surprising and ironic. Conflict of interest in the investigation of the crash of another Transport department King Air had been one of the cases cited to show that the CASB should be independent of the Transport Department.

* * *

The most serious case under consideration at the time I joined the board concerned a helicopter that had crashed shortly after leaving an oil rig anchored offshore from Argentia, Newfoundland — about 180 miles south of Gander. The Bell 214ST lifted from the rig on the evening of March 13, 1985, and disappeared into a low cloud cover. A search-and-rescue station heard the first signal from the flight about an hour later; it was the beeping of the emergency locator transmitter or ELT (a radio distress signalling device activated by crash forces). The helicopter had crashed into the icy ocean, killing the two crew members and all four passengers.

Six months following the Argentia crash, the investigators still had no hypothesis about the cause. Worried about the safety of continued operation, the operator decided to carry out tests with its remaining Bell 214ST. On October 23, 1985, this second machine crashed while trying to duplicate the flight conditions at Argentia. Eye-witnesses saw the helicopter break apart in flight. Both pilots were killed.

Bell Helicopter Textron Inc. of Fort Worth, Texas, had developed the 214ST, the largest of its family of two-bladed helicopters, in the late 1970s, to the specifications of the Shah of Iran. One unusual feature of the high-performance military transport was its powerful Stability and Control Augmentation System'' in which computers augmented the pilot's movement of the controls according to their own internal logic. Failure of the SCAS would render the machine uncontrollable—but in military operations the increased risk was deemed a fair trade for higher performance.

The sale of the 214ST fell through in 1979, after the forces of the Ayatollah Khomeini toppled the Shah. Bell decided to recover its investment by offering the 214ST for commercial use, even though such heavy reliance on flight control computers was unprecedented in civilian operations. The 214ST reminded me of a string of crashes of the highly computerized UH-60 Blackhawk military helicopters, eventually attributed to problems with the computerized control system. Incredibly, the investigators' report on the Argentia crash didn't comment on the extraordinary features of the system, let alone study the possibility of a subtle malfunction.

Rather, the report speculated on intangible human factors. "On the day of the accident, another aircraft operated by Universal Helicopters entered the positive control zone without advising ATC [air traffic control]," it stated. Guilt by association led to a finding that "operational control in the company was inadequate." Another finding stated that "the operator assigned the captain to multiple duties, elements of which conflicted with his flying duties and may have affected the pilot's decision-making".[34]

I was amazed at the subjective speculation about intangibles at the expense of a detailed technical investigation. My experience with complex computer programs had made me well aware of—and occasionally the victim of—the "logic bomb" errors that can lurk undetected in a computer's innards until triggered by unusual circumstances. I remembered that the first launch of the space shuttle *Columbia* in 1981 had been scrubbed at the last minute because an insidious software bug had disabled all five flight control computers. To me, the loss of two of these highly computerized helicopters, under similar circumstances and with no apparent mechanical failure, called for a careful evaluation of the automatic flight control system.

Dave Mussallem and Norman Bobbitt, the two other aeronautical engineers on the board, appreciated my concerns, but the staff who were available to answer our questions scoffed. They noted that the 214ST had received a certificate of airworthiness from the FAA and to them this made the control systems beyond reproach. I recalled the problems with the space shuttle computers and the crash of the *Challenger*, pointing out that the most celebrated design team in the world had fatally miscalculated the odds of a tragic malfunction.

Chairman Deschênes was obviously irritated by such remarks from his new board member. I was coming to realize that boardroom presentations were tightly scripted to support the desired conclusions. Staff members answering questions were not technical specialists, but part of the administration, and conclusions that might have been reached after considerable debate among the specialists were presented by these "investigators" the way information is doled out to the media by public relations staff.

By April 1987, when the investigators submitted their report on the Gander crash, I was well aware that Deschênes would attempt to gain quick approval by his usual methods. All the same, I wasn't about to let the report on the worst aviation disaster in Canadian history pass without a thorough technical review. But when I and like-minded members of the Board embarked on our doomed attempt to force a review of the Gander crash, we had no way to foresee the boobytraps that lay in our path. To understand the twists, turns and dead-ends we would meet, it's necessary to look back at how the good intentions behind the founding of the Canadian Aviation Safety Board went so far astray.

BOOK 1

POLITICAL AND BUREAUCRATIC IMPERATIVES

BOOK 1

POLITICAL AND BUREAUCRATIC IMPERATIVES

Part 1

Roots

In the general election of May 26, 1979, Canadian voters signalled their displeasure with the Liberal Party, which had formed their government for the past sixteen years, by electing the minority government of Joe Clark, new leader of the Progressive Conservative Party. With the Conservatives out of office for so long, Clark had few experienced hands to form a Cabinet. One widely applauded appointment was that of 43-year-old Don Mazankowski as Minister of Transport. Mazankowski had been a popular MP from Vegreville, Alberta, for the past eleven years. As opposition Transport critic he had often scored solid body-punches on the increasingly* weary Liberal government, and in the months leading to the election, he had waxed particularly indignant about the state of aviation safety. The affable "Maz", a private pilot with good connections in the aviation community, had worked up a network of sources in the Transport department. Leaked reports put lead in his gloves for pummelling the Liberal government.

Mazankowski charged that Transport doctored aviation accident reports to cover up government incompetence. He had been particularly vocal about the investigation of the crash of a Pacific Western Airlines Boeing 737 at Cranbrook, British Columbia, on February 11, 1978—the first major airline accident in Canada in three years, the worst in eight. The 737 crashed and burned while trying to abort a landing after the crew noticed a snow sweeper on the runway. Thirty-nine passengers and four crew members died, and five passengers were seriously injured.

Like many of Canada's smaller airports, Cranbrook had no air traffic controllers. A radio operator provided information to arriving aircraft, but there was no control tower and the radio operator couldn't see the entire runway. Controllers at nearby Calgary looked after the airspace.

The Pacific Western 737 arrived from Calgary after a 23-minute flight. The Calgary controller had notified the Cranbrook radio operator when the flight was on its way, but the 737 arrived ten minutes early. Snow was falling and the runway was being cleared for the incoming flight. There was little doubt that miscued communications between aircraft and ground played some role in the accident.

The possibility of a cover-up came to public notice when the Transport department refused to give a coroner's jury accurate transcripts of the radio conversations. As the department saw it, such transcripts "might do more harm than good in the hands of the untrained jury".[1] Six months after the accident, with the investigation still in progress, Pacific Western sued the Transport department for negligence.

As Transport critic, Mazankowski knew that an internal study produced about a year before the Cranbrook crash had examined problems of air-to-ground communications at outlying airports.[2] The report had recommended measures to prevent conflicts between aircraft and ground vehicles at airports without a control tower. Officials claimed that if these recommendations had been applied at Cranbrook, they would probably have prevented the accident. It seemed that this information was being suppressed because it was embarrassing and harmful to the government's defence against Pacific Western's suit.

During the months leading up to the dissolution of Parliament in the spring of 1979, Mazankowski repeatedly accused the government of sweeping air safety problems under the carpet. In particular, he accused the Minister of Transport of distorting the causes of the Cranbrook crash to protect the government. His litany of horror stories prompted the Conservatives to make aviation safety an election issue. The Transport department released its report on the Cranbrook accident on March 29, three days after Parliament was dissolved, and the Conservatives promised that, if elected, they would hold a public inquiry into aviation safety and bring in legislation to clean up the mess.

On August 3, Mazankowski began to fulfil those promises by announcing the appointment of Charles Dubin, a justice on the Ontario Court of Appeal, as a one-man commission to study the safety of Canada's civil air transport system. The minister's press release said that Dubin would "inquire into the investigation of accidents and the reporting and investigation of incidents involving aircraft." Aiming a shot at the previous government, it noted that "a public inquiry, rather than an internal investigation is consistent with the government's policy of responding to concerns of the public in an open manner."

The Ottawa bureaucrats didn't welcome Mazankowski's plan for exposing the inner workings of a government department. As journalist Stevie Cameron explains, "In Ottawa, the most important people are not so much the politicians; rather they are the top public servants — the Clerk of the Privy Council and the deputy ministers — who, along with the heads of agencies, boards and commissions, run the place like feudal barons''[3]. In the normal course of events, the feudal barons knew how to deflect a new minister with delusions of rearranging the furniture. Only the fact that Mazankowski's blueprint for a public inquiry had been hatched in opposition, and the turmoil brought by the ascendancy of a new governing party, gave Dubin the chance to set up his commission without prior constraints from Ottawa insiders.

Dubin chose Toronto lawyer John Sopinka as his chief counsel. To conform with practical and political realities in Canada, Quebec lawyer Gary Q. Ouellet became Sopinka's French-speaking associate counsel. Sopinka and Ouellet would play key roles in writing the recommendations that led to the Canadian Aviation Safety Board. Both these high-horsepower lawyers would attain bigger and better roles under a future Conservative government; each would reappear for a cameo part in the collapse of the ill-fated board they had helped create.

As the inquiry held hearings across Canada, it provided a form of primal-scream therapy for critics of the Transport department. The commissioner encouraged the department's footsoldiers to spill the beans, and publicly berated senior officers when he got wind of reprisals.

On December 13, 1979, just as Dubin's hearings were gathering momentum, the Clark government fell.* A triumphant Pierre Trudeau, enticed out of recent retirement, led the Liberal party back from the wilderness to a majority win in the general election of February 18, 1980. But the Dubin Commission had

---

*The Canadian political system follows the British parliamentary tradition, which calls for the government to cede power if defeated on certain important issues.

already cleared the launch pad, and it continued to enliven the news with hair-raising stories of special favors for the state-owned airline, Air Canada; cover-ups of Transport department bungling; and nonchalant enforcement of safety regulations.

The new Transport minister, Jean-Luc Pepin, tabled the first instalment of Justice Dubin's report in the House of Commons on May 27, 1981, two years after the commission was set up. The report fired a devastating salvo at the Transport department, and some of its most telling shots tore into senior managers who watered down accident reports to avoid criticism. He pinpointed the intransigent resistance of managers to criticism as a major obstacle to effective investigations. The Transport department should be stripped of all its air safety investigation tasks, he said, because these tasks created conflicts of interest—and he emphasized that he was talking not about the appearance of conflict, but about actual conflict.

John Sopinka summarized the dilemma of an individual investigator: "Since the investigator is dependent on the employer for his livelihood, he is placed in a position where his interest and his duty conflict." When the investigator's ultimate employer, the Crown, defended itself in legal suits, "it may appear that the investigator will seek factors supporting the Crown's position or turn a blind eye to those which put the Crown in jeopardy."[4]

To cure these ills, Dubin recommended that the responsibility for investigating aviation accidents should be handed over to a new tribunal, independent of all departments of government.

Roots of the NTSB

The problem Sopinka had identified was not new. In fact, conflicts of interest reappeared whenever a sensational aviation accident brought an investigation into the public eye. The dilemma had been identified some forty-five years earlier in the United States.

On May 6, 1935, five people, including Senator Bronson M. Cutting of New Mexico, died in the crash of a Transcontinental and Western Air DC-2 near Kirksville, Missouri. The Bureau of Air Commerce, which investigated aviation accidents at the time, took less than a month to find a cause. "The probable direct cause of this accident," it concluded, with the kind of insightful inference that characterizes accident reports to this day, "was an unintentional collision with the ground while the airplane was being maneuvered at a very low altitude in fog and darkness". The investigators found "no evidence that the established Department of Commerce navigation aids were not functioning properly." They did, however, find that the airline had "initiated or permitted certain irregularities which were inexcusable."[5] In short, the investigators' employer was blameless; other parties were culpable.

These findings outraged the airline, not to mention Cutting's senatorial colleagues. In those early, budget-balancing days of the Roosevelt administration, they suspected that slashed funding for navigation aids might have been a factor. Could it be that the investigators didn't look hard enough—turned a blind eye, even — at their employer's responsibilities? "Can a government agency sit in judgement upon an activity in which it itself participates?", asked the journal *Aero Digest*, anticipating Counsel Sopinka's soliloquy by some thirty-five years.

The United States Senate mounted its own probe of the "Cutting Air Crash", calling into question not just the investigation's findings but the policy behind it.

The Senate investigation found more fault with the government than with the airline. In particular, the senators claimed that the accident could have been avoided if the government had lived up to its responsibilities in providing aids to air navigation.

Repercussions ultimately led to an Air Safety Board to investigate accidents, determine probable cause and recommend prevention measures, independently of the new Civil Aeronautics Authority. The FAA *Historical Fact Book* tells us that "Though the Federal aviation role would be shuffled and reshuffled in subsequent years . . . accident investigation and determination of probable cause was never again placed in the hands of the agency responsible for establishing and maintaining the Federal airways."[6] One of the subsequent reshuffles, the Department of Transportation Act of 1966, gave American aircraft accident investigators their present administrative home, the National Transportation Safety Board, or NTSB.

Since the Canadian air transport system is modelled closely on the American one, the value of an independent accident investigation agency was apparent in Canada well before Don Mazankowski started needling the government in the spring of 1979. But the inertial self-protection of the civil service was able to stave off such an agency for decades. When the CASB Act was finally introduced in 1983, that same self-interest ensured that the much-vaunted independence of the new board became little more than a façade to preserve the comfortable status quo.

Earlier DC-8 Crashes

On November 29, 1963, when the news all over the world was still dominated by the assassination of President John F. Kennedy in Dallas a week earlier, a DC-8 belonging to Trans-Canada Airlines* crashed into a wooded lot near the town of Ste-Thérèse, Quebec. Flight 831, a "businessman's special", had left Montreal's Dorval airport some five minutes earlier. The crash destroyed the aircraft and killed all 118 occupants. The Ste-Thérèse crash was the worst air disaster in Canada at the time and would hold this record for another twenty-two years—until the death tally was more than doubled in the crash of another DC-8, at Gander.

The precise cause of the Ste-Thérèse crash was never pinned down, but the investigation highlighted jurisdictional disputes and conflicts of interest. A great deal of the wreckage was buried in the ground. Protracted salvage operations using heavy equipment and up to 1,500 workers continued for about five months. Some 26,000 cubic yards of earth were removed and screened and some 80 per cent — over 105,000 pounds — of the aircraft's total weight was recovered. Since the airline paid for the recovery operations, "some senior TCA officials took the position that it was a TCA investigation",[7] causing confusion and friction between the airline and the Transport Department.** The conflicts induced the government to commission J.R. Booth, a retired Assistant Deputy Minister of Transport, to study how such fiascos could be headed off in future. The National Transportation Safety Board then being formed in the United States suggested the possibility of an similar independent

---

*The name was changed to Air Canada on January 1, 1965, while the investigation into the Ste-Thérèse crash was still in progress.
**The Department of Transport (DoT) is also called the Ministry of Transport (MoT or MOT) or transport Canada. I use "Transport department" except in direct quotes.

agency, but the Canadian bureaucracy was not keen on the idea.

"In Canada, something like the U.S. system of a separate organization would seem at the present time to be impracticable or at least undesirable," wrote Booth. Instead, he recommended that accident investigation be separated from the regulatory arm but stay under the umbrella of the Transport department.* But even this mild shift in the status quo proved too radical for the Ottawa establishment. The Booth study was studied and restudied without action until it was overtaken by a second catastrophic DC-8 crash.

On July 5, 1970, Air Canada Flight 621 was en route to Los Angeles from Montreal with a brief stop at Toronto. The approach was routine until almost the point of touchdown. Then, at about sixty feet, the DC-8 suddenly lost lift and thumped down hard. Not realizing that the right outboard engine had broken off, the crew decided to abort the landing. The plane climbed to about 3,000 feet, where explosions ripped off most of the right wing. It plunged to the ground and all 109 occupants were killed.

The cockpit voice recorder, which was quickly recovered, showed that the hard landing had been caused by an inadvertent extension of the ground spoilers just before touchdown.** As was well known, a lockout device released when the landing gear struts began to

---

*The Aeronautics Act was amended to allow the Minister of Transport to appoint a board of inquiry with judicial powers to look into accidents. Previously the full Cabinet had to do it under the Inquiries Act.

**Ground spoilers, also called "lift dumpers", are hinged plates on top of the wings that can be flipped up to "spoil" the wings' lift once the aircraft is firmly on the ground. With the wings' lift spoiled, the brakes slow the aircraft more efficiently.

compress could prevent such premature deployment of the ground spoilers. But the FAA didn't require a ground spoiler lockout on the DC-8, and the Canadian Transport department was not in the habit of questioning American airworthiness decisions.

Don Jamieson, Minister of Transport at the time, decided to forestall controversy by appointing a judicial inquiry to consider the circumstances of the accident after the completion of the department's technical investigation.

Justice Hugh F. Gibson of the Exchequer Court of Canada, who conducted the inquiry, found that "the Captain's method of arming the spoilers was not according to the manual and he insisted that the first officer follow his wishes." We may safely assume that an investigation conducted solely by the Transport department would have found this pilot error — and declared the case closed. But while Justice Gibson noted the crew errors, he also concluded that "a design defect" in the DC-8 allowed the ground spoilers to be deployed in flight. He reported that the flight manuals contained "equivocation, inaccuracies and misinformation", and that contributing circumstances included "the acceptance and approval of the ground spoiler system" by the Transport department as well as the department's "failure to detect the deficiencies and misinformation in the flight manual.[8] He recommended that the department "strengthen its capacity to approve design of aircraft imported into Canada" and do a better job of monitoring flight procedures.

Justice Gibson's report rekindled interest in the idea of an independent board. Retired Judge Advocate Brigadier-General Harold A. McLearn was commissioned to conduct yet another study, and finished it in 1973. Mclearn recommended the creation of an independent board along the lines of the NTSB—"the only effective means of eliminating, or substantially reducing, the existing risks of conflict of interest . . . ." McLearn pointed out that his proposal required new legislation and amendment of existing statutes. "As this would doubtless be a slow process," he noted, "it is desirable that an interim step be considered." He suggested that the Minister of Transport could, on his own authority, establish an Air Accident Review Board consisting of a chairman "drawn from a department other than the Ministry of Transport" and two officials from the department who are "not involved in accident investigation and safety".[9] But by the time McLearn submitted his report in 1973, the commotion had died down.

The tranquillity didn't last long. On February 27, 1974, a Rockwell Saberliner executive jet crashed into a mountainside on Baffin Island in Frobisher Bay, killing its nine occupants. The Transport investigators refused to turn over their results to a coroner's inquest. "Such action leads to the speculation that there may have been something in the report which was damaging to the MoT, and which the MoT did not want to make public," said an editorial in the November issue of *Canadian Aviation*. "Some kind of independent authority" to investigate aviation accidents "is long overdue."

In the face of such criticism, the government declared its intention to establish a Transportation Accident Investigation Board. But before it had time to act on this good intention, another accident raised even greater concerns. On October 30, 1974, a Lockheed Electra owned by Panarctic Oils Ltd. crashed into the ice near Rae Point in the Northwest Territories, killing thirty-one on board; one of the three survivors died en route to hospital. The crew were not wearing shoulder harnesses, and there was evidence that the captain had been incapacitated by some physical problem at the time of the crash.[10]

In the wake of yet more accusations of conflict of interest, a Cabinet decision dated April 17, 1975 approved in principle a new agency to investigate transportation accidents along the lines recommended by McLearn. Introducing new legislation was indeed a slow process. It would take another nine years to create the Canadian Aviation Safety Board—and then five years to get rid of it.

The AARB

The Cabinet's approval of an independent Transportation Accident Investigation Board provided at least a temporary out for those who preferred the status quo. As McLearn had recommended earlier, the Minister of Transport set up an advisory board under his own jurisdiction, as an interim measure pending the introduction of legislation.

Good work takes time—so it took more than a year to set up the interim board. On February 18, 1976, Walter M. McLeish—the Transport department's air administrator (the title used to designate the senior aviation official at the time), invited H. R. Footit to be chairman of an interim Aircraft Accident Review Board. Footit was a 62-year-old retired air commodore and senior civil servant, well used to the ways of Ottawa. McLeish made it very clear who was calling the shots. "The Air Administrator," he wrote, "recommended that you be appointed Chairman of the Review Board and Mr. Lang [the Minister of Transport at the time] was pleased to approve your appointment." The procedures given to Footit stipulated that the board would review only those accidents designated by officials of the department.

Two other members were also appointed on the recommendation of Walter McLeish — Ron Baker, a retired Air Canada captain, and Frank R. Thurston, who was at the time director of the National Aeronautical Establishment.

The board, dubbed the AARB, was supposed to protect the Minister of Transport from the conflicts of interest inherent in having both the investigatory agency and the regulatory agency under his authority. But practice fell far short of theory. The AARB revealed its inadequacy in the laconic five-page report submitted to the minister after a year of unremarkable operation, which said in essence that the AARB had established provisional procedures and reviewed some accident reports. Short and uninformative as it was, the report was revealing in noting that the procedures it had established for itself had been sent to Mr. McLeish and his staff "to ensure that the board would carry out its mission as stated."

In other words, the AARB needed the approval of the very officials it was supposed to monitor for conflict of interest. It took a low profile and remained virtually unknown outside the Transport department. It resolved conflicts of interest, not by pointing them out to the minister, but by jawboning behind closed doors with officials and involved parties. This penchant for backroom dealing was so natural in the Ottawa environment as to be altogether unremarkable.

In March 1979, shortly before Parliament dissolved for the general election that would bring in the short-lived Clark government and set the Dubin Commission in motion, two new members were added—again on the recommendation of Walter McLeish. Max T. Friedl was the first of a line of retired air force generals with supporting roles in the creation and destruction of the CASB. The other new appointee would play an even bigger role; he was Montreal lawyer Bernard M. Deschênes. Deschênes kept a low profile within the low-profile AARB — "sitting off to the side in a haze of cigarette smoke", according to a staffer who frequently attended meetings.

By the fiscal year 1977-78, when Friedl and Deschênes joined the AARB, the annual report had shrunk to four pages listing seven "Major Accident Reports reviewed and audited by the Board for release as public information". Three of these would serve Justice Dubin as case studies to highlight the problem areas in the investigation of aviation accidents, and these case studies would expose the AARB as utterly ineffective in preventing conflicts of interest between accident investigation and the Transport department's other functions.

Justice Dubin's Case Studies

The first case Dubin considered was the crash of a Hawker-Siddeley HS-125 at Churchill Falls in December 1977, the first major Canadian air crash after the AARB was set up. The corporate jet crashed into the ice two miles short of the runway during its final approach. A helicopter searching for the downed aircraft flew over the accident site within fifteen minutes, but poor visibility prevented the rescuers from spotting the wreck. By the time the wreckage was found two days later, all eight occupants had died. Two of the victims had lived for some hours and could have survived had they received medical attention. Both of the crashed jet's ELTs (emergency locator transmitters) had been removed because of an elaborate bureaucratic foul-up in the Transport Department.

The ELT story went back to a notorious crash five years earlier. On December 9, 1972, a Canadian Forces aircraft flying over the Arctic had picked up a weak signal from the ELT of a downed Beech 18. The pilot, Martin Hartwell, a 46-year-old ex-Luftwaffe officer, was the sole survivor of a flight that had crashed in foul weather thirty-two days before. His passengers — Judith Hill, a 27-year-old British-born nurse, 14-year-old David Kootook and Neeme Nulliayak, who was pregnant—were all dead. The crash site was some 180 miles off course, well outside the original search area. In the subsequent inquest, Hartwell described his month-long ordeal in the frigid weather, which included having to eat the bodies of his dead passengers to stay alive. The delay in finding the downed aircraft was, in part, because of an inadequate ELT.

Sensational publicity led to various committees and studies and eventually, in July 1974, to a regulation requiring standardized ELTs on Canadian aircraft. The Hartwell case emphasized the need for special consideration of the intense cold in the Arctic, so—in a rare break with American standards—the Canadian regulation specified superior cold-weather performance. Conventional batteries were inadequate, but fortunately, or so it seemed at the time, a suitable battery — the so-called lithium battery — had recently become available. Some 17,000 lithium-battery ELTs were sold in Canada during the next several years.

Unfortunately, the lithium batteries also had drawbacks. Sulphur dioxide generated in the units tended to corrode electrical contacts. More ominously, such corrosion could produce an internal short circuit, which could, in turn, cause an explosion. In 1977, after various incidents including one where a lithium battery exploded in flight, the Transport department ordered the removal of lithium-battery ELTs. While the department pondered its next move, aircraft were allowed to fly without them — with tragic results for least two of the occupants of the HS-125 that crashed at Churchill Falls that December.

In the draft report on the Churchill Falls crash the Transport department's investigators concluded that "The removal of the aircraft's emergency locator transmitters seriously delayed the search and rescue activity and may have caused unnecessary loss of life."[11] Walter McLeish had not only approved the removal of the ELTs, but had recommended the members of the AARB to the minister; moreover, he was the investigators' boss. He judged the part about "unnecessary loss of life" a gratuitous editorial comment. A revised report fudged the conclusion to state that the rescue effort might have been "expedited" if there had been a serviceable ELT on board, and if various other assumptions were true.

But there was still a problem. The aircraft's owners and other interested parties already had copies of the draft, along with the offending conclusion. Censoring the reference to "unnecessary loss of life" showed the Transport department whitewashing accident reports and demonstrated the impotence of the AARB. The facts were exposed in an inquest conducted by the Province of Newfoundland, and the publicity provoked a lot of anguish among field investigators, Transport Canada management, the AARB and the minister's office. The upshot of "public comment on the tragedy" was that the minister "felt it advisable" to release the original version of the report.[12]

Dubin's second case history involved one of the Transport department's own aircraft. Two department employees were killed in May 1979 when a wing broke off a Beechcraft King Air 90 during a routine flight to check out a radar antenna at Montreal's Dorval airport. It was quickly established that the fitting that held the right wing had given way because cracks had not been detected during maintenance. Evidently the fitting had been either badly designed or improperly inspected in the Transport department's own shop. The investigators' report claimed that Transport mechanics had omitted the inspection which should have detected the cracks because the FAA-approved directives were ambiguous. The manufacturer and the FAA begged to differ; they tartly suggested that the Transport department was dodging responsibility for its own negligence.

Dubin declared himself satisfied that the investigators had "arrived at their conclusions with expertise and in good faith and . . . without any intention on their part to favor" their employer, but he also believed that the investigators "were in an untenable position by reason of their conflict of interest", "The investigation," he said, "therefore, became ineffectual."*

Dubin's third case study illustrated yet another way in which the investigators' conflicts of interest could

_____

*Neither the Transport department nor Dubin noted another way of looking at the accident: the department had certified and allowed its inspectors to work with an aircraft where the failure of single highly stressed part could cause a catastrophe. Ironically, a modification was available for the Beech 90 which would have allowed it to fly to a safe landing even aftr a wing fitting had broken. Many private operators had adopted this modification, but endorsement by the manufacturer and the certificating authorities—might have been taken as admission of design fault.

obscure the causes of an accident. In this case the worst fears were realized: while the search for the cause of one accident was being delayed by such conflicts, a second accident resulted from the same cause.

The de Havilland Twin Otter, or DHC-6, was designed, manufactured and certified in Canada. It was powered by Pratt & Whitney PT6 engines that were likewise designed, manufactured and certified in Canada. Thus the all-Canadian DHC-6 was a great source of national pride, to individual Canadians, the de Havilland Company, and the government. The float-plane version of the DHC-6, which served various communities on the west coast of British Columbia was especially symbolic of the rugged bush-plane traditions of the Canadian aviation industry.

On September 3, 1978, a float-equipped DHC-6 was approaching Vancouver's Coal Harbour with two crew members and a party of Japanese tourists on a sightseeing flight. About 175 feet above the surface, the aircraft suddenly rolled and crashed into the water. The two crew members and nine of the eleven passengers were killed, and the other two passengers were seriously injured. "The investigation as to the cause of this accident", still under way while Dubin was conducting his public inquiry, "became a lengthy, complicated and highly technical inquiry as well as a controversial one."[13]

"Shortly after the investigation was commenced," Dubin explained, a senior manager in the Transport department "began interfering with the investigation." In light of "the particular importance that Twin Otters played in the Canadian manufacturing industry", the department formed a committee to study Twin Otter accidents, including the one at Coal Harbour. Dubin allowed that the committee "was formed in good faith and with the best of intentions" but added that "it was never made clear to all the participants what the role of the ad hoc committee was and how this role interrelated with the normal investigation process." Lorne Tapp, the investigator in charge, summarized it more succinctly, saying that "no one knew exactly who was running what."

A more worrisome effect of superimposing this committee was that it included representatives of de Havilland and manufacturers of the engines and propellers. This, Dubin noted, added another source of conflict, by making them "part of an inquiry into matters which may have found them at fault." According to the accident investigators, the de Havilland Company became aggressively defensive. "It seemed as though the purpose" of the company was not to find the cause of the crash but "to refute any preliminary analysis" that might indicate that "their product was deficient."[14]

Relations stretched to the breaking point when the laboratory suggested that one of the wing flaps on the DHC-6 had retracted suddenly — presumably due to failure of a rod holding it in place. The rods holding the opposite flap were corroded, and inspection of other Twin Otters found further examples of such "stress corrosion". Although the flap rod presumed broken on the crashed plane had not been recovered, engineers in Transport Canada's laboratory inferred that it must have failed in flight. "Because de Havilland did not accept the position that their product was deficient and other critical components similarly manufactured were also susceptible to stress corrosion, nothing was done to rectify the overall problem."

In fact, the company had been aware of the problem for some time. Nine years earlier, in November 1970, it had issued a technical advisory bulletin that aileron control rods similar to the suspect flap rod "have proven to be vulnerable to corrosion." But the company denied that corrosion could be a factor in the Coal Harbour crash. On September 30, 1979, another float-equipped DHC-6 crashed into the water at Sechelt, off the coast of British Columbia. This crash, which killed the pilot and one passenger, was unmistakably due to stress corrosion failure of an aileron control rod.

The company continued to resist the notion that the Coal Harbour accident might have had a similar cause, and senior management in the Transport department sided with de Havilland against the department's own investigators. The company carried out some tests and appealed directly to the AARB. The AARB attempted to conciliate, without success. Eventually, renewed dredging of Coal Harbour turned up the missing flap rod and established beyond reasonable doubt that it had failed in flight from stress corrosion.*

The Siddon Connection

Dubin's decision to use the Twin Otter crashes as case studies may have been the first step to my eventual appointment to the Canadian Aviation Safety Board. Both pilots had been constituents of Conservative MP Thomas Siddon. In addition to being a Member of Parliament, Siddon was at the time a professor of Mechanical Engineering, on leave from the University of British Columbia. He had taken a great interest in the investigation of the two crashes, and was the only MP to testify at the Dubin Commission. His testimony outlined some of the conflicts of interest, along lines

---

*These crashes show yet another type of conflict of interest: no agency, as far as I know, has ever attributed a crash to delay, incompetence or indeed any failing in a previous investigation.

eventually noted by Dubin, and urged the creation of an independent accident investigation agency.

Ten years before Tom Siddon testified at the Dubin Commission, he and I had been fellow graduate students and aviation enthusiasts at the University of Toronto Institute for Aerospace Studies. We discussed the Twin Otter accidents and the need for an independent investigation agency numerous times before and after his testimony. Later, we shared a keen interest in the implementation of Dubin's recommendations. After the Conservatives formed a majority government in 1984, Siddon became a Cabinet minister. Later still, he recommended me for an appointment to the CASB.

Actually, I should say I have every reason to believe he recommended me. On his suggestion I sent a letter expressing interest, along with a curriculum vitae, to the Minister of Transport John Crosbie. One of Crosbie's aides got in touch and after some discussions of the conditions, in which nothing political was mentioned, I was told that Mr. Crosbie had recommended my appointment to the Cabinet.

Later, with the controversy over the Gander investigation in full swing, I naively told this story to a reporter. The result was a syndicated column about my "patronage appointment", which dwelt on my friendship with a politician and left out the rest. The column greatly delighted my detractors, and kept a lot of photocopiers warm as they made a hobby of mailing out copies of this proof of my incompetence and sycophancy.

* * *

Dubin's first three case studies provided ample high-octane fuel for critics of the Transport department's way of investigating air crashes. But he saved his best shots for the PWA 737 crash at Cranbrook — the case of the snow sweeper on the runway, the case that gave Mazankowski conniptions just before the 1979 general election, and the one cited by Mazankowski when he launched the Dubin Commission.

Dubin's analysis exposed gaping philosophical gulfs among accident investigators, between investigators and senior management, as well as between investigators and Justice Department lawyers. It exposed deeply held principles, stubborn arrogance and a sensational story of withheld information and shredded evidence. It showed the AARB useless in preventing conflicts of interest, and in fact helping to hide them by giving out a misleading aura of independence.

The Minister of Transport, Otto Lang, released the final report on the Cranbrook investigation to the public on March 29, 1979. (Lang, just one of the long line of Transport ministers who duck in and out of the CASB story, would lose his seat in the election that had been called three days earlier.) The part of the report that drew Dubin's most pointed attention was an obscure editorial interjection potted on page 32, like a small flower in a bed of gravel: "The failure to report on final approach and the unnecessary talk on company frequency "represent an unacceptable standard of cockpit practice and discipline."

The Justice Department lawyers who put Dubin on to this comment pointed out that the report's findings said nothing about unnecessary talk or unacceptable discipline. Would Mr. Justice Dubin care to find out why not? The lawyers also volunteered that they were disturbed by the "pattern of conduct" they had encountered while trying to defend the government against the Pacific Western Airlines suit. A memorandum from W.J. Hobson, senior defence counsel for the Cranbrook case, suggested that "evidence to support the real cause of the accident has been destroyed and might not be made available by fresh investigation. . . . Allegations of cover-up, non-disclosure and deceit might well be made against the Department of Transport as the litigation proceeds."

This was not an anonymous source leaking to the media or a member of the opposition; this was an attorney making a statement on the record. The problem was that certain investigators were withholding information on their own initiative — not to protect their employer, but because they felt honour-bound to keep their sources confidential, "especially so for information of an embarrassing or inculpating nature."[15]

In this case, the "information of an embarrassing or inculpating nature" was the testimony of unidentified pilots who claimed they had heard the captain of PWA Flight 314 engage in "unnecessary talk on company frequency", rumored to concern the stock market and a beautiful female dispatcher. The government lawyers realized, of course, that proof of the captain's preoccupation with irrelevant gossip would reduce the government's liability. But none of the investigators was willing to defame the dead pilot by admitting knowledge of such rumors.

Transport department managers claimed they had demanded their subordinates' full cooperation. Since the suit against the government named many of these managers, their eagerness to cooperate with the defence may have had a twinge of self-interest. In any event, the blame for withholding information about the alleged conversation fell on the individuals who had drafted the report.

Dr. François Dubé, a physician from the Department of Health and Welfare, became the principal scapegoat. A specialist in aviation medicine and a frequent participant in accident investigations, Frank Dubé was an expert in "human factors"—that is, in finding the medical and psychological factors that bore on an accident. Medical factors include such things as combustion products in the tissues of the victims; psychological factors include possible distraction by an irrelevant conversation at a critical time.

But Dr. Dubé was not about to turn sensitive medical stuff over to a bunch of lawyers. He muttered that "you can't subpoena what doesn't exist", and by the time the lawyers got really serious, his sensitive Cranbrook files had been shredded.

While defending his actions, Dr. Dubé mentioned that there were "certain things which could not be put in an accident report because they would raise questions which might lead to confidential sources of information." Apparently miffed by the lawyers' continued badgering, Dubé professed surprise "that nobody had questioned what was the real cause of the accident and which they had buried in the Accident Investigation Report."[16]

The suggestion that the "real cause" had been buried piqued the lawyers' curiosity. Dube pointed to the statement about "unnecessary talk", claiming that the others on the investigation team shared his appraisal, and then added a truly remarkable postscript. "Dr. Dubé said that much pressure was brought by the company and would be brought by any carrier in these circumstances to ensure that this type of comment would not be made as contributing to the cause of the accident."

A highly experienced investigator, Dubé accepted such pressure to distort accident findings as a normal part of business. Indeed, it might be naive to expect otherwise. But what is really interesting about his comment is its tacit admission that investigators yield to such pressure according to their own whims.

Dubin deplored the destruction of evidence and the omission of a key finding, but he didn't pursue Dubé's casual admission that a party with a vested interest had successfully interfered with the investigation. He berated the AARB for not ensuring that the cryptic reference to "unnecessary talk on the company frequency" was backed up by witness statements but he was silent on the successful attempt to influence the investigation. And the remedies he proposed for eliminating conflicts of interest omitted an important element — the influence of vested interests while gathering information was being gathered.

Dubin's analysis of the Cranbrook investigation greatly influenced my approach to reviewing the CASB investigators' reports. I was very aware of the need to verify sources and the need to make explicit findings. But my attempts to do so ran headlong into Bernard Deschênes' concept of the "exclusive authority" of the director of investigations. The exclusive franchise he claimed not only fulfilled the bureaucratic dream of no external yardsticks to measure performance, but also removed the threat of embarrassment from the kind of questions that Dubin had asked about the Cranbrook crash. No one on the outside would ever know if pressures had been brought to bear, in CASB cases, let alone if anyone had capitulated. Somehow, between diagnosis and inoculation, Dubin's prescription had lost all its potency.

Part 2

Bureaucracy and Politics

Independent Tribunal

With the Conservatives again relegated to the opposition, Dubin had to hand his recommendation for an independent tribunal for accident investigation to the newly reinstalled Liberals. But public interest ensured that the report could not be swept under the rug, and when Transport minister Jean-Luc Pepin tabled Volume I of the report in the House of Commons on May 7, 1981, he announced plans to introduce legislation to set up the Canadian Aviation Safety Board.

This time, it seemed that the idea of an independent agency had enough momentum to go the distance. Nevertheless, heavy duty damage control operations were immediately apparent. Pepin simultaneously announced that he had set up a "Minister's Advisory Committee" to advise him on implementing Dubin's recommendations. One of the members, Walter McLeish, had candidly warned that Dubin's recommendations "may be quite unacceptable to us as professional aviation people".[17] It was not unreasonable to suppose that other members of the committee — who included a senior assistant deputy Transport minister, a former deputy Transport minister and Bernard Deschênes, described as "a former transport department legal adviser" — would be sympathetic to McLeish's point of view. Opposition critics and editorial writers condemned the committee as "the wrong men for the job", who would "undermine confidence that Dubin's message has been heeded".[18]

Jack Ellis, the Conservative transportation critic, said the minister had "put a fox in charge of the chicken coop".[19] Canadian aviation associations, including the Air Transport Association and the Canadian Air Line Pilots Association, in absolute agreement for the first time ever, called on the minister to change the committee.[20] The minister resisted, and the committee carried on.

In the meantime, the Transport bureaucracy geared up to meet the challenge of the impending legislation. Don Button, director of the Aviation Safety Bureau (the unit responsible for accident investigations), had testified about problems with senior management; he was replaced by Kenneth Johnson, an administrator from the department's west coast office. Johnson, a former military and commercial pilot, was trained as an accountant, without experience in the practical side of accident investigation. But as a defendant in PWA's suit over the Cranbrook crash he got a good education in the strategical side, and made the right connections, not only in Transport but in the Justice department and the Privy Council Office.

Johnson proved the ideal choice to oversee the bureaucratic birth pangs of the new agency — arranging new office space; writing job descriptions; dividing files and data bases; reassuring unions, auditors and other watchdogs. Naturally, the departments who had supplied medical and legal support for accident investigation were loath to release funds and staff to a new organization. But Johnson stickhandled deftly through such minefields. The organizational charts that appeared after the dust settled owed much more to Johnson's bureaucratic talents than to the wisdom of Justice Dubin or the legislators. And the slots on these charts were not filled with names of those who had spoken to Dubin with less than complete enthusiasm about the department's policies.

By the time the Canadian Aviation Safety Board Act was passed in October 1983, Johnson's organization was ready. There was, however, one fly that might drop into the ointment at the last moment. The chairman and board would be filled by "order-in-council" appointments, which would be made at the unfettered discretion of the Cabinet. What if the Cabinet appointed a chairman and board who wanted a different type of organization?

The Minister's Advisory Committee worked mightily to ensure that the legislation would rule out interference from the political appointees. But they were told that the legislation could not overtly remove "executive responsibility for the day to day operation of the investigative staff of the board" from "the Chairman of the CASB or his several colleagues." The bureaucracy's talent for skirting inconvenient legislation with administrative sidesteps is an enduring source of marvel or rage, but it is rare to find as candid a description as that written by the AARB to the Minister's Advisory Committee. "Having regard for the practical politics of the situation as it now exists," chairman Max Friedl wrote, "the AARB has concluded that there is no point in suggesting fundamental changes to the draft Act, but wishes to recommend relatively innocuous additions that might either (and preferably) be slipped into the Act or implemented by purely administrative actions".[21]

Since Bernard Deschênes was a member of both the AARB and the Minister's Advisory Committee, he was, in effect, advising himself to slip innocuous additions into the act or, failing that, to get around it by administrative actions.

* * *

Saturday, July 23, 1983. Air Canada Flight 143, a Boeing 767 carrying sixty-five passengers and a crew of eight, took off from Ottawa for a scheduled flight to Winnipeg. Cruising at 41,000 feet at about the half-way point, the aircraft ran out of fuel. Both engines flamed out and the crew had little choice but to try to glide to an abandoned military airport in Gimli, Manitoba, some seventy-five miles away. The field had been converted to a drag-racing strip and drivers and their families had set up camp just beyond the end of the stripland. Fortunately the crew made a sensational deadstick landing and brought the aircraft to a halt on the runway. Aside from a collapsed nosewheel, there was no damage and no injuries — thanks to a combination of the crew's cool heads and skill, and incredible luck.

The legislation to create the Canadian Aviation Safety Board had been introduced the month before, but it was not yet in effect. The Aviation Safety Bureau of Transport Canada — acting as an autonomous unit under director Kenneth Johnson in anticipation of the new act—launched an investigation. Two weeks later, Lloyd Axworthy replaced Jean-Luc Pepin as Transport minister. Flight 143 had been headed for Axworthy's home riding of Winnipeg, and his constituents took a great interest in the incident. On October 13, Axworthy appointed Mr. Justice George H. Lockwood to conduct a public inquiry.

Ken Johnson testified at length about the ongoing investigation. "The investigation procedures are basically an interpretation and an application of the scientific method where we would go out and gather data," he told the hearing, "and the data-gathering involves the collection of the various records involved in the operation of the aircraft. . . . It includes measuring and verifying the conditions of the aircraft . . . interviews of the crews and any other witnesses . . . analysis of the cockpit voice recordings. . . . "

The investigators introduced a document titled "Joint testimony of the two Air Canada technicians who refuelled the B-767. . . . Testimony recorded on cassettes." The document had the appearance of a verbatim transcript of tape-recorded interviews, and everyone assumed that that was what is was. It reported a series of questions and answers between the Ministry of Transport investigator and the mechanics.

So the Air Canada lawyers were considerably surprised when the mechanics disowned the words attributed to them. It turned out that the document was not a transcript but, as one lawyer said, "a figment of an overheated imagination". The investigators called it a "summary" written in the form of an interview.

The *Winnipeg Free Press* suggested that the investigators could have concocted a more believable story. "The department could have said, for example, that its investigators were budding playwrights and the document was the script for a play, in which there was a clear hero (the ministry) and a clear villain (Air Canada's policies)."[22]

Judge Lockwood was not amused. He adjourned the hearings until an accurate transcript could be produced, but the verbatim transcript revealed a confusion of two investigators talking to two witnesses at the same time. "The tapes, themselves, were of poor quality. The recording equipment was less than adequate," Lockwood reported. "To make matters worse, they were jointly interviewed by two investigators. . . . Instead of one investigator asking questions and then allowing the witness to answer, both investigators asked questions from time to time and sometimes at the same time. There were times when both interviewers and both witnesses were talking at the same time with one of the investigators getting excited."[23]

In his final report Lockwood called the doctored tapes "a very serious lapse in Transport Canada's methods of investigation". By the time his report was published in April 1985, the Canadian Aviation Safety Board had taken over. "The Board should review procedures formerly followed by the Aviation Safety Bureau," he wrote, "and if it has not already done so, correct them. It should train its accident investigators in the technique of conducting interviews. Such training should also include instruction on comportment before tribunals."[24]

Lockwood didn't realize, of course, that there was no audience for his recommendations. Bernard Deschênes' concept of the "exclusive authority" of the director of investigations, "slipped into the Act or implemented by pure administrative actions", made sure of that.

Bureaucracy and Politics

The Canadian Aviation Safety Board Act passed into law when it received unanimous consent on third reading in the House of Commons on October 24, 1983.

Despite the unanimous consent, certain expectations had not been fulfilled. The Minister of Transport was made responsible for the new board, for example; Dubin had urged that some other minister be responsible to avoid a conflict of interest. The headquarters for aircraft accident investigation were to be moved from Ottawa, where they had always been and where the Transport department had just built a new aviation safety laboratory. The laboratory, which provided essential technical support to investigations, was located next to the aeronautical laboratories of the National Research Council and the Department of Transport at the Ottawa airport.

The site of the board's proposed new headquarters, Montreal, was, as it happened, also the home town of its newly appointed chairman, Bernard Deschênes. Deschênes' most notable qualifications for the job were, as the news media gleefully noted, his political connections. He was chairman of one of the Liberal Party's key committees and was, even then, working on a study that would eventually recommend an expansion of his new empire into the investigation of marine accidents.

As luck would have it, the go-ahead for the CASB coincided with what everyone in Canada knew to be the final countdown for another general election. Pierre Trudeau, who had returned from retirement to upset the Clark government, would not be seeking reelection. The Liberals chose John Turner as their new leader. Meanwhile the outgoing Prime Minister had a surprise for his successor: before leaving office on June 30, he intended to make hundreds of political appointments to reward favored courtiers. One of those so favored was former speechwriter and defeated Liberal politician William MacEachern. MacEachern's sudden appearance at the CASB surprised even its recently appointed chairman.[25]

Trudeau's intended appointments included so many sitting Liberal MPs that the new leader would be left without a majority in the House of Commons; Turner would inherit a minority government and lose control over the timing of the election call. So Turner agreed to make the remaining appointments himself. He was sworn in as prime minister on June 30, and when he called an election on July 7 he also announced nineteen patronage appointments to federal boards and agencies — including seventeen sitting Members of Parliament. "The appointments," he said, "except one, were in furtherance of a commitment to Mr. Trudeau, my predecessor, or had been discussed and agreed to at Cabinet meetings of the previous administration."

The one exception was Arthur Portelance, Member of Parliament for the "safe" Liberal riding of Montreal-Gamelin, which he had won by a plurality of about 26,000 in both 1979 and 1980. Turner appointed Portelance to the CASB to make way for a "star candidate" of his personal choice, Lise Thibault. Though Portelance had been an MP for some fifteen years, he had shown scant interest in aviation matters.

The aviation press denounced the appointment as "merely another patronage plum to be handed to the Liberal party faithful". Moving the board to Montreal was seen as "a political maneuver" with "no basis in logic or common sense".[26] And the derision and cynicism that greeted the news that Turner was continuing the pork-barrel politics he had sworn to eliminate sent the election campaign into a nose-dive and ultimately made it the worst defeat in Canadian history. Even Montreal-Gamelin turned to the Conservatives.

Don Mazankowski's prior stint as Transport minister had taught him something about dealing with the bureaucracy. He arrived back at his old department with a long list of directives for the deputy minister. The list—dated September 18—directed, for example, that "The name of the previous Minister of Transport is to be removed from all signs, documents, billboards, advertisement, publications, etc. by Friday, September 21st at the latest." To make sure the bureaucrats knew he wasn't kidding, Mazankowski added, "There will be no exceptions permitted with respect to the timing of this directive." He also instructed that the relocation of the CASB to Montreal be "put on hold". The implementation of the Safety Board Act went ahead on October 4, as scheduled. On November 5, Benoît Bouchard, a sort of junior minister to Mazankowski, announced that "The new Canadian Aviation Safety Board will have its head office in Hull, not in Montreal as proposed by the previous government."[27]

Over the next two years, a series of blunders, faux pas, scandals and forced ministerial resignations would teach Mazankowski and his colleagues that the cooperation of the civil service was essential to their survival. There would be no more reversals of the bureaucracy's plans for the CASB.

BOOK 2

CORPUS DELECTI

Part 1

Gander: Where Is the Airplane?

Detection is, or ought to be an exact science.
Sherlock Holmes[1]

Fans of detective fiction know that murder scenes are cordoned off, objects are photographed and dusted for fingerprints, investigators chalk victims' outlines on sidewalks before bodies are taken to the morgue. As countless whodunnits reiterate, a seemingly trivial or irrelevant observation often holds the key that unlocks the mystery. The great Sherlock Holmes invariably spotted vital clues that eluded Scotland Yard's competent but unimaginative Inspector Lestrade.

In real life as well as in fiction, tough cases can hinge on minute details — the torn end of a single strand or a latent print on an insignificant object — not just in criminal investigation, but in probing the causes of airplane crashes. But Sherlock Holmes will not be around to pick up clues that bypass crash investigators. They must collect all the evidence on their own and preserve it until the case is closed. As Bill Muncey of the CASB said of his work on the Gander investigation, "It's a detective job. The key is to gather more information than you'll ever need so you won't find out too late you've missed something important".[2] No irony was apparent at the time.

The irony of this remark became only too obvious after the investigators had submitted their report — a report that vaulted over a thicket of clues, striking circumstantial evidence and conspicuous leads. Some of these leads pointed to an in-flight explosion, but suggested any number of sources: contraband cargo, sabotage, dangerous souvenirs, or booby-trapped luggage. The range of possibilities was bewildering, but the clues were conspicuous. What mysterious myopia could have induced the investigators to march like lemmings on the dead-end trail to ice on the wings? Before we can venture an answer, we must look at the array of confounding and self-contradictory clues left by the Gander crash, much like the mystifying scenes that confronted Dr. Watson in Sherlock Holmes' most difficult cases. Like Watson, we will have only one certainty — that the official investigators followed the wrong track.

At the Crash Scene

The first indication of the devastation from the crash was a single broken treetop 2,975 feet beyond the far end and 720 feet to the right of Runway 22, the runway from which the Arrow Air DC-8 had attempted its doomed takeoff. Falling terrain let the crippled jet fly farther than it could have over level ground, but then the plunging aircraft's tail brushed the top of the slender 40-foot tree on the downslope at a point about 150 feet below the end of the runway. From there, the treetops were clipped at successively lower points. The scar through the woods was angled some 20° to the right of the DC-8's original heading.

The knife-like slice through the treetops soon became a disorderly trail of uprooted and broken trees increasingly overlaid with debris. The path of destruction bisected a service road and led to the final scene of carnage—later called "the pit."

Fire-fighters arrived some seven minutes after the crash. They saw no sign of life. Pieces of wreckage and human remains littered an oblong of charred ruination some quarter of a mile long. Some bodies were still strapped to the seats; others had "come out of their green Army fatigues and were lying in the snow"[3]. An upturned undercarriage bogey, a piece of the tail cone without fins and a twisted, burned-out segment of fuselage bearing a scorched stars and stripes were all that remained to suggest the origin of the field of shrapnel.

Constable Frank Ward of the RCMP arrived just after the fire trucks. He heard a "continuous chatter of minor explosions that appeared to be small arms ammunition burning." A series of larger explosions, some "strong enough to lift large mounds of rubble several feet into the air", completed the image of a battlefield.[4]

The devastation left an unforgettable impression on witnesses. To Greg Seaward, a Gander photographer who arrived at the site an hour after the crash, "the scene was very surreal, sort of like a bad movie." Seaward described bodies "frozen stiff like manikins" and he too heard what sounded like "bullets going off in the woods".[5]

One vivid impression stood out in the memory of early observers: the completeness of the devastation — the total fragmentation of wreckage, the random scatter of debris that seem to belie a sequential breakup. Harvey Day, one of the first fire-fighters, told me, "My first reaction was: Where is the airplane?" "There was strikingly little wreckage," reported correspondent Michael Rose in the December 23

Maclean's. "The plane had, in effect, disintegrated." Larry Hudson, the only television reporter to gain access to the site, broadcasted that the aircraft "had obviously disintegrated into small pieces. . . . It's almost what you'd expect if you threw a grenade and the pieces flew everywhere."

When an aircraft flying through cloud slams into a mountaintop, or when an out-of-control airplane dives into the ground at high speed, the fuel tanks disintegrate. When fuel driven from the ruptured tanks in a fine mist mixes with the surrounding air, it can detonate in a giant fireball. But aircraft move relatively slowly at takeoff; the impact of a crash may split the fuselage and cause a fire, but as a rule the hull structure retains its general shape. So the total disintegration of the fuselage puzzled many investigators who pored over the photographs and eye-witness reports. The slender trees at the crash site, which should have helped cushion the impact and dissipate the forward momentum, were scattered like matchsticks. The total destruction and the absence of survivors were not what is expected when a shortage of thrust or lift keeps an aircraft from climbing after takeoff.

Consider another DC-8 military charter that crashed and exploded while attempting to takeoff in freezing drizzle. The Capitol International Airways jet was taking off from Anchorage on November 27, 1970, bound for Cam Ranh Bay, South Vietnam, with a load of 219 servicemen and dependents and a crew of ten. The DC-8 failed to accelerate and crashed just off the airport. Forty-three of the occupants died; 186 survived and the aircraft did not disintegrate. Investigators from the National Transportation Safety board attrib-

uted the DC-8's lack of acceleration to a dragging hand
brake.

Even more significant comparisons can be made
with crashes clearly due to the wing's failure to generate sufficient lift. Take, for example, the first fatal
crash of a Boeing 747, on November 20, 1974. The
Lufthansa jetliner felt sluggish after lifting off from
Nairobi but the crew reacted immediately. The first
officer, who was at the controls, held the nose down
to gain airspeed. The captain raised the gear to reduce
drag. It was no use; the 747 sank back to the ground,
dragging its tail until it hit an elevated road some 3,000
feet past the runway. The impact broke the fuselage
and swung it around, and an intense fire broke out.
But unlike the field of shards at Gander, the burned
wreck of the Lufthansa 747 could be easily identified
as the remnants of an airplane. Fifty-nine of the 156
occupants died; many escaped unhurt.

The Lufthansa crash is particularly relevant because
investigators quickly identified the cause; the leading
edge slats of the 747 had failed to extend. This meant
that the angle of stall was drastically reduced and the
wing could no longer generate the required lift —
much as if it had been covered with a layer of ice.
Moreover, in lowering the nose and quickly raising
the landing gear the crew had reacted with the instinct
of experienced airmen. There was no reason to believe
that the crew of the Arrow Air DC-8 would have done
differently in similar circumstances. Captain John
Griffin was, after all, a DC-8 training captain and
check pilot.

Witnesses saw the Arrow Air DC-8 level off after
lift-off, as if the crew were attempting to hold the nose
down to gain speed, but — inexplicably — the landing
gear was found locked down. Experienced airline cap-

tains such as Ross Stevenson and Dave Mussallem suspected an in-flight emergency. If all engines were producing full thrust, the DC-8 should have been able to fly — even with the gear down and small amounts of ice on the wings.*

In the weeks and months following the crash, no official source drew attention to the pattern of debris from the Arrow Air DC-8. By contrast, just two days after an Avianca Airlines 727 crashed after takeoff from Bogotá, Colombia, on November 27, 1989, American investigators were telling newsmen that "wreckage is strewn over a two-mile area in a manner consistent with a bomb explosion".[6]

The CASB's submission "to assist the Indian Inquiry into the crash involving Air India Flight 182 on 23 June 1985" includes sections on "Wreckage Mapping and Surveying" and "Wreckage Distribution" as well as an appendix with a plot of the final resting-place of the pieces of debris; the "distinct distribution patterns" of these pieces under 6,000 feet of ocean were deemed noteworthy despite the 31,000 foot fall and the unknown action of winds, tides and currents. By contrast, the pattern of wreckage from the Arrow Air DC-8 was never analysed. All traces of havoc from the 160 tons of aircraft and contents were eradicated well before the investigators submitted their first draft report. The remaining trees were cleared away, and bulldozers covered the scars in the ground with tons of topsoil. Six months after the crash, visitors to the site saw only a sloping meadow with a small stone memorial.**

---

*Computer simulations later bore out this supposition.

**A more elaborate memorial was erected during the summer of 1990.

The hasty bulldozing of the crash site would eventually fuel suspicions of a less literal cover-up, suspicions that the causes and circumstances of the crash had been deliberately concealed. What would we think, after all, if Inspector Lestrade had the floor scrubbed, the furniture replaced and the victim embalmed before Sherlock Holmes had even been notified?

The American Team

The closest official U.S. military presence to Gander is at the naval base in Argentia, Newfoundland, about 180 miles to the south — the northernmost end of the top secret North Atlantic surveillance and command system.

At the first word of the Arrow Air crash, Captain Joseph Payne, commander of the Argentia base, assembled a group of forty volunteers and rushed to help. His team arrived around noon but were not allowed on the site; the Department of the Army in Washington had dispatched a more specialized team headed by Major-General John S. Crosby, Deputy Chief of Staff for Personnel.[7]

Crosby, a craggy-faced 53-year-old with a Master's degree in mechanical engineering, had spent his entire adult life in the U.S. Army. As fate or design would have it, he was a former commander of the 101st Airborne — the same division as the crash victims. A number of observers felt that the steely-eyed Crosby, who strode about the accident site in battle fatigues, seemed entirely too lean and mean to be a professional paper-pusher.

Be that as it may, Crosby and his 43-member team arrived at about three p.m., some eight hours after the crash. Crosby was given the base commander's office as temporary headquarters and, according to internal army correspondence, "assumed duties as on scene commander".[8] Publicly, General Crosby described his role more modestly. "We are here to help in any way we can," he told reporters—and with great diplomacy, he deferred to the Canadian investigators.

At about 1:30 in the morning of the 13th, another twenty Americans arrived in a U.S. Army C-141. On the 14th, the U.S. Army Chief of Staff, General John A. Wickham, Jr., joined General Crosby for a brief unannounced visit "to inspect the crash site and the temporary morgue".[9] Wickham, who had recently set up a classified Technology Management Office to provide oversight over all "black" programs in special operations and deception, left before long.[10] After his departure, General Crosby was able to describe his role a little more fully: "We have . . . logistical personnel from the 101st Airborne here to gather and secure military equipment carried aboard the aircraft."[11]

Crosby did, in fact, take possession of "the badly damaged personal weapons of the troops", which had been gathered from the site by the RCMP. It was later disclosed that a padlocked briefcase recovered from the site had also been turned over to General Crosby, who had cut it open and found three 45-caliber pistols.* Captain Tom Badcock, a volunteer from the Canadian Armed Forces, said that personnel on the site were asked to be on the lookout for a large quantity of cash and that a large amount was indeed recovered. It's not clear if the money was in the same briefcase,

---

*The Transcript (p.1478) says two pistols, but the tape clearly says three.

but its presence would become one of the minor mysteries that fuelled speculation about the role of the American troops in the Sinai.*

Although the Americans kept a low profile, rumors began to circulate about their intentions. One of these was recorded later by Michael Mendez, director of maintenance for Arrow Air, who had been sent to help the investigators in Gander. In a report dated February 24, 1986, Mendez described his frustration at having to cool his heels while American military officials toured the crash site with the investigators. Mendez claimed that one of these officials told him that "Major-General Crosby wanted to bulldoze over the crash site immediately.".[12] This seemingly bizarre request appears to have been corroborated by McDonnell-Douglas representative Frank Wiggers, who reported that selected parts of the wreckage were being transferred to the CASB laboratory in Ottawa, adding the note: "Remainder to be plowed under — Army request."[13]

Suspicions about what seemed like inordinate haste to obliterate the evidence leaked into the public domain during the CASB's public hearing in April 1986. The transcript and video record show a strained exchange between Allan Markham**, a Washington

---

*On Oct. 23, 1989, the U.S. Army's Deputy Chief of Public Affairs wrote ABC Television producer Victor Naufeld that the money in the briefcase "was the mid-month payroll for the soldiers in the plane." This was later denied by the Deputy Assistant Secretary, who claimed in a letter to Congressman Robin Tallon that the briefcase contained personal cheques and vouchers and only $469.79 in currency.

**Mr. Markham also represents former CIA "proprietary" Southern Air Transport, which was heavily involved in the arms deliveries to the Contras and to Iran in 1985 and 1986.

attorney representing Arrow Air, and chief investigator Peter Boag.

After a few preliminaries, Markham says ". . . it has come to my attention from several sources that early in the investigation a high-ranking U.S. military officer who was at the site in Gander was urging the very quick bulldozing over of the crash site within the first, as I understand it, two or three days. Can you comment on that for me please, sir?"

Boag replies assuredly, "No. I cannot. Not to my knowledge. That was not proposed."

"You are totally unaware of any such request on the part of anybody from the United States?" asks the incredulous Markham.

Boag leans forward and begins confidently: "I am totally unaware that was made as a request, yes. I think it may have been discussed at some point, certainly not in the terms you have mentioned — two or three days after the accident. It was discussed in the context of when the investigation was complete. So that probably. . . ."

At this point the chief investigator falters, seeming to lose his train of thought. He hesitates, shifts on his chair, leans forward, crosses his arms on the table and continues more slowly, ". . . not probably — in deference to the concern for the families and the next of kin of those killed. That people would not be rummaging through any debris or possible remains that might be left."

Markham persists: "Then, if I understand you, the investigation team did not have any pressure put upon it to conclude the on-site investigation and cover up the scene?"

Chairman Deschênes reacts instantly to the words "cover up". "Mr. Markham, I will not allow such a question. The fact as explained by the investigators is that they completed their investigation on the site and that no such bulldozing took place and the evidence was protected throughout."

Markham gets the message. "Very well, Mr. Chairman," he says, "I have no further questions."[14] Three years later he was still outraged. "It came to me that the army wanted the site bulldozed as soon as possible. I got it directly from people who heard the conversations. I have never, ever witnessed a more poorly handled, inept investigation than this one. It appalled me."

The bulldozing issue festered behind the scenes until the CASB released its report. When contacted by a reporter in early 1989, General Crosby denied any desire to destroy evidence. There might have been some talk about bulldozing, he allowed, "but certainly not in December."[15]

Notwithstanding this recollection, Crosby's contemporaneous notes show that the disposition of the crash site was very much on his mind at the time. His untimely concern about tidying up seems especially out of place in view of his apparent lack of curiosity about the cause of the crash. On December 20, 1986, for example, he wrote a "Memorandum for the Record" that discussed "arrangements concerning the closeout of the crash site" with Canadian officials "in Gander and Ottawa". These officials included the Commissioner of the RCMP and the Deputy Minister of Transport. Crosby was most anxious that "clearance and restoration work" begin as soon as possible and that "a representative from the Department of the Army is present" at all times.

Apparently the investigators didn't move fast enough; in another memorandum dated December

23, Crosby referred peevishly to a telephone conversation with Peter Boag, who "reiterated that the Safety board will not turn over the site for cleanup until they have all the information they need for their investigation."[16] The inference — if it needs to be spelled out — is that, within ten days of the crash, General Crosby was pressing for disposal of the wreckage, with no apparent concern for the technical investigation.

In fairness to Crosby it should be pointed out that nothing in the record indicates sinister motives in his wanting to dispose of discomfiting reminders that so many Americans had perished while under the care of the U.S. Army. To all outward appearances the investigators and the Canadian and American governments were blind to the circumstantial evidence suggesting that the Arrow Air DC-8 might have been sabotaged. Yet the evidence was more than circumstantial. Investigators were uncovering physical evidence suggesting that the crash had been the result of an in-flight explosion.

The Public Inquiry

The only readily accessible public record related to what went on at the crash site—indeed the only readily accessible public record of the whole investigation — is the 2,000-page transcript of the public inquiry held by the CASB in Hull between April 8 and 16, 1986.

Chairman Bernard Deschênes announced the decision to hold an inquiry on the day of the crash. "In the case of such an air disaster, a full public inquiry is necessary to serve the public interest in aviation safety," he said.[17] The board's press release explained that the inquiry "will be fact-finding in nature, aimed at determining the contributing factors and causes that led to the accident.'' Deschênes added that ''This inquiry represents an opportunity taken by the board itself to ensure that the investigation team is provided with the factual information that will be required to analyze the causes of the accident. It also ensures the facts available to the investigators are available to the public.''[18]

The board set the date of the inquiry on January 15, 1986. The minutes record that ''it was also proposed'' that board members Pultz, Stevenson, Thurston and Roger Lacroix ''be designated to assist chairman Deschênes who will be acting as Presiding Officer, as persons to conduct the inquiry.'' The convoluted, ambiguous wording was far from accidental.

Board Members recall that around this time minutes of meetings were produced several months after the event. The chairman would distribute outstanding minutes in bunches, a short time before a scheduled meeting, and they were accepted as routine house-keeping, with little attention to the wording. Office staff who helped type and duplicate the minutes say they went through numerous drafts according to the dictates of chairman Deschênes and Executive Director Ken Johnson. Neither the board's press releases nor its annual report even named those who conducted the inquiry with Deschênes. In hindsight, it's easy to see that the minutes were a carefully contrived plan of ''historical engineering'' to build a precedent for the chairman's absolute power.

The minutes record, for example, that on February 14, 1986, ''the board agreed to recommend to the Presiding Officer that the persons listed in Annex 1 be designated as participants.'' Those at the meeting recall that a list was produced and there was some vague discussion without formal vote or approval. The implication that the board had no executive authority, that it could only "recommend" to the chairman, seemed trivial—"Bernard putting on airs," one of them thought.

* * *

While planning for the public inquiry, Deschênes shared a suite of offices with Ken Johnson, who at this time styled himself "Associate Deputy Head" in some internal correspondence. Ordinary board members were relegated to offices in a different building and seldom saw Deschênes outside of scheduled meetings —except for member Frank Thurston.

A wily old bureaucrat, seventy-one years old at the time, Frank Thurston had a special entrée. Thurston, who had retired six years earlier after a long career as director of the Aeronautical Laboratories of the National Research Council of Canada, had kept busy dishing out backroom advice on various advisory committees and boards, including the CASB's predecessor, the AARB. When Deschênes was named chairman of the newly formed CASB in 1984, he lobbied strenuously to get his fellow AARB member appointed to his board. Thurston's long government career and credentials as an aeronautical expert would lend the Montreal lawyer credibility in his role as head of a technically oriented agency. Perhaps more important, Thurston's evident qualifications could counteract the ridicule heaped upon the patronage appointments of Bill MacEachern and Arthur Portelance. Deschênes' efforts succeeded and Frank Thurston was appointed as a part-time member in June 1984 —for a one-year term.

Frank Thurston, an erstwhile dabbler in amateur theatre, had perfected the role of elder statesman-scientist. Tall, thin and ascetic, he could affect an air of great erudition. Regardless of the topic at hand, he could always come up with a seemingly profound question to show his understanding. His expertise as a bureaucrat shone brightly whenever there was a disputed fact or conclusion. Where others might ask for more data or analysis, Frank could always avoid delay by coming up with a convoluted but at first blush plausible wording that would cover all possible interpretations.

Ordinary board members didn't see much of Frank Thurston between formal meetings, but from time to time he would be seen visiting Deschênes or Johnson in their executive suite. At the board meetings, he served as *de facto* deputy chairman. Deschênes and Thurston, heavy smokers, usually retired into an adjacent room during coffee breaks to puff away and discuss strategy.

All the same, at the time of the Gander crash and the public inquiry Frank Thurston was an interloper. His one-year appointment had lapsed in June of 1985 and frantic behind-the-scenes lobbying by Deschênes and Johnson had thus far failed to secure his reappointment. (He was finally reappointed for a three-year term in June 1986.) In the meantime, however, he continued to attend board meetings and take part in decisions.

Roger Lacroix recalls being aware that Thurston's appointment had lapsed. He can't remember how he got the impression, but somehow he believed that Thurston had been reappointed soon thereafter, the appointment being backdated to provide continuity. Norm Bobbitt, who was only appointed to the board in October 1985, recalls that Thurston was introduced as a board member and that he had no reason to question his credentials until the web of intrigue and impropriety overtook the CASB's operations.

And so this *eminence grise* sat at the right hand of the Chairman throughout the public inquiry, with no one else aware that he was not a member of the board. Keeping board members' names off official documentation avoided an overt record of the deception. Thurston eventually sent a bill to Executive Director Ken Johnson requesting $275 a day for services rendered during the public inquiry, and authorization for payment was justified on the basis that Thurston had provided advice and guidance, presumably to Johnson.[19]

We discovered Thurston's deception much later, after he had taken a lead role in defending the ice-on-the-wing theory proposed by Ken Johnson's staff. Until then, I could never understand how Thurston could fail to see the scientific and logical fallacies in the theory. "Why is he doing this?" I kept asking myself. "He's got to know better." But when I belatedly discovered that he had not even been a board member during the public inquiry, I could see how the honoured scientist might have been trapped into promoting a scientific absurdity. Deschênes and Johnson had had full confidence that they could manipulate the process of political appointment. They had led Frank to believe he would be reappointed retroactively. It didn't work out and it was too late to back out gracefully. The schemers had to sink or swim together.

Whatever role Frank Thurston may have had in advising the chairman, other board members had no say in the preparations for the public inquiry. According to the board's records, board members were "briefed on the arrangements" and a list of potential witnesses was distributed on March 18, 1986. Stevenson and Bobbitt both recall that there was neither consultation nor discussion.

By the time the witness list became an issue, Deschênes had resigned in mid-crisis over his stewardship of the board. His successor, Ken Thorneycroft, would eventually explain, "We had taken evidence from 300 witnesses, 200 to 300 witnesses. Obviously we can't call them all to appear at the public hearing. So let's have a round table discussion, decide what evidence we want to have come out at the public hearing, and then select a group of witnesses who will provide that evidence."[20]

* * *

However the witnesses may have been chosen for the public inquiry, investigator-in-charge Peter Boag became lead witness and overall star. Boag played a double role of witness and interrogator. As the CASB's only witness, he summarized the investigators' findings to date and was the sole source of wisdom about the technical investigation. After this initial appearance, Boag took the lead in examining other witnesses, occasionally relinquishing this role to juniors in the style of the veteran trial attorney.*

Boag's testimony was self-confident, well rehearsed —no hems, no haws, no clearings of the throat. Reading from a three-ring binder, with his arms crossed on

---

*Lawyers protested Boag's dual participation as witness and counsel. For example, Alfonsus E. Faour, Lawyer for Newfoundland's Justice Department, later wrote to Deschênes labelling Boag's role "inappropriate".

the table in front, the chief investigator summarized the history of the flight, calling attention to the fact that "the aircraft was not de-iced, nor was this requested by the flight crew".[21] He noted that some parts were shipped to the laboratory in Ottawa for examination and that the flight recorder data was not good.

His summary didn't, however, mention the gigantic explosion that had signalled the crash for miles around. In fact, he didn't even mention the word "explosion"; he merely noted that "a severe post-impact fuel-fed fire ensued following the impact"[22] and that "the aircraft was destroyed by impact forces and subsequent ground fire".[23] "All wreckage was examined in the field for signs of pre-impact distress, failure or fire," he reported, adding that "to date, investigators have found no evidence of any pre-impact fire, explosion or failure to the aircraft structure".[24] When Boag had finished his summary, the chairman allowed representatives of the interested parties to pose questions — reminding them that "all examination is in the nature of examination in chief and no cross-examination is allowed."[25]

Boag's self-assurance persisted. When lawyer Stuart Goldstein, who represented the next of kin of the flight crew, began by asking "how the board determined that the master fire warning light was not illuminated", Boag took the opportunity to let everyone know who was in charge: "First of all, I would emphasize that it was not the board that determined that. It was the *investigators of the board* that determined that." Having dispensed of that misconception, the chief investigator could modestly admit that "I am not an expert on light-bulb analysis," before giving the unenlightened a little lecture on the art.

Eventually Goldstein got onto the subject of the wreckage. "What exactly is the basis for the conclusion, if it is a conclusion at this point, that there was no pre-impact fire or explosion?"

"We have, and our investigators have, conducted an extensive examination of the wreckage that was left," said Peter Boag. "Considerable amounts of that wreckage sustained extensive fire damage and limited useful information could be determined. However, from that wreckage that we did examine, and were able to examine, we have found no evidence that could be consistent with an airborne or pre-impact failure. All of the damage that we have looked at has been considered as the result of impact and the post-impact fire."[26]

Goldstein wanted to make sure he had it clear. "As I understand it, you are not excluding, then, pre-impact explosion or fire?"

Boag was patient. "I'm not necessarily excluding that. But I am saying that we have, at this point, found no evidence that would be consistent with that."

But the lawyer still didn't seem to get the point. "That would not be unusual in view of the condition of the aircraft, the way you found it, would it?

The chief investigator's patience was getting noticeably thinner. "I would have difficulty putting a 'usual' or 'unusual' on anything in that case," he said. "In accident investigation, and particularly examination of the wreckage, it is difficult to use those types of terms, Mr. Goldstein."

Goldstein persisted, "Then is it fair to say that you are not certain that there was no pre-impact fire or explosion?"

Boag could no longer contain his sarcasm. "I was not there in the aircraft at the time," he said. "So you are right, we cannot be *certain* of what happened." With the tone of a long-suffering kindergarten teacher, he tried again. "Certainly, we can reach reasonable conclusions based on the evidence we have examined. At this point we have seen no evidence that we would think would be consistent with a pre-impact failure, fire or explosion."

The lawyer gave up and moved to another subject. Boag easily handled the rest of the questions, and that was the last word at the public inquiry about the CASB's examination of the wreckage. But the evidence was considerably more ambiguous, and the opinion of the CASB's investigators considerably more divided, than anyone could have divined from the chief investigator's testimony.

The Explosives Experts

John Garstang, an engineer at the CASB's laboratory, was the only member of the official investigation team with specialized training in forensic science. In the months preceding the Arrow Air crash, Garstang's work on the Air India investigation immersed him in the arcane arts of detecting traces of explosives and helped him form close professional relationships with the world's foremost specialists. Later, he would be invited to join American explosives experts Walter Korsgaard and Tom Thurman to search for traces of an explosion in the wreckage of Pan Am Flight 103 at Lockerbie, Scotland.

When describing his work at Lockerbie, Garstang speaks with animation and self-assurance. He delights in expounding on the difficulty of detecting "the very subtle differences between soot from explosion and post-crash fire." "Fracture damage typical of explosives," he explained to us, "is not much different than break-up damage and can only be differentiated through minute examination. Detecting evidence of an aircraft bombing is very difficult even when you know what you're looking for." He told me later that, as he saw it, finding the suitcase fragment that conclusively established that a bomb had exploded on board Pan Am Flight 103 had been "incredible luck".

But when pressed about the Gander accident, Garstang turned cautious and reticent. "I didn't look at Gander in the same way," he explained. "My job was site survey and security."

Who, I wondered, looked for "the very subtle differences between soot from explosion and post-crash fire"? "You would have to refer to the RCMP for that," he said curtly.

So, in spite of his special expertise, John Garstang's assigned role in the Gander investigation was limited to an inventory of the wreckage. He arrived at the crash site around noon on the day of the accident. While others were recovering the human remains, Garstang and his team started to catalogue the twisted and charred bits and pieces that remained of the DC-8. Their site survey was one of the few proficient documents to back the investigators' final report. It lists Item Number 39.1 as "Composite material from cargo liner".[27]

But Garstang's original description of this seemingly insignificant item was more detailed. He marked the object with Stake Number 29 and described it in his personal notebook as a "6 foot by 3 foot piece of composite material cargo liner exhibiting soot and burn marks and small punctures." He carefully noted that the piece was "found in area where ground not burned" at the extreme right side of the crash trail.

In an interview in 1990 Garstang re-emphasized his task as that of cataloguing the wreckage with no responsibility for interpretation. He recalled that he had felt that the location of the charred piece in an unburned area was of potential significance to finding the cause of the crash.

He also allowed that the puncture marks could have been signs of an onboard explosion. Accordingly, the report he turned in to Peter Boag included all the details and observations recorded in his personal notebook. Garstang would not venture an opinion as to why his description was abbreviated in the public version.

In any event, once Garstang had drawn attention to the suspicious piece, it was sent to the RCMP forensic laboratory. A one-page report describes the work done:

> Ex[hibit] 1,000,683 was examined visually. It was crumpled out of shape and had several holes punctured in it. The fiberglass was charred in the areas around the puncture holes. The areas surrounding the puncture holes were examined microscopically for foreign material, with negative results. Spot tests for explosive residue were applied to areas surrounding the holes, with negative results.[28]

None of the documentation I was able to examine explained why this item was sent to the lab or how the burn and puncture marks might have been produced, or even mentioned that the burned piece came to rest in an area remote from other fire damage.

Garstang said that the RCMP's tests would have detected residue from conventional high explosives—if these had not been washed off by snow and rain. He didn't need to add that the RCMP tests would not have detected residue from sophisticated chemical explosives such as the Czechoslovakian-manufactured Semtex responsible for the destruction of Pan Am Flight 103. "To this day, I do not have a good feeling about what caused those holes," he said.

While sabotage of the Arrow Air DC-8 would have created sensational headlines, it would also have had a big influence on insurance claims. Coverage for the DC-8 was provided by a consortium of companies, Associated Aviation Underwriters (AAU). As usual in commercial operations, the airline was covered for hazards associated with mechanical and human failures; civil liability insurance doesn't cover deliberate sabotage. It was evident to the AAU's attorney, Mark Dombroff, that if there was any substance to rumors of evidence of an onboard explosion, his clients would be able to save some of the $200 million eventually paid out in settlements. Dombroff advised AAU to consult an independent expert.

I. Irving Pinkel had retired as director of NASA's Aerospace Safety Research and Data Institute in 1972. Since then, he had been working as an independent investigator and expert witness on aviation and industrial accidents. Pinkel's testimony as an expert witness on crash and fire had greatly impressed Dombroff during the civil litigation arising out of the crash of a Pan American 707 in Pago Pago in 1974. Pinkel was just the man to help with Arrow Air.

In 1967, while still at NASA, Pinkel had served on the team that investigated the launch-pad fire that killed America's prime astronaut team and threatened to derail President Kennedy's plan to land an American on the moon before the end of the decade. Pinkel's report on the origin and development of the fire was one of the ingredients that, in the words of back-up astronaut Walter Cunningham, helped ''to take the inadequate junk represented by the blackened pyre on Pad 34 and transform it into the brilliant machine that carried us to the moon.''[29] Pinkel had also participated in the investigation of the second serious accident to mar the Apollo program, the explosion aboard Apollo 13.

As associate chief of the Physics Division at NASA's Lewis Research Laboratories, Irving Pinkel had initiated a massive research program to study how fires break out after takeoff accidents. In addition to his experience with fire and explosion, his work at NASA had included research on aircraft icing and the drafting standards for the design of aircraft icing protection systems. In short, Pinkel was uniquely qualified to assess whether fire or ice was the most probable cause of the Gander accident.

By the time Pinkel arrived in Gander in May 1986, the wreckage had been cleaned from the accident site, so he was conducted to the piles of twisted, burned metal in Hangar No. 22 at Gander airport. As it turned out, he did not get to see several key items. Nevertheless, some of what he did see aroused his curiosity. His attention was particularly drawn to an oblong hole about a foot in diameter in a piece of fuselage apparently ripped from the forward right-hand side. The edges of the hole had a suspicious outward curl. Seeing that hole ''almost knocked me off a pile of rubble,'' Pinkel recalled. ''How could they have missed it?''[30]

''The missing fuselage piece that occupied the hole,'' Pinkel wrote in his report to the insurer, ''appears to have been punched out by an object that struck at high speed. The fuselage section showed no further damage, but a partial window frame just above this hole was distorted outward, as if from an internal blast." The hole was located just above the line where the fuselage skin had been riveted to the cabin floor. As Pinkel had perceptively observed, the skin had retained its original contour. You can crumple a sheet of paper with your fist, but you can't punch a hole through it unless the edges are supported. Through similar reasoning, Pinkel deduced that an explosion must have driven an object through the airplane's side while the skin was still held taut by the cabin floor—that is, *before* the fuselage was ripped apart in the crash.

In theory the hole could have been produced by a secondary explosion during the moments while the airplane was breaking apart but before the fuselage skin was ripped away, but no one could imagine how this might have occurred. A more plausible hypothesis was that of an explosion before the crash.

Pinkel searched diligently for additional clues. He believed he had found some when he examined the remains of the inlet guide vanes, or fixed struts, from the mouth of the inboard right engine. Three consecutive vanes showed "a slight flattening on the leading edge." Could these slight dents be signs of an object cast into the engine while the vanes were still in place? "Examination of these vanes at moderate magnification," he wrote, "showed that the middle one had a faint marking of red orange color on the leading edge." It looked as if some reddish paint might have rubbed off whatever had struck the vane.

The mouth of the inboard right engine seemed to be lined up with the suspicious hole in the right side of the fuselage. An object driven sideways through the hole by an explosion could have entered that engine and dented the inlet guide vanes. Forensic examination of the tiny paint smear might have given a clue to its origin.

But Irving Pinkel's findings didn't impress the CASB's investigators. They did ask the RCMP to test the area surrounding the suspicious hole for evidence of explosive residue, but the RCMP laconically reported that "No explosive residue was detected."[31] This would not have surprised Pinkel, who had written that "In view of the extensive weathering of the airplane debris, it is likely that nitrate detection for explosives may prove fruitless." But he did make a suggestion about where to look for additional evidence. "Also, the absence of bomb fragment impingement . . . may be due to the capture of these fragments by the panel insulation and other surfaces that lie between the fuselage skin and the explosive."

In drawing attention to the panel insulation, Pinkel was underlining the potential for finding clues in bits of upholstery and other soft components—a potential enshrined in a textbook example of explosives detection some twenty years earlier. An in-flight explosion of a BEA Comet over the Mediterranean Ocean on October 12, 1967, would have remained a mystery but for a seat cushion. British accident investigator Eric Newton detected a number of tiny tunnel-like holes in the material, and fine wires passed through these holes came to a common point that would have been at floor level under the seat. From this insight, Newton established that the holes were the tracks of fragments from a bomb that had exploded under the seat. (It was speculated, but never proved, that the bomb had been planted in an assassination attempt on Archbishop Makarios of Cyprus.)

Since the Comet investigation, explosives experts pay careful attention to non-metallic materials. To help them, they have catalogued signatures characteristic of high explosives through controlled experiments. An explosion may be indicated, for example, by highly shredded and teased tears in woven fabrics. Single strands of nylon-type materials may show explosive heat fracture in the form of globules without discoloration or gas bubbles. As the International Civil Aviation Organization's (ICAO) Manual of Aircraft Accident Investigation says; "these features may not readily indicate the characteristic evidence of an explosive event, but when viewed at high magnification, much valuable information may ensue."[32] Expert examination can also distinguish subtle differences remaining after the detonation of high explosives from those left after the explosion of aviation fuels. High explosives can, for example, accelerate tiny particles to supersonic velocities. Larger particles are sometimes embedded in metal components. The smallest can penetrate deeply into soft material such as seat cushions or human tissue.

In the words of the ICAO manual, "Evidence as to the probable cause will be anything but clear during the early stages of the investigation . . . . Examine everything. Document everything."

Photographs of the Gander accident show that some seat cushions and other soft materials survived the fire, but there is no indication that anyone paid any attention. The investigators were not only devoid of initiative to search for evidence of an explosion, they ignored specific leads. The sum total of their recorded thoughts about Pinkel's hypotheses is contained in two short memoranda from the engineering laboratory to Peter Boag. The first concedes that the hole in the fuselage "appears to have been formed by an object being forced through from the inside to the outside''.[33] No thoughts are offered on what this object might have been, but, on the basis of the arguable assumption that ''there were no witness observations of any explosions during take-off'' the memo concludes that the fuselage section ''received the observed damage from impact.'' Pinkel's deduction about the stiffening from the floor is ignored.

The second short memo brushes off Pinkel's hypothesis about the inlet guide vanes.[34] There seems to have been a slight mixup in the lab—a bookkeeping error discovered after Pinkel had left. The dented guide vanes really came from an *outboard* engine. A part blown from the aircraft could not have reached that far. ''Nevertheless,'' the memo says, ''an examination as to the possibility of pre-impact damage was carried out with negative results.''

The red paint? ''It was noted that this piece of wreckage had been subject to intense heat, and any coloring material would most likely have been burnt off,'' the memo states. ''Therefore, it is reasonable to assume that the color deposits occurred after the crash rather than before or during the accident.'' It's possible, the memo suggests, that there was ''paint transfer from the equipment'' moving the wreckage. But there was no attempt to determine if *possible* was also *probable*—nor was there an attempt to account for the flattening of *successive* blades, which suggests that something entered the inlet before the crash.

The investigators' nonchalance in running down clues was hardly what one would expect in the investigation of a major air disaster. To put it in context, consider the response of Indian investigators to a suspicious gouge on a piece of fuselage recovered from the wreck of the Air India 747. It was possible that the gouge indicated in-flight damage but it was also possible that it had been produced by the bridle cable used to winch the piece from the ocean floor.

In this case, the investigators proved the skeptics wrong. They reported that "a sample bridle cable was obtained from the [recovery ship] and gouge marks were produced by pressing this cable to an aluminum sheet. The gouge marks thus produced . . . appear to be different from those observed" on the piece in question.[35] In short, the investigators determined that while it was *possible* the cable had produced the gouge, it was not *probable*.

When I got to look at Irving Pinkel's report, some two years after the CASB's investigators had rejected it, I was at first puzzled and then enraged by the off-hand, almost mocking rejection of an apparently well reasoned hypothesis. Irving Pinkel's vast experience and professional background stand in marked contrast to those of the men who so cavalierly dismissed his observations. Neither the director of Investigations, Tom Hinton, nor the chief investigator, Peter Boag, had ever investigated a major airliner crash. John Garstang, the only CASB employee with training in forensic investigations, had been banished to the sidelines. By default, the responsibility for investigating the possibility of an in-flight explosion had fallen to the RCMP.

The RCMP Investigation

. . . it is impossible for me to say whether the present case is an instance of crime or not, but the course of events is certainly among the most singular that I have ever listened to.

Sherlock Holmes

When air traffic controller Glen Blandford cleared Arrow Air "Big A 950" for takeoff on the morning of

December 12, 1985, "It was sort of a heavy-type morning . . . nothing appreciable had fallen around the tower area that I could see—and nothing on the windows of the tower cab".[36] As soon as Blandford saw the Arrow Air jet lift off Runway 22, he diverted his attention to an inbound flight. Out of the corner of his eye he saw the DC-8 level off before disappearing from view. "At that point," Blandford said, "I saw a glow, which proceeded along the lakeshore . . . and then a mushrooming fireball".[37] The incoming pilot saw it too; "Tower, it looks like we have some sort of explosion off to the west," he reported.

Glen Blandford picked up a special hotline telephone and triggered a well-rehearsed emergency plan —his call simultaneously alerting the fire hall and the local office of the Royal Canadian Mounted Police.

As Canada's national police force, the RCMP operates central forensic laboratories and is responsible for counter-terrorism and domestic intelligence — functions analogous to those of the FBI in the United States. But unlike the FBI, the RCMP also polices Canada's international airports and provides general police services in most provinces, including Newfoundland. Thus the RCMP had several levels of responsibility.

The RCMP's immediate reaction to Blandford's call followed established procedures. The airport duty officer relayed word of the crash to Constable Frank Ward. A few minutes earlier Ward had been parked near the ramp, watching the troops board. Everything had seemed routine. There had been "very light snow that didn't even warrant turning on the windshield wipers." When the doors were closed up and the DC-8 was ready to taxi, Ward drove off on patrol.

Before Constable Ward had time to react to the news of the crash, the radio call was interrupted by the voice of his superior officer. Staff-Sergeant Vincent O'Donnell had seen the fireball from his home near the airport. Ward sped to O'Donnell's home and took him to the airport before continuing to the crash site. He reached the site at 7:05 a.m., shortly after the first fire trucks. The fierce blaze precluded an immediate search for survivors. Ward began to take names of witnesses and to direct traffic to clear the way for emergency vehicles.

Meanwhile, O'Donnell carried on from the airport office. Helicopters played a key part in the emergency plan — searching for survivors, evacuating the wounded and ferrying rescue workers to the crash site — two of the many calls summoned RCMP helicopter pilots Don Turner and Gerald Greening.

Turner arrived at the airport at around 7:20, some thirty-five minutes after the crash. A drizzle was falling, but the "drizzle was not freezing on the helicopter bubble while parked on the tarmac".[38] He took off, flew to the accident site and began to search for survivors. After circling the site several times, he noticed light ice forming on the helicopter. After a few more minutes he returned to the hangar for de-icing. He made several more flights during the day, having to return for de-icing from time to time.

Gerald Greening didn't reach the airport until 8:15. By this time, conditions had deteriorated so that he could fly for only ten minutes before having to return to de-ice. On the previous day, freezing rain had forced him to cancel a scheduled flight; "There was icing the day prior to the accident and after the accident," he recalled.[39]

The investigators who arrived later were not immediately aware that no freezing rain had fallen during the hour and fifteen minutes that the Arrow Air DC-8 had spent on the ground. A detailed account of the precipitation would emerge later from the testimony of the official weather observer and witnesses such as Don Turner and Gerald Greening, but by then the investigators were already committed to the ice-on-the-wing theory.

In the meantime, the RCMP took charge of the emergency. Its role was outlined at the CASB's public hearing by Staff-Sergeant William Ronald Fraser, head of the RCMP's fifteen-member detachment at Gander airport. Fraser had spent his entire twenty-year career with the RCMP doing general police duties in Newfoundland, and he was put in charge of the RCMP's investigation.*

First and foremost, the RCMP carried out the normal police duties needed in the wake of any disaster. Officers cordoned off the area, established control over traffic routes to the site, notified hospital and other emergency services and carried out a myriad of related tasks laid out in a detailed checklist. They began to organize the search for survivors, stringing a rope grid so the location of bodies and pieces of wreckage could be recorded. They also began to interview witnesses. Verbatim transcripts of these interviews would prove essential in correcting misinterpretations that

---

*The RCMP won't allow interviews, nor will it release document relating to the investigation. Section 16(3) of the Access to Information Act exempts from disclosure ''information obtained by the RCMP while performing policing services for a province or municipality. . . . '' The details of RCMP activities have to be obtained from Sergeant Fraser's testimony and written report.

emerged later from the short interview summaries kept by the CASB investigators.

As well, the RCMP anticipated the specialized needs of the aircraft accident investigation. Sergeant Fraser explained that after the immediate emergency response, "Our continued investigation into the site itself involved photographing and examination and assistance to the aviation investigators in the examination of the wreckage and the transportation of the items for forensic studies."[40] The cockpit voice recorder was recovered within an hour and the flight data recorder by mid-afternoon; both were rushed to Ottawa and handed over to Bernard Caiger, the National Research Council's chief analyst, by nine o'clock that evening.

As with any accident causing death, the police were responsible for recovering the victims' remains. To deal with the horrible carnage, personnel from an identification unit from Grand Falls and from the mortuary division of the U.S. Army augmented the RCMP's Gander detachment. In total, twenty RCMP officers and about forty-three U.S. Army personnel were involved in searching the site. Such police work was routine in scope, though unprecedented in scale.

The RCMP was additionally responsible for probing the possibility of a crime. Sergeant Fraser explained that, in Newfoundland, the RCMP is responsible for investigating all deaths not due to natural causes. "Also, in relation to this incident, it was of primary importance to determine if, in fact, we could support any evidence of a bomb of any sort which may have contributed to the crash of this aircraft."[41]

Traces of high explosives in the wreckage of commercial aircraft are usually construed as evidence of sabotage. But the Arrow Air DC-8 was carrying soldiers and their military gear; high explosives could have detonated as the unfortunate result of military ordinance placed, wittingly or unwittingly, onboard. As it happened, neither the RCMP nor the CASB turned up evidence of high explosives. They were thereby spared the embarrassment of having to resolve whether the crash involved a criminal act to be investigated by the RCMP or a misadventure under the investigational jurisdiction of the CASB.

Since the search for signs of high explosives called for outside expertise, the RCMP's Gander unit was assisted by "some explosives people who were at the site" and "a scientist from the chemistry department at the forensic laboratory in Ottawa." The scientist was later named as Dr. B. W. Richardson, chief chemical scientist in the RCMP's Central Forensic Laboratory.

Dr. Richardson is a laboratory chemist trained to test suspicious items that field investigators have picked from the rubble, but the RCMP had no one actually on the site who claimed to be qualified to detect such items. In fact, Sergeant Fraser's testimony carefully distanced the RCMP from this aspect of the investigation: "Well, the investigators from the Aviation Safety Board had done a complete examination of all the wreckage on the site," he said, adding that the RCMP "had assisted in doing the photographic work and examining the wreckage."[42]

The "explosives people" who conducted this complete examination have never been identified. John Garstang, the CASB's acknowledged forensic expert, explicitly denied that he played a part. Peter Boag would later defer all questions concerning the search for evidence of an explosion to the RCMP. The RCMP continues to counter it only assisted the CASB.

According to Fraser's testimony, the "complete examination" of the wreckage, however it was conducted, turned up only two items of interest. One of these was the piece of cargo liner that John Garstang found near Stake 29. When questioned directly by lawyer Allan Markham, Fraser said he didn't know what had caused the punctures or whether the piece had been found inside or outside the area of the fire.[43]

The second item singled out for special examination was another piece of fiberglass panel similar to the first piece, but with a different distribution of punctures. It's not recorded who found this item or what aroused suspicions about it, but it too was sent to the lab for examination and judged innocuous.[44]

Sergeant Fraser did explain at some length that, aside from three non-explosive mortar shells used in training, no live or expended ammunition was found on the site. The RCMP's search for explosives was completed by two members of the Gander Airport Police Dog Service Section. Constable Russ Mirosty and his companion "did a complete search of all the aircraft parts . . . at Hangar 21 to determine if there was any evidence of explosives that P.D. [police dog] STRAPS might detect."[45] "Throughout that search," Sergeant Fraser said, "they had not detected or showed any indication of any explosive substance within the wreckage".[46]

No Canadian investigators noted the section of fuselage with the hole that Irving Pinkel described as important direct evidence. With most of the wreckage discarded, no one can give a reassuring response to

Pinkel's rhetorical question, "How much other important evidence did they overlook?"

* * *

Had there been an in-flight explosion, it would be reasonable to hope that a diligent search of the wreckage would turn up positive indications. Of course, the absence of direct evidence neither rules out the possibility nor precludes determining that an explosion did occur.

The mid-air explosion of a Korean Air 707 off the coast of Burma in November 1987 left no accessible wreckage at all, but Korean investigators "analyzed the overall situation and concluded that there was a great possibility the plane had been blown up in the air by terrorists."[47] In the absence of direct evidence, conventional police methods suggest a search for motive, means and opportunity for foul play. Using such methods, Korean investigators identified and arrested a pair of suspects. A confession from 26-year-old Kim Hyon-hui provided full details of the plot to sabotage the flight.

While he avoided putting it in such terms, Peter Boag prompted Sergeant Fraser to comment on the search for indirect evidence of sabotage. "Now that we have discussed the on-scene aspects of the investigation," Boag said, "did other investigative activities take place with respect to the RCMP that you are aware of?" When Fraser, who seemed baffled by the question, hesitated Boag added, "Did you receive assistance from RCMP headquarters in Ottawa?"

Up to this time Sergeant Fraser had testified in an earnest monotone, arms crossed impassively on the table in front. His measured demeanor, along with his grey jacket, striped blue tie and metal-rimmed glasses, gave him the appearance of an owlish accountant rather than a police officer. When the question turned from regular RCMP duties to the cooperation with headquarters, his tone became pained and his syntax even more fractured.

"Well, the following day of the air incident," he answered slowly, "a report had been published, or made aware of a report, in which the terrorist group in Beirut, the Islamic Jihad, had claimed responsibility for planting a bomb on this flight and that they had claimed responsibility for this disaster."[48] He seemed unaware that CASB staffer John Issenman had taken a call from Reuters on the day of the accident.

"From this outset," Fraser went on, "we contacted our national headquarters in Ottawa and asked them if they could follow up this release or determine any further information relevant to it. As a result, they had made contact with foreign sources and determined or advised that they were unable to get any support, determine any evidence or proof to verify whether or not this here particular Islamic Jihad had any involvement in this incident. There was no evidence to support it from the inquiries that they were able to make."

Although many people went away with the impression that Fraser had said the RCMP or its "foreign sources" had determined that the calls about the Islamic Jihad had been a hoax, he had said nothing of the kind. All he said was that they were unable to tell, one way or the other. When asked for details about its investigation of the claim by the Islamic Jihad, the RCMP will not clarify Fraser's murky statement.

Boag also prompted Sergeant Fraser to specify the foreign sources that had assisted the RCMP. "Earlier in your testimony you mentioned that you worked with other investigative agencies or another investigative agency. What agency was that?"

"The FBI has provided assistance to us in this investigation. On the outset, the FBI had arrived at Gander and had some very brief discussions in relation to their arrival; but other than that, there was no investigation and contact with that agency, except that we had made a request of them to conduct interviews with the previous crew who had handled this flight."

"Has that agency or any other law enforcement agency provided you with any information or evidence that would indicate the loss of this aircraft was the result of the detonation of any kind of explosive device?"

"No there has not been."[49]

The FBI Investigation

The wave of terrorist attacks against Americans in 1985 spurred the United States Congress to introduce the so-called "long arm" law that empowers the FBI to investigate suspected terrorist acts outside the U.S. But the law was not enacted until August 1986.[50] At the time of the Gander crash the FBI could have launched an investigation under another statute if it suspected that "the crash was caused by an explosive or destructive device, or other criminal act."[51] It did not.

The FBI did provide what it termed "routine assistance" to the RCMP by interviewing the crew that had flown the Arrow Air DC-8 on the initial (Cairo to Cologne) leg of the fatal flight. It was convenient to conduct the interviews in the United States, where the RCMP had no jurisdiction.

While testifying at the CASB's public inquiry, Sergeant William Fraser de-emphasized his part in the

FBI interviews. Asked if he was aware of the FBI's line of questioning, he replied, "Yes sir. I do have copies of those statements from the crew which they had obtained."[52] But, later, in response to a direct question, Fraser admitted that he had been present during these interviews.

The statements obtained by the FBI were not placed in evidence at the CASB's public hearing. Summaries were eventually obtained through the American Freedom of Information Act. The summary of Captain Art Schoppaul's statement makes strange reading in light of Fraser's denial of having received "information or evidence that would indicate the loss of this aircraft was the result of the detonation of any kind of explosive device."

"Captain Schoppaul theorized that the crash of N950JW at Gander was caused by either structural failure of the aircraft at take-off, or explosion of a bomb that may have been placed in the aircraft in Cologne." The summary includes a copy of Schoppaul's sketch showing the hypothetical location of the bomb in the belly compartment, close to the leading edge of the right wing.[53]

Schoppaul's theorizing may not have been reliable or even credible information, but it was certainly information "that would indicate the loss of this aircraft was the result of . . . detonation." It's not clear whether FBI agents interviewed Schoppaul on more than one occasion, so—in light of Fraser's sworn testimony—the summary may refer to an interview other than the one that Fraser witnessed. If so, we are left to wonder why the FBI didn't share its concerns with the RCMP.

Captain Steve Saunders, a DC-8 pilot for Arrow Air, claimed that in February 1986 a man identifying himself as Inspector William Scarborough of the Homestead office of the FBI interviewed him with respect to the Arrow Air investigation.* The man posed a number of hypothetical questions about in-flight explosions on board a DC-8. "If an explosion occurred in 'B' pit, what areas would be most adversely affected structurally, and if hydraulic and flight control systems were adjacent would irretrievable catastrophic failure occur? What seat-row occupants adjacent to these cargo areas would be most likely struck by shrapnel" from such an explosion?[54]

Mark Dombroff, the Washington attorney who recruited Irving Pinkel, tried to track down the details of the FBI's investigation. An internal FBI memorandum sparked by Dombroff's inquiry states that a special agent (whose name is blacked out) from the "Explosives Unit, Laboratory Division, FBIHQ," advised that he was a member of the forensics team that traveled to Gander, Newfoundland, to assist the Canadians." The same memo states that the FBI's legal attaché in Ottawa had "coordinated the travel of various FBIHQ forensic experts to Gander, Newfoundland, shortly after the crash. However, the Canadians declined this offer of assistance."[55]

Yet another heavily censored memorandum to the "Fugitive/General Government Crimes Unit" refers to "results of investigation conducted by Special Agent [blacked out] at San José, California, concerning captioned matter"—the caption being "Arrow Air DC8-63, Gander, Newfoundland." Uncensored portions of other documents refer to related interviews conducted by the "Miami and San Francisco Divi-

---

*Schoppaul had some problems with his pilot's licence and refused to give interviews after the controversy broke out.

sions" and "results of our investigations. . . . disseminated to the Defence Intelligence Agency."

Despite these indications, spokesmen for the FBI have steadfastly maintained that "the FBI has no investigative interest in the Arrow Air crash and has not conducted active investigation regarding this matter."[56] The senior FBI official responsible for investigating terrorists acts reaffirmed this lack of interest in an interview with Terry Taylor of Pittsburgh's KDKA Television in the spring of 1990. Associate Deputy Director Oliver "Buck" Revell said the FBI had "no indication that it was a terrorist incident" and repeated that the agency had not conducted an investigation. "We helped only with the identification of the deceased through our identification division and our disaster team. There certainly was no basis for us to proceed on an investigation of a terrorist act or sabotage." Revell said.

Saunders may have been interviewed by an impostor—indeed the FBI suggested this in a letter to Mark Dombroff — but if so, the bureau showed little concern. The denials of official interest also seem strange in view of the indications of activity gleaned from partially censored documents obtained under the Freedom of Information Act. Above all, the FBI's apparent lack of interest in the sudden death of 256 Americans in a foreign country seems inexplicable in light of bureau's duties and its reaction to the loss of a Pan Am 747 at Lockerbie, Scotland.

It's worth noting that the FBI's investigation of the Lockerbie crash was launched by none other than Buck Revell. Revell swung into immediate action after catching an early bulletin on Cable Network News about an hour and a half after the crash. By the time the FBI received official notification a few minutes later, Revell had already alerted Director William Sessions and both the criminal investigation and intelligence divisions.

The first sketchy report of the Lockerbie crash mobilized a gamut of counter-terrorist interests at the Justice department and other agencies. Computer operators at the CIA counter-terrorism center and the State Department Office for Combating Terrorism started to comb databases on bombs, bomb-makers and bomb threats. Experts at the Federal Aviation Administration "pulled their own files on bombs and bomb threats and began going through them for clues."[57] The U.S. Army Criminal Investigative Division became involved because military personnel had been on the flight.

It's important to note that the agencies swung into action on no other basis than early informal reports of a mid-air disintegration of an American aircraft. Immediate speculation about the cause raised several possibilities: mid-air collision, catastrophic mechanical failure or bomb. No matter; to cover the possibility of a bomb the bureau dispatched Special Agent James T. Thurman, one of its best explosives technicians to Scotland immediately.

The FAA reacted with equal alacrity; Walter Korsgaard took off for Scotland within ninety minutes and the CASB's John Garstang joined him soon after. Through diligent search of the wreckage and incredible luck, the team of explosives experts would find conclusive evidence that a bomb had brought down the jet.

FBI director William Sessions publicly explained the FBI's early involvement a week after the crash, and stressed that the bureau had no suspicions of criminal or terrorist involvement. The director cited Title 18,

U.S. Code 32, the Destruction of Aircraft or Motor Vehicles statute, which authorized the bureau to investigate any possibility of "interference with or destruction of United States aircraft."[58]

Similar reasoning would have applied, *a fortiori*, to the crash of the Arrow Air DC-8 at Gander. So what are we to make of the FBI's dogged insistence that they didn't conduct an investigation? The claim has been repeated so often by so many officials that we must accept it as literally correct.

But the senior FBI officials who rushed a team of explosives experts to Lockerbie must have at least considered doing the same when they got word of the Gander crash—"to cover the possibility" of a bomb. In fact, it's clear they did just that. FBI documents that have been made public make at least two references to the "forensics team that travelled to Gander." According to one description, this team returned to the U.S. because "FBI personnel were not given access to the crash site, and therefore, no FBI investigation was conducted."[59]

This statement, made in an internal memo to the Assistant Director for Criminal Investigations, undoubtedly tells the literal truth. But it's beyond belief that Canadian authorities would have kept FBI forensic experts from the crash site while welcoming the assistance of dozens of U.S. military personnel. The quote attributed to the FBI's legal attaché in Ottawa that "the Canadians declined this offer of assistance" seems to be a diplomatic nicety to allow the FBI team a reasonably graceful withdrawal from the turf of the army.

The crash of the Arrow Air military charter was, in essence, a purely military affair. If a military transport had been used, the U.S. Department of Defense would have conducted the investigation under a bilateral agreement — notwithstanding that the crash had occurred in Canada. It's understandable, then, that the U.S. Army should claim control over the investigation. But to do so, it would have to turn off the attentions of other American agencies — the FBI, for example — and bureaucratic battles would have to be fought behind the scenes. Canada, on the other hand, would have found it legally and politically impossible simply to turn the investigation over to the Americans. Canadian law called for several overlapping investigations: an investigation by the CASB to determine the causes and contributing circumstances, a criminal investigation by the RCMP; and a sort of coroner's inquest by a judge of the Province of Newfoundland. They would at least have to go through the motions of investigation.

Although the disintegration of the aircraft suggested an explosion, the U.S. Army was much more interested in cleaning up the site than in encouraging forensic examination. Civilian forensic experts were not called in, as at Lockerbie. In fact, readily available expertise was shunted to the side. Yet, in contrast to the army's seeming apathy toward the remains of the aircraft, it showed a keen, almost obsessive interest in the human remains.

The Human Remains

As General John Crosby wrote worried memoradum about disposal of the wreckage during the final days of December 1985, an even greater hurdle had already been cleared: it had been quietly agreed that the victims' remains would be promptly transferred to the United States.

The victims, all American citizens, consisted of eight crew members from Arrow Air and the U.S. troops. Emergency medics quickly determined that there were no survivors, and at 8:10 a.m. (about an hour and a half after the crash) the James Patton Hospital in Gander called off preparations for a major emergency. At around two p.m. rescue workers began to remove the remains to a temporary morgue set up in Hangar 21.

That evening, Dr. A. Shapter, Gander District Hospital medical director, instructed the pathologists to prepare for autopsies in the morning. But General Crosby, arguing on behalf of the U.S. government, insisted that the autopsies be conducted in the United States. A heated, behind-the-scenes debate ensued. The RCMP and the hospital's medical director wanted the autopsies conducted in Canada, while the CASB reportedly backed the American demand on the grounds that the autopsies should be ''completed within a reasonable time frame . . . by pathologists trained in aviation pathology''.[60] The local pathologists, who ''were not enthusiastic about the enormous task'', were quite willing to leave the grisly work to their American counterparts.

Crosby's insistence on American autopsies undoubtedly stemmed from both practical and patriotic reasons. But there were also bureaucratic concerns. The U.S. Army had good reason to be sensitive to problems with the identification of mutilated human remains; during the Vietnam war the identification of soldiers missing in action had become one of the army's most contentious public relations problems. Well-publicized cases such as that of James Cowan, Jr. — one of the prisoners of war released by the North Vietnamese in 1973 — had subjected the army to accusations of incompetence and fraud: Cowan's grieving parents had buried the "positively identified" remains of their son three years earlier.[61]

Controversy had erupted again just a few months before the Gander accident. In July 1985, the army claimed to have "positively identified" thirteen victims of a C-130 that had been shot down in southern Laos in 1972. The army had based the identifications on a total of three pounds of charred bone fragments. Skeptical families consulted outside experts who could substantiate only two of the thirteen identifications. A Pentagon evaluation team was eventually forced to admit "mistakes of a blatant nature".[62]

Anne Hart, widow — or possibly wife — of Lieutenant-Colonel Thomas T. Hart, one of the crew members on the C-130, claimed to have intelligence documents proving that her husband had survived the crash. Mrs. Hart filed a lawsuit against the Pentagon. Katherine Fanning also harbored doubts about the identification of her husband, who had been shot down over North Vietnam in 1967. On hearing of the Thomas Hart case, Mrs. Fanning had the remains buried as her husband disinterred. Independent forensic examination concluded that positive identification was impossible, and Mrs. Fanning also sued. The lawsuits were pending in U.S. District court at the time of the Gander crash.* In the context of the unfolding scandal, the U.S. Army was not eager to turn over the task of identifying bodies to foreign doctors in a godforsaken Canadian town no one had ever heard of.

According to a report by Emergency Preparedness Canada, the dispute over the autopsies was referred

---

*Scarborough denied that he had interviewed Saunders.

to federal and provincial Cabinet ministers, who quickly ruled in favor of the United States.[63] The decision to ship the remains to the U.S. was recorded in the form of a "Memorandum of Understanding" signed by various officials on December 14. The most significant of its sixteen clauses stated that "Pathological examinations will be conducted by the Armed Forces Institute of Pathology at Dover, Delaware."*

The other provisions were window dressing to minimize the appearance that the United States was taking control. For example, the leading item stated that the pathological examinations "will be conducted under the control and supervision of the Canadian Aviation Safety Board." Another clause stated that "Death Certificates will be signed by a qualified physician licensed to practise medicine in the Province of Newfoundland," and without further ado the Newfoundland College of Physicians granted a licence to Dr. Robert McMeekin, director of the Armed Forces Institute of Pathology. Death certificates issued by Dr. McMeekin in Delaware would satisfy the letter of the agreement.[64]

General Crosby signed the memorandum on behalf of the United States Army. Other signatories included Newfoundland's Deputy Minister of Justice and the Chief Superintendent of the RCMP. The Canadian Aviation Safety Board's chairman, Bernard Deschênes, got around the formality of involving his board members by delegating the signature to Peter Boag; the CASB neither discussed nor documented this decision to contract out a major part of its most important investigation.

---

*Ann Hart was awarded $632,000 for her severe emotional distress in October 1988.

In allowing the autopsies to be conducted in the U.S., the Province of Newfoundland tacitly agreed to forgo an inquest. Later, next of kin would argue that the province had not met the specific conditions for dispensing with such an inquest. At the time, no one questioned the decision.

The elaborate charade of the Memorandum of Understanding was necessary, at least in part, because "Some of those involved expected the decision to cause a reaction and that the media would ask about Canadian sovereignty."[65] As it turned out, the Canadian media were not interested, while it didn't even occur to the American media that there could be any alternative.

* * *

An early precursor of the uncertainties that surfaced later about the autopsies concerned the seemingly straightforward issue of the total number of victims.* Initial news accounts placed the death toll at 258: 8 crew members and 250 passengers—a full load. These accounts were based on the official flight manifest Captain John Griffin had left in Gander. Captain Art Schoppaul had marked the same total on the manifest for the first leg of the flight, between Cairo and Cologne. Early news reports reinforced the picture of a full load of homecoming passengers. The Associated Press reported, for example, that Frank Brady of Kansas City was greatly relieved to get a call from his son Mark, who was supposed to have been on the Arrow

---

*The Dover mortuary had been the site of the autopsies after the bombing of the Marine barracks at Beirut airport in 1983, the mass murder/suicide at Jonestown, Guyana, in 1978. The remains of the *Challenger* astronauts would also be moved to Dover.

Air flight but had "volunteered to wait behind when there wasn't enough room for all those who wanted to go."[66] A telephone call to Karen Thomas in Detroit told her that her brother, Michael Thomas, had not made the flight even though he was listed on the manifest.*

But apparently the impression of a full flight was not quite correct. The day after the accident, General Crosby announced that the number of military victims had been revised from 250 to 248 "because we just didn't have the manifest".[67] It seemed that two soldiers hadn't made the flight at the last minute: Chris Carlin had decided to visit his girlfriend in Israel and Eric Harrington had misplaced his passport and was not allowed to board. But why had the flight attendants not noticed the two empty seats and corrected the official manifest? Why, for that matter, had the vacancies not been filled by soldiers on the waiting list?

Although the revision of the reported number of victims passed without comment, the army was sensitive to the issue. Major P. Shearston, who observed the airlift of the remains from Gander to the Dover mortuary, noted that "A great deal of importance was placed on transferring exactly 256 cases." As Shearston explained, "We were told the press was counting, with the implication that we did not want to convey the impression that any remains were left behind. As it turned out . . . the final group of transfer cases con-

---

*With 258 victims, the Gander crash would have ranked as the eighth-worst aviation disaster of all time — just one ahead of the number of victims claimed when an Air New Zealand DC-10 slammed into the slopes of Mt. Erebus in the Antarctic in November 1979.

tained various body parts, none of which could have classified as a 'body', i.e. — having as a minimum a head and torso.''[68]

These precautions proved unnecessary. All the bodies had not been found and "The U.S. Army were still interested in having the crash site searched and examined a second time".[69] A team of forty-two army personnel came back to Gander on January 6, 1985, hired local help and proceeded to clear the accident site of trees, including the ones that showed the initial swath prior to the point of impact. Shelters were constructed to allow searchers to melt the snow and collect every remaining fragment. On January 17, they found a body that had been buried by the blast and another on January 20. By the time the second search was called off on February 6, some three hundred more anatomical portions and four and a half tons of personal effects had also been recovered.

An information paper issued by the U.S. Army on February 10 said that Dr. McMeekin "estimates that there may be a number of soldiers (as many as ten) that will not be matched to a set of remains."[70] The paper added that "a decision memorandum" discussing potential sites for a group burial would be forwarded to the Secretary of Defense. But Dr. McMeekin's pessimism seems to have been unwarranted; just two weeks later, the army announced that all 248 victims had been positively identified.

By this time the army (at least as personified by Dr. McMeekin) had already dismissed any fleeting notion that the Arrow Air DC-8 might have been sabotaged. McMeekin had been issuing Province of Newfoundland death certificates from the Dover mortuary since late December, and in each case he showed the immediate cause of death as "multiple traumatic injuries" with the antecedent cause of "aircraft accident". He certified the approximate interval between onset and death as "instantaneous" and on the line asking whether the cause of death was "accident, suicide, homicide or undetermined" he marked "accident". To the question asking, "May further information relating to the cause of death be available later?", he confidently and, as it turned out, incorrectly answered, "No."

Three years later, the deaths of many of the victims would be reassessed as neither instantaneous nor due to multiple traumatic injuries. Even the claim of 100 per cent identification was challenged. Dr. Samuel Dunlop, a former employee of the army's Central Identification Laboratory, estimated that up to 35 per cent of the Arrow Air victims might have been misidentified.[71]

* * *

The revised number of victims caught my attention soon after the CASB board received the investigators' draft report in the spring of 1987. By this time, the Air India disaster had been attributed to a bomb in the suitcase of a passenger who had not boarded. I wondered about the baggage of the two soldiers who had not made the flight; could their duffel bags have remained on board? No one seemed to know exactly how the CASB had learned of the error in the documented number of passengers. The official files contained only a list of 248 passengers, provided by the U.S. Army some weeks after the accident, while the passenger manifest signed by Captain Griffin declared 250. The crew of the first leg believed that the aircraft had left Cairo with a full load of 250. Both the pilot,

Art Schoppaul, and the senior flight attendant, Mona Ogelsby, said so in sworn testimony.[72]

By this time, too, details of the Iran-Contra affair were emerging. The story of the clandestine arms-for-hostages negotiations broke in November 1986 after the Lebanese magazine *Al Shiraa* reported that the United States had been secretly supplying arms to Iran and that former National Security adviser Robert McFarlane had made a secret trip to Tehran for negotiations with the Ayatollah Khomeini's representatives. At first President Reagan said the story had no foundation, but he was forced to admit that there had, in fact, been a secret diplomatic initiative to Iran after Ali Akbar Rafsanjani, speaker of the Iranian Parliament, confirmed the *Al Shiraa* story.

McFarlane and a delegation including Lieutenant-Colonel Oliver North had visited Iran in May 1986. These latter-day Magi travelled with falsified Irish passports on a Southern Air Transport Boeing 707 ''combi'', bearing a Bible annotated by President Reagan. Other gifts included a pallet of spare parts for HAWK missiles, a key-shaped cake to symbolize the sought-for opening to Iran and a number of Colt duelling pistols.

But secret forays by top-deckers such as McFarlane and North have to be supported by precursor and follow-up trips by lesser officials — and their trips wouldn't merit a specially chartered 707. There must, moreover, be a limit to the number of forged Irish passports that the CIA can provide on short order. Suppose, for the sake of argument, that the two missing passengers had been replaced at the last moment by taciturn strangers wearing Ray-Bans and carrying khaki duffel bags. Had the flight proceeded normally, the replacement passengers would have blended unobtrusively with the troops and slipped quietly away on arrival at Fort Campbell.

But what kind of official reaction could we expect if such mysterious strangers were killed in a tragic accident?

Richard A. Gabriel, a former army intelligence officer and now a professor of Political Science has written that "the U.S. military has a long history of covering up casualties taken in 'black' operations. It was normal practice in Vietnam to assign different causes and places to the death of servicemen killed on missions in Laos, Cambodia and even North Vietnam." Gabriel reported that twelve soldiers assigned to a "black" unit and killed in four separate helicopter training accidents in 1983 were listed in army records as deaths in automobile accidents. The army was forced to admit that it had deliberately disguised the death of helicopter pilot Captain Keith Lucas, who was shot down during the invasion of Grenada in 1983, because his unit was in support of a "black" operation.[73] Gabriel states that there is reason to believe at least two other deaths associated with the Grenada invasion were "laundered" to make them appear to be accidents at another time and another place.

While keeping such musings to myself, I tried to establish how the documented figure of 250 passengers had come to be revised to 248. Questions in the boardroom were invariably deflected. Eventually, in February 1988, I asked Chairman Deschênes in a memo "where in the files or elsewhere I can find a record of how and to whom the message on the revised figures was communicated." But by the time of his precipitous resignation in April, Deschênes had not replied.

I pressed the issue with the new chairman, Ken Thorneycroft, and he provided an answer of sorts on May 6, 1988. Thorneycroft wrote that "the Investigation staff have researched your question", discovering that "United States military personnel on the scene at Gander provided the investigators with a manifest containing the names of 248 passengers." He added that "the exact manifest could not be identified" because only a copy received some weeks later "remains following reorganization of the files in order to facilitate the review process." So I never did get an answer, but Thorneycroft unintentionally confirmed that the investigation files had been purged.

The puzzling discrepancy between the passenger manifest provided by the U.S. Army and the official documentation took an unexpected twist just before the report was released. Dr. Richard Shepherd, a lecturer in Forensic Medicine at the University of London, England, was struck by an incongruity between the lists of autopsies and laboratory tests. The laboratory seemed to have toxicological results for two victims not corresponding to any of the official autopsies. In a memo to Arrow Air's lawyer dated November 18, 1988, Shepherd dryly remarks that "Given the degree of care in removing, marking, storing and transportation of the remains, I find this somewhat perplexing."

Initially the investigators simply ignored the inconsistency, but the fuss forced them to address the issue in their final report, albeit obliquely. The report notes that 250 passengers were scheduled originally but "In the days immediately preceding the flight" this was reduced to 249, apparently without changing the paperwork, and finally one passenger was not permitted to board in Cairo because he had misplaced his passport. The report notes without comment that "The load sheet prepared by the flight crew in Cairo and carried over to the departures from Cologne and Gander showed a load of 250 passengers."[74]

One of the findings in the draft report had stated that "documentation regarding the weight of passengers and cargo" was inadequate. The concession made to clear up the mystery of the passenger list was to add two words and refer to the inadequacy of the "documentation regarding the *number* and weight of passengers".[75]

The investigators' final report also notes that "the personal baggage" of Eric Harrington, the soldier who was not allowed to board in Cairo, "remained on the aircraft". But apparently the thought of a bomb in the baggage was too preposterous to investigate.

Part 2

Egypt: A Bomb in the Baggage?

Circumstantial evidence is a very tricky thing.
Sherlock Holmes[76]

During the first weeks and months after the Gander crash, no one without full access to the facts could have credibly challenged Peter Boag's disavowal of physical evidence consistent with an in-flight explosion. Circumstantial evidence pointing to sabotage was more accessible, if not completely so.

Terrorist activity had soared over the past year. On November 27, 1985 the U.S. Transportation Department warned airlines of a pending attack on a civil aircraft by Iranian terrorists. According to a *New York Times*/CBS poll in April 1986, Americans considered terrorism to be the gravest threat facing the nation. And a variety of overt and hidden motives gave plausibility to the Islamic Jihad's claim to have brought down the Arrow Air flight.

It was also evident that the enemies of the U.S. had more than strong motives: they undoubtedly commanded the means to strike down a DC-8. The bombings of the American embassy and Marine barracks in Beirut attested to ample sources of explosives and expertise, and fanatic determination. But even these may not suffice to bring down an airliner through a single device stashed in the baggage.

Deliberate destruction of commercial flights by hidden bombs dates back to the sabotage of two DC-3s in 1949. On May 3 of that year, a crude timebomb exploded aboard a Philippine Airlines flight, killing all thirteen occupants, including the husband of the instigator's mistress. Several months later a Canadian Pacific DC-3 met a similar fate while en route from Quebec City to Baie Comeau. All twenty-three aboard were killed when a timing device made from an alarm clock detonated several sticks of dynamite in an Air Express parcel. Police investigations uncovered a plot by Montreal jeweller J. Albert Guay to kill his wife and collect the insurance, and Guay was tried and hanged.

Over the years, the technology of terrorism progressed. Alarm clocks and dynamite gave way to micro-electronics and Semtex, barometric triggers replaced simple timers, political terrorism supplanted personal motives. On February 21, 1970, barometer-triggered bombs concealed in transistor radios exploded in the baggage compartments of two separate flights from Europe to Israel. Both explosions were premature and both flights had time to make emergency calls. Swissair Flight 330 crashed during an emergency approach near Wurenlingen, Switzerland, killing all forty-seven occupants, but the Austrian Airlines flight was able to make an emergency landing in Frankfurt without injuries.

As airport authorities introduced counter-measures, the terrorists improved their technology. Combined barometric/time-delay triggers defeated airport decompression chambers designed to harmlessly set off simple barometric devices. Planting bombs on unwitting passengers got around restrictions on unaccompanied baggage.

In August 1982 a bomb exploded under a seat in a Pan Am 747 en route from Tokyo to Honolulu, but the explosion didn't breach the fuselage or disrupt vital controls, though it did kill the passenger in the seat and injure fifteen others. But in June 1985 a single bomb in the baggage compartment of Air India Flight 182 destroyed the huge aircraft and its 329 occupants at a stroke.

There could be little doubt that those with a motive to strike the Arrow Air DC-8 had the means to do so. Did they also have the opportunity?

* * *

First Officer Hans Bertelsen landed the Arrow Air DC-8 at Cairo International Airport at 5:34 p.m. local time, December 11, 1985, some twelve hours later than scheduled.* The sun had gone down, but a number of Russian and Egyptian military aircraft could still be discerned across the runway from the DC-8's parking spot in the civilian area on the north side of the terminal building.

The ageing DC-8 had brought its load of 250 American soldiers almost halfway around the globe, with refuelling stops at Gander and Cologne. The nineteen-hour journey had not been pleasant; disembarking troops complained of cramped seating, cold coffee, dripping water, peeling panels, opaque windows and leaking duct tape around the doors. "I would rather swim back to the States than fly Arrow Air again," one of them remarked.[77]

Members of the 101st Airborne who had finished a six-month tour of duty were standing by to go home.

---

*Problems with starters on two of the engines had delayed departure from McChord Air Force Base, near Tacoma, Washington.

Their return flight would complete the second of the three round trips needed to exchange 750 American troops; the first had been made a week earlier, using the same route, airplane and crews. The manifest was adjusted to get as many of the married soldiers on this flight as possible — to give them more time with their families during the Christmas holiday. Colonel Frank Partlow, the U.S. Army's Chief of Staff in the Sinai, ruefully remarked after the crash, "At the time, this seemed like a good idea."[78]

Prior to this cycle of three flights, the exchange of incoming and outgoing soldiers had always taken place at Ras Nasrani airport in the southern Sinai. This time, the MFO had instructed Arrow Air to pick up the troops in Cairo because construction on the main runway at Ras Nasrani had made the airport unfit for large aircraft. The troops would be shuttled between Cairo and the Sinai on smaller Boeing 737s chartered from Egyptair. Their baggage would go overland by truck.*

These new arrangements provided additional transfer points for the baggage and correspondingly expanded opportunities for someone to tamper with the cargo. With both motive and means at hand, it's reasonable to wonder if terrorists may have sought such opportunities. But there is another reason to suspect there may have been explosives in the soldiers' baggage.

---

*A memo referring to an Egyptian NOTAM (Notice to Airmen) closing Runway 04L/22R is the only reference in the files to the closure. Apparently the adequacy of the runway was discussed in an exchange of telexes between the MFO and Arrow Air in Jult 1985, but I have not been able to find a documented reason for the closure.

Americans, it has often been remarked, are steeped in the culture of individual weaponry. Passengers frequently try to carry weapons or explosives aboard aircraft without criminal intent. During the first decade of anti-hijacking screening at American airports more than 25,000 weapons were detected and confiscated.[79] While the CASB investigators were drafting their report on the Gander accident in 1986-87, security screening at U.S. airports yielded 3,361 firearms and 12 explosive/incendiary devices.[80]

A pair of "nunchakus" or "Ninja sticks", among the more usual weapons picked up from the crash site, reminds us that young soldiers are even more enamored of weaponry than the average American. Other soldiers have shown a remarkable penchant for stashing much more dangerous souvenirs of their trade in their baggage.* As we shall see, military authorities occasionally wink at rules against the transport of explosives on passenger flights. In short, military explosives may have been secreted on board, quite aside from the possibility of sabotage. But the fact remains that baggage inspection in the Sinai and the loading in Cairo presented points of vulnerability.

Inadmissible Evidence

On March 24, 1986 the CASB held a "pre-hearing conference" to establish an agenda for the public inquiry that was to open in a few weeks. The participants included representatives of government agencies, Arrow Air, Douglas Aircraft, Pratt & Whitney and the families of the flight crew. These "interested

---

*The smuggling of weapons became a minor scandal for the U.S. Army after the invasion of Grenada in 1983.

parties" were called together to comment on a list of proposed topics and witnesses.

Chairman Bernard Deschênes presided. Board members Lacroix, Pultz, Stevenson and Thurston were members of the tribunal, but Lacroix and Stevenson recall it as the chairman's show. In so far as he consulted his board members at all, Deschênes presented decisions for *pro forma* ratification when it was already too late for changes. Members of the board had no opportunity to review witness statements or to preview the investigators' preliminary findings. To complete their isolation, they were excluded from the consideration that led to a choice of topics and witnesses. Stevenson and Lacroix were not happy but, much to their later chagrin, they went along with the chairman's wishes.

Deschênes proposed six areas to be examined at the public inquiry:
- history of the flight
- potential for airframe icing
- weight of the aircraft and method of calculation
- pre-impact state and serviceability of the aircraft
- aircraft takeoff performance
- flight crew performance

The list telegraphed the investigators' tentative, and as it turned out, final conclusions.[81]

Allan Markham, Arrow Air's attorney, had no objection to examining the potential for airframe icing, but he argued that the inquiry should also take evidence on the potential for an explosive device. The lawyer had some recent news he considered highly relevant. A month earlier, an unpublicized incident at Norton Air Force Base had sent waves of apprehension through operators of military charters. On February 26 a group of U.S. Marines were boarding a DC-8 chartered from Transamerica Airlines for a flight to Cherry Point, North Carolina. "While loading baggage, one of the loaders noticed a strange item (later identified as blasting caps) had fallen out of a bag belonging to . . . a passenger on the flight. Search of the bag brought forth four rolls of detonation wire."[82]

Carrying such mind-boggling hazards on a passenger flight contravened the terms of the charter, standing orders and all common sense. The loader notified the airline and the troops were told that all baggage would be searched, but the commander promised amnesty to anyone who voluntarily turned over other forbidden items.

The seventy-one additional dangerous items that were sheepishly trotted out included six rolls of detonator cords, a 40 mm shell, nine slap flares, a smoke bomb and a signal flare. A badly shaken Transamerica representative described the incident as "a catastrophe waiting to happen". "What I am very much concerned with," he wrote to the Military Traffic Management Command, "is the potential loss of lives . . . and this just after the Arrow Airways crash with Airborne troops returning from foreign duty."

The reference to Arrow was highly suggestive. Markham sought to introduce the Transamerica letter as an exhibit, Deschênes wouldn't hear of it.* He was adamant that the six suggested topics would not be expanded. He ruled that "pre-impact state and serviceability of the aircraft" could cover the possibility of an onboard explosion.

But Deschênes did make one concession: he agreed that witnesses could be questioned on security meas-

---

*By contrast, Deschênes did allow "seven articles on icing" to be tabled by Douglas Aircraft.

ures and customs clearance procedures in Egypt.[83] Testimony would be provided on behalf of the Multinational Force and Observers, who were, in any event, to give evidence on the contractual arrangements and loading procedures. James Williams, the MFO's Director for Contingents, Aviation and Support Services, was a witness on Deschênes' list.

But there was a hitch. Citing "international privileges and immunities", the MFO wouldn't allow Williams to testify. After some official correspondence and behind-the-scenes dickering, Deschênes agreed that Williams could provide a written statement.[84] Although the MFO would not submit to direct questions, it did send a lawyer to question other witnesses. U.S. Army personnel formerly associated with the MFO testified in person.

When the CASB convened its public inquiry in Hull between April 8 and 16, chief investigator Peter Boag read Williams' statement into the record. Military witnesses with knowledge of the security measures and customs inspection, who emphasized that they were not representing the MFO, answered questions from Boag or one of his assistants. Lawyers for the interested parties could question the witnesses, subject to being ruled out of order should they come "dangerously close to cross-examination".

The CASB investigators' interest in the baggage focused on the possibility that the DC-8 might have been overloaded; testimony about security arrangements was incidental. Lawyers who tried to probe this area were hampered by a process that prevented a preview of the evidence, contentious questions or calling additional witnesses. They didn't make much headway.

Testimony about the baggage-handling would become highly significant when sensational allegations of contraband cargo emerged later on. One theory claimed, for example, that light anti-tank weapons had been smuggled on board with the connivance of senior officers. According to another theory, the flight was secretly returning the bodies of U.S. soldiers killed in a failed attempt to rescue American hostages from Lebanon.[85] Such allegations were dismissed by allusions to stringent security measures in the Sinai.

Customs Inspection in the Sinai

Most of the troops of the 101st Airborne travelling home on the Arrow Air DC-8 carried laundry bags containing equipment harnesses, flashlights, ammunition pouches, canteens and other items of their trade. They all had individual weapons — some dismantled in the laundry bag, others wrapped in paper. Most also carried a small overnight bag. "Many had cameras. Less than half carried an attaché case, a zippered leather case or a cardboard personal record holder," testified Captain Schoppaul. Other carry-on items were similar to what might be seen in the lounge of any international airport. One thing particularly caught the attention of senior flight attendant Mona Ogelsby. "Oh, they had huge, big radios," she recalled.

Two kinds of baggage were loaded into the DC-8's belly cargo compartments: the remainder of the troops' personal belongings, packed in duffel bags and other, less easily classified cargo that belonged to the battalion as a whole — "unit-owned equipment". At the time, the distinction between personal and unit-owned baggage seemed unimportant.

Testimony about the inspection of baggage began with the written statement submitted by James Williams on behalf of the MFO. On the basis of reports from others, Williams believed that "approximately half of the duffel bags and all unit stores and equipment were inspected by Egyptian Customs and/or U.S. military customs personnel in the Sinai".[86] By "unit stores and equipment" Williams seems to mean the baggage stored in the cargo compartment, other than the duffel bags containing the soldiers' personal gear. The MFO statement said nothing about carry-on baggage.

A representative of the U.S. State Department supported Williams' written statement with a letter that "was cleared by the Department of the Army". The letter explained that "In accordance with standard U.S. customs inspection procedures . . . all cargo and 50-60% of the personal baggage was subjected to a physical inspection."[87] It's not clear whether "personal baggage" includes the carry-on items.

Leaving aside for the moment the fact that nothing was claimed about the thoroughness of the inspection, neither the written statement nor the supporting letter from the State Department jibed with the direct testimony from Lieutenant Colonel James Kelly, who "acted as an expediter and co-ordinator of the movements of troops to and from the MFO out of Fort Bragg, North Carolina." Kelly had not been on the scene in the Sinai either. But at least he was a step closer to the action than those who had submitted written statements; he was also available to answer questions.

Kelly explained that certain MFO troops were empowered to inspect baggage in the Sinai on behalf of U.S. Customs. When the MFO flights had begun two years earlier, all the baggage had had to be inspected, but the process was painfully slow. U.S. Customs had agreed to expedite procedures by checking only a sample of the baggage. On this flight, Kelly testified, "the captain ordered about a 50 to 60 per cent check of cargo and 100 per cent of carry-on"[88] — that is, the reverse of the ratios cited as being "In accordance with standard U.S. customs inspection procedures."

Captain Gerald De Porter, accredited as a customs supervisor for the MFO, was the only witness able to give first-hand testimony about how the baggage had actually been inspected, as opposed to how it should have been inspected. De Porter had travelled to the Sinai a week earlier, with the first batch of troops in the planned three-flight rotation. He testified that the customs inspectors were mainly interested in contraband narcotics and proscribed agricultural products such as "camel saddles . . . stuffed with all kinds of foreign matter that is prohibited by U.S. Agriculture". The Americans inspected "probably 10 per cent, maybe 15 per cent of the baggage. Then the Egyptian Customs people come in after that and inspect what they consider to be a reasonable amount . . . .They just at random pick baggage out for inspection purposes. They looked at about 40 per cent."

> Q: For this particular flight, the flight that crashed, can you tell me exactly or estimate how much baggage was inspected in detail?
> A: We looked at between 60 to 70 per cent . . . .The hold baggage is what I'm talking about now.
> Q: Only hold baggage?
> A: Only hold baggage.

In answering questions from the lawyers, De Porter clarified that the 10 or 15 per cent inspected by U.S. officials was actually 10 or 15 per cent of the baggage that evoked a response from a dog trained to sniff out narcotics. He also said the dog wasn't trained to detect explosives. The Egyptians then inspected 40 per cent chosen at random—''it may have been the same bags, it may not have been.'' Finally, the Americans inspected a further 10 per cent of the baggage that had passed the dog.

De Porter couldn't specify what was in the cargo. ''We looked at some tool kits — what kind of tool kits they were, I have no idea — a couple of resuscitation dolls for training purposes, things of that nature. Any more specific than that, I really cannot give you.''

Stuart Goldstein, the lawyer representing the next of kin of the flight crew, questioned De Porter on the contradiction with the 50 to 60 per cent inspection of personal baggage mentioned in the written submission on behalf of the MFO. De Porter said he could not explain the difference. ''The percentages are really hard to figure, other than just an estimate. We do not keep an actual bag-by-bag count of what we inspect. We look at rough figures . . . .''

De Porter said that after the inspection, the hold baggage was loaded onto trucks by hand, ''then the back gate is locked, and then a lead wire seal is put on it.'' He was not asked whether he personally observed the loading. He did say he left instructions that all carry-on bags were to be inspected, but couldn't testify first hand. ''I feel certain that the lieutenant would have followed my orders,'' he said. He admitted that he had not actually seen the baggage being inspected, and could not confirm that, in fact, the carry-on baggage had been inspected.[89]

The CASB made no attempt to reconcile, or even comment on, the inconsistencies, but from this haze of imprecise and contradictory testimony the investigators eventually distilled the following:

> All personal effects carried on board the aircraft were subject to a rigorous pre-flight inspection by United States Military Customs Inspectors and Egyptian Customs officials. Approximately 60 per cent of the baggage placed in the cargo pits of the aircraft was inspected. Bags were selected at random, emptied and the contents examined. One hundred per cent of the carry-on baggage was inspected.[90]

What happened to the baggage after this haphazard inspection? As Captain De Porter testified, after the inspection at Ras Nasrani the hold baggage was loaded onto trucks that were then sealed. The trucks left for Cairo, where they were to be guarded by designated troops. De Porter rode with the trucks but, again, he had no personal knowledge of the security arrangements.

Q: . . . You cannot personally verify that . . . the truck or the trucks were guarded?

A: . . . Major Moore assured me that the cargo would be guarded and was guarded . . . .

Q: Do you know if the people who guarded the trucks stayed there while the bags were being loaded?

A: Again, I have no knowledge of that.

The U.S. Army liaison officer in Cairo, Major Ronald William Carpenter, may have known a little more. Carpenter, a clean-cut Eagle Scout type who could have passed for Oliver North's younger brother, met the trucks at Cairo airport at around two p.m. on

December 10, the day before departure. He escorted the trucks to the Customs area where they were guarded by shifts of two American soldiers.

The next day, Carpenter was on hand to observe the transfer of the truck's contents to the DC-8. He did not know who was authorized to break the customs seals. But he remembered that the guards had not stayed with the trucks while six or eight employees of ZAS Aviation (the Egyptian cargo firm who did the ground servicing) transferred the baggage to the DC-8.

On December 11 the soldiers left the base for Cairo Airport on board two Egyptair Boeing 737s. Each 737 carried 121 troops, and for this leg of the trip their carry-on baggage was stowed in the baggage holds. The other eight men (or seven, if the manifest had already been reduced to 249) scheduled for the flight apparently travelled by ground transport, possibly with the baggage trucks. But no evidence was taken on such details.

The first 737 arrived in Cairo around eleven a.m. and the second around two p.m. The troops disembarked and waited on the ramp to claim their bags as they were unloaded. As the Arrow Air flight would not be leaving until 10:35 p.m., the soldiers then took a five-mile bus ride to the Hyatt El Salam Hotel, where they rested and had something to eat. "To my knowledge, the hand baggage was kept with the individual the entire time," Carpenter testified, "I was not with the soldiers after they departed the Cairo airport . . . but, when they departed the airport and when they returned, their hand baggage . . . was with each person." Carpenter didn't know who would have counted to see if the same number of bags went into the Arrow Air DC-8 as had been taken from the Egyptair 737s.

No one could have known if one of the soldiers had picked up a good bargain at the Hyatt — on a Toshiba "Boombeat" radio-cassette player, say — a model not too well known at the time, but one that would receive some notoriety after the Lockerbie crash.

Loading the Arrow Air DC-8

As soon as the incoming Arrow Air DC-8 had parked in Space Number 7 in the civilian area, Egyptian ground personnel rolled a ramp to the forward passenger door. The door was opened and the inbound soldiers filed off.

The baggage handling started with a glitch when the ground handlers couldn't open the hatch on the forward cargo compartment. They proceeded to the other three compartments while Flight Engineer Charles Alonso and First Officer Bertelsen struggled with, and eventually opened, the front hatch. As darkness fell, it was discovered that the baggage compartment lights weren't working. A ground power unit was brought out.

The turnaround was expected to take two and a half hours, but it took two hours just to unload the incoming baggage. Loading then began for the return flight. The trucks that had brought the boxes and duffel bags from the Sinai were backed near the DC-8. Ground handlers put the baggage on a conveyor belt and then into one of the cargo compartments. Bertelsen said the job was accomplished by "15, maybe 20 Egyptians . . . milling around the airplane", some dressed in western clothing, some in traditional Egyptian robes.[91]

One young Egyptian soldier with a rifle and a civilian topcoat was on guard at the bottom of the steps near the front of the aircraft. Captain Schoppaul later recalled that the guard "was mostly catching Coca Colas while he sat at the bottom of the steps. He wasn't very alert."[92]

Julius Graber, Arrow Air's European manager, travelled in the jumpseat on the return trip between Cairo and Cologne. He witnessed the loading and recalled that "there was a period of time when there was no light, because the ground power unit had been pulled."[93] The conveyor belt broke down "a couple of times." Captain Schoppaul "noticed that the Egyptian baggage workers got into a fist fight outside the aircraft during the loading procedure" and "thought this was an extremely unusual event."[94] Graber, who also witnessed the fight, added that the altercation involved a number of people and took place "in front of belly number one and number two" — that is, in front of the two forward cargo compartments.

Graber was struck by "the general lack of security", and by something a good deal odder. Asked whether he observed anything else unusual, he replied, "The unusual was to me that a considerable number of wooden boxes were among . . . the baggage."[95]

The Mysterious Wooden Boxes

As Julius Graber watched the boxes going into the baggage hold, he began to suspect that there wouldn't be enough room for the soldiers' gear. There wasn't. When the baggage holds were filled, over forty duffel bags were left over.

According to Graber, Colonel Marvin Jeffcoat, the battalion commander, was "rather upset" and "requested that all bags be loaded — period." "The troop commander was very concerned about those boxes," Graber recalled. "In particular, the troop commander requested that if nothing more could be loaded, that we swap bags — that we would remove bags already loaded from the bellies in order to accommodate the wooden boxes" that "the troop commander said contained very important military material."

Graber saw between ten and twenty of these boxes — footlockers, or whatever they were — most of them two feet wide by three feet long by two and a half feet high. He objected to the term crates: "They were definitely boxes which were closed on all six surfaces, you could not see into them." It turned out that one of them had not been shipped on the trucks, but had been brought on one of the Egyptair 737s. Major Carpenter estimated the size of this box as about two by three by five feet. He said the soldiers tried to take this box into the aircraft cabin but the flight attendants wouldn't allow it, and the box went into the cargo hold along with the others. Carpenter speculated that it might have contained "last-minute records or something that was in a footlocker for convenience", and assumed it must have been inspected with the hand baggage. No one thought to ask where this box had been kept, or who had guarded it while the soldiers were resting in the Hyatt El Salam.

The commander's insistence notwithstanding, there was just not enough room for all the cargo, and Captain Schoppaul was called to mediate. He had all four baggage compartments reopened — to little avail. Flight Engineer Alonso crawled into the compartments and found room for two or three more duffel bags. After he had done his best, forty-one bags remained.

"The bags to be left behind were then laid on the ground under the floodlights and the colonel took the name from each bag," Schoppaul recalled in an *aide-mémoire* he wrote shortly after the accident. Colonel Jeffcoat "then used the aircraft P.A. system and said, 'Any man whose name I call out stand up and tell me if you have any classified or restricted articles or weapons in these bags. If you do, come forward.' He then read out the names." No one acknowledged any forbidden items. Schoppaul suggested that the colonel's question about contraband proved that the baggage had not been inspected.

Colonel Jeffcoat's insistence on taking the boxes in preference to his troops' personal gear surprised a number of observers. Colonel James Kelly, chief of the MFO division at Fort Bragg, testified that the purpose of the charter flights was to move the troops and their personal gear. Other cargo was allowed if "no troops' bags get bumped".[96] "My big concern," he explained, "is that the troops' bags go on first." General Burton Patrick, who commanded the 101st Airborne at the time echoed the sentiment. Patrick didn't know about the duffel bags that had been left behind, but said, "It would be unusual to separate a soldier from his equipment."[97]

Despite the army's well-known penchant for record-keeping, little could be learned about the contents of the boxes. Captain De Porter explained that "we do not do a detailed indication of the equipment on the manifest."[98] The manifest was filled out after the customs inspection in the Sinai on December 9, 1985. It consists of two pages. The first page has one entry: 481 bags—the duffel bags containing the soldiers' personal belongings, 41 of which were left behind. The second page has fourteen short entries, adding up to 48 pieces. The list includes 11 "tool boxes", 6 boxes of "medical records", 5 boxes of "multipack commo/

equipment", 11 "footlockers" with unspecified contents, 2 boxes of "legal forms", a box of training aids, a box of charts and other miscellaneous items.

The mannequins or "resuscitation dolls" that Captain Gerald De Porter noticed while strolling through the customs inspection in the Sinai were presumably part of the unspecified contents of the eleven footlockers. The inert "practice type" mortar rounds found among the wreckage were also, we must suppose, part of the "very important military material" that took precedence over the troops' personal belongings at the insistence of the company commander.

In November 1989 the National Security Subcommittee of the U.S. Congress asked the Comptroller-General of the United States to direct the General Accounting Office (GAO) to "determine to the extent possible the identification of the cargo loaded aboard [the] Arrow Air [flight] prior to its departure from Cairo."[99] The GAO reported back ten months later — with the same generalized list.[100] Despite numerous other requests — by private individuals under the Freedom of Information Act, by Members of Congress directly to the army — no itemized description of the cargo was ever released.

Neither the testimony taken at the public inquiry nor anything else obtained by the CASB sheds light on which items were inspected by Customs officials and which, if any, went with the troops on the 737s. In any event, it's hard to imagine any of this "unit-owned equipment" meriting the unusual separation of soldiers from their equipment. It's also difficult to reconcile the manifest with the recollections of those who observed the loading.

Major Carpenter described about a dozen "foot-lockers in cardboard boxes". These were "standard U.S. Army footlockers, perhaps two feet, three feet across . . . , perhaps five feet in length—wooden foot-lockers." He said he didn't know exactly what was in the footlockers, but suggested "medical records or tool kits or whatever."[101]

Captain Schoppaul gave a slightly different, more detailed description. In his *aide-mémoire* he described "a few, not over 20" bulky cardboard boxes carried by the troops. He supposed they contained souvenirs or Christmas presents. Schoppaul also saw "possibly 4 wooden boxes", which were put into either the first or second cargo compartment. He noted that it took two men to lift each, guessed the weight as 160 pounds and the dimensions as "6 feet by 2 feet by 14 inches", just big enough for a life-size resuscitation doll—that is, about the size of a coffin.

Weapons and Ammunition

If the vague flight manifest and the chaotic customs inspection fail to give pause about what might have been loaded into the Arrow Air DC-8, consider the testimony about weapons and ammunition carried on the flight.

Colonel Kelly, the MFO chief at Fort Bragg, testified that troops were directed to carry their individually assigned weapons, a 45-caliber pistol or an M-16 light assault rifle, unloaded and inconspicuous. Other weapons such as machineguns and mortars were not to be carried with the troops.

> Q: Is that normal procedure that all other machine guns would be pre-positioned?
> A: That is our procedure for the Sinai.

> Q: Is that adhered to?
> A: Yes sir.[102]

Notwithstanding this testimony, twenty-four M-203 grenade launchers and two M-60 machine guns were "believed to be on board the aircraft at the time of the accident".[103] A U.S. Army list obtained under the Freedom of Information Act lists 26 M-203 grenade launchers and various other differences from the lists reported by the CASB.[104]

The apparent contradictions did not pique the investigators' attention. Neither did an even greater discrepancy: senior officers had winked at violations of the regulations about the transport of live ammunition. Lieutenant Kelly succinctly summarized the rules: "You do not carry live ammunition on an aircraft with personnel unless you are in a combat situation."[105]

Major Carpenter explained that one breach of this rule was exposed as the soldiers were boarding one of the Egyptair 737s at Ras Nastrani. A metal detector triggered an alarm and an Egyptian security guard found that one of the troops was packing a loaded 38-caliber snub-nosed revolver, and became "very upset about it". There were many witnesses.

Some quick diplomacy convinced Egyptian authorities to permit the pistol and ammunition to go along in custody of the pilot. Major Carpenter tried to smooth things over in Cairo. He gave the unloaded pistol to one of the officers telling him to "get the weapon back to the individual, but you can not bring ammunition through the Cairo airport."

But the battalion commander put up an argument. Colonel Jeffcoat collared Carpenter in the airport departure lounge. "He told me that he could not see what the big deal was about taking ammunition," Carpenter recalled.[106] To make his point, Jeffcoat said that "there had been ammunition on the 4th of December flight," that is, on the rotation a week earlier. The major turned over the ammunition.

To defuse possible accusations of a cover-up, the State Department "requested the Army to make inquiries in response to questions concerning the possibility of ammunition or hazardous cargo being transported on the airliner which crashed" at Gander. The army's answer is not known, but a State Department representative sent a carefully phrased letter to the Secretary-General of the MFO. The letter says, without specific attribution, what "the Army reports". Deschênes contrived to include it in the record of the CASB's public inquiry and, after Peter Boag had read the letter submitted by James Williams in lieu of MFO witnesses, Deschênes remarked with studied casualness, "I guess you should also read the letter annexed to the file."

Boag read the letter. It adopted the classic diversionary tactic of muddling what should have happened with what did happen. "It is not the Army's practice to transport ammunition or explosives on passenger flights," it asserted, and then described standard operating procedures for the inspection of baggage. It went on to note, almost off-handedly, that "The only ammunition thought to be on board the aircraft was carried by a CID inspector, who was required to be armed because he was transporting evidence to be used in a criminal trial, and by the Battalion Commander. Both individuals are reported to have carried one clip of ammunition for their sidearms."[107] There was no mention of who had done the thinking and reporting.

The story was fleshed out only slightly during questioning of the live witnesses by the lawyers. Jack Lange, the lawyer representing the MFO, asked Major Carpenter what CID meant, and Carpenter explained that it was the U.S. Army's Criminal Investigation Division, and that basically it meant undercover military policeman.

> Q: Major, do you happen to know what evidence the CID inspector was transporting?
>
> A: What he told me that he had on his person was contraband evidence, drug evidence, marijuana or some sort of drug evidence that he was in custody of.[108]

The CID officer can be identified from the list of victims distributed by the Pentagon as Chief Warrant Officer 2, Dirk Miller. Evidently, Miller's mission to the Sinai was of some importance. The army posthumously awarded him the Legion of Merit, the highest military honor, and a building at Fort Campbell was named in his honor in 1986.[109] But no details were ever released of Miller's mission to the Sinai or of the evidence that required him to carry a loaded gun contrary to regulations.*

Colonel Jeffcoat died in the crash, but the officers and men who had travelled on the Arrow Air flight on December 4 were available for interviews. There is no indication in the record that any of these soldiers was asked about Jeffcoat's contention that live ammunition had been carried. Although Frank Ward, the RCMP officer who arrived on the accident site just after the fire trucks, reported hearing "explosions that

---

*Post mortem examinations did not reveal any traces of marijuana or other illegal drugs in the soldiers' remains.

appeared to be small ammunition burning", no trace of the two clips of ammunition was ever found.

Despite the "catastrophe waiting to happen" discovered at Norton Air Force Base on February 26, 1986, Chairman Deschênes wouldn't allow evidence of the contraband explosives discovered on the Transamerica charter to be introduced at the CASB's public inquiry. Ironically, yet another alarming incident occurred a week after the inquiry.

A security bulletin put out by the Federal Aviation Administration described the situation:

> On April 19, 1986, the airport police at Oklahoma City received information that unidentified passengers on a Transamerica Airlines B-747 aircraft were in possession of explosives. The Transamerica Airlines flight had been chartered by the U.S. Army to transport military personnel. . . . An appropriate search of the passengers and their baggage, both checked and carry-on, resulted in the recovery of various items of military ordnance which were being transported without authorization as souvenirs . . . [one] item, if triggered, would have resulted in a severe fire and probable crash of the aircraft. Among the items recovered was a trip flare with the triggering pin loosened, rendering it extremely dangerous. If the trip flare would have been [*sic*] set off a magnesium fire would have resulted.[109]

While Bernard Deschênes deemed the Transamerica incidents unworthy of note, the authorities responsible for the military charter flights were sufficiently alarmed to send a message to all operators of military charters. Arrow Air received the telex, headed "Hazardous Materials Prohibited on Pax [passenger] Aircraft", on July 28, 1986. It is apparently dated March 10, 1986 — before the incident at Oklahoma City — yet it refers to hazardous items on "two MTMC [Military

Traffic Management Command] arranged commercial air movements that have moved in the last 60 days." The communiqué refers specifically to the February incident at Norton. Without specifying the other incident, it says, "In light of the fatal air crash of a DC-8 in Gander, Newfoundland on 12 Dec 85, request that all concerned with the movement of troops aboard commercial aircraft . . . do their part to ensure that hazardous items do not get aboard aircraft."[110]

The two Transamerica incidents bracketed the CASB's public inquiry. Internal correspondence shows that the relevance of these incidents to the Gander crash was perfectly obvious to military authorities. While the incidents were not common knowledge, they had certainly been brought to the investigators' attention. Nevertheless, the investigators continued to focus on the ice theory. Their refusal even to consider the possibility of an onboard fire was all the more peculiar in view of the sensational direct evidence that had already emerged from the post-mortem examinations.

Part 3
Gander: The Big Chill

The Postmortem Examinations

> The investigator . . . must not draw
> conclusions too quickly lest he make
> irreparable mistakes. There may just be things
> out there in the world that he hasn't seen yet.
> *Colonel Robert R. McMeekin*[111]

Bernard Deschênes once mentioned that he had discussed the decision to conduct the autopsies in the United States with Newfoundland's Deputy Minister of Justice. He wouldn't say what was discussed or name others he had consulted. But the CASB chairman certainly did not discuss the issue with members of his own board.

A "Memorandum of Understanding" signed by chief investigator Peter Boag provided a semblance of due process. Apologists could point to a clause stating that the post-mortem examinations were to be conducted "under the control and supervision" of the CASB to argue that the board had not ceded its authority.

It was later claimed that this "control and supervision" was exercised on behalf of the board by Dr. David Elcombe. At the time of the Gander accident, Elcombe was in the process of signing on as the board's director of Safety medicine. A large, informal man affecting horn-rimmed glasses and tweed jackets, Elcombe has the benevolent, disorganized look of a perpetually harried country doctor. For the previous eleven years, he had been chief of the Health department's Civil Aviation Medicine Unit in Toronto.

Elcombe did, in fact, observe some of the autopsies, which were conducted simultaneously by six pairs of pathologists. When I pressed him in a board meeting in January 1988, he admitted that no one but himself *could* have exercised control and supervision on behalf of the CASB. Yet, notwithstanding the Memorandum of Understanding, Elcombe balks at assuming the responsibility. "I prefer the words liaison and co-ordination," he says.

Dr. Elcombe's modesty is understandable. There is no room to doubt that control and supervision over the post-mortems remained with the U.S. Army—and the representative of the army was, as far as any outsider can determine, Colonel Robert R. McMeekin— the doctor who had been licensed to issue Newfoundland death certificates from the Dover mortuary, and who had found all the deaths due to "accident".

On April 15, 1986, Colonel McMeekin was on hand to testify at the CASB's public hearing. With his close-cropped hair, metal-rimmed glasses and immaculately pressed uniform, he looked more like a missile launch officer than a doctor, and he responded to Dave Elcombe's long-winded interrogatories with laconic, often monosyllabic answers.

McMeekin was not a man to waste words. He had joined the army after Yale Medical School and a surgery internship at the University of North Carolina, and the army had trained him in forensic pathology. He wasn't asked, and he didn't volunteer, his recent role in a highly sensitive task — supervision of the autopsies of the astronauts killed in the explosion of the space shuttle *Challenger*.

About five weeks earlier, U.S. Navy divers had found the *Challenger*'s crew capsule about fifteen miles north-east of the launch site, under ninety feet of water. The fragmented crew compartment was held in one piece by an embedded web of electrical wire, much as fragments of an egg's shattered shell can be held together by the inner membrane. The remains of the astronauts were soon retrieved, albeit in poor condition after six weeks in the ocean. Nevertheless, it was clear that, contrary to early reports, the initial blast had not vaporized the crew. The astronauts had activated their emergency oxygen packs and survived two and a half minutes in free fall, but were crushed by the force of the capsule's 200 mph impact on the ocean surface.

The USS *Preserver* took the remains of the crew to Port Canaveral. Florida law requires autopsies to be conducted under local authority, in this case under the authority of the Brevard County medical examiner, Dr. Ronald Wright, but NASA dreaded a reversal of the neat, tidy account of instantaneous death.* The astronauts' remains were stuffed into thirty-gallon plastic garbage cans, loaded on a blue-grey, open-bed pickup truck and spirited to Patrick Air Force Base,

---

*History was repeating itself. NASA had also announced that Gus Grissom, Roger Chaffee and Ed White, the astronauts who perished in the *Apollo I* launch-pad fire in 1967, had died an instantaneous death. Tape recordings were later obtained that showed the astronauts' desperate efforts to open the hatch and escape the flames.

twenty-five miles to the south. The base, including the hospital and morgue, are under exclusive federal jurisdiction.

Investigative reporter Dennis Powell, who uncovered the story, wrote, "NASA wanted the autopsies to be done by military doctors from the Armed Forces Institute of Pathology. This way, the space agency could maintain control over what information would be released."[112]. Dr. Wright later told Powell, "What NASA did was illegal. Against the law. I don't know how to make it any plainer or what good it will do. The whole thing stinks."

The astronauts' death certificates were not filed with any state registry, but were kept at the Johnson Space Center in Houston. NASA maintained that the cause of death "cannot possibly be determined."[113] Clearly, Dr. McMeekin, who had supervised the autopsies, was a man who could be trusted with a sensitive task.

Dr. McMeekin was the only witness to testify on post-mortem examinations and other medical evidence at the CASB's public hearing. His testimony conveyed a seamless, simple account of instantaneous death — just like the initial accounts of the *Challenger* deaths. He .said that, by dint of hard work and cooperation from other agencies, all the victims had been positively identified, but he was not so confident that the post-mortems could shed light on the cause of the crash.

Q: Would an analysis of your results through what is described as injury pattern analysis be contributory to the understanding of this accident?

A: Perhaps. . . .

Q: Would this apply possibly to the existence of pre-impact or post-impact fire?

A: It could, yes.

Q: And the possibility of an explosive device of any type?

A: Perhaps.

Q: Would you have any particular experience in that area of expertise that you could relate to us?

A: In injury patterns?

Q: Injury patterns and particularly explosive device analysis.

A: I have worked on a number of aircraft accidents in which explosives were a consideration. . . .

Q: To date, has there been any evidence from that area, the X-ray group, to indicate any form of fragmentation, projectile, missile device?

A: As from an explosion?

Q: Yes.

A: No, there has not.

Elcombe didn't ask what proportion of the X-rays had been completed "to date" or if the many caveats that Dr. McMeekin outlined in his technical papers had been observed. But Elcombe did make sure that the record was complete in another respect:

Q: Just to clarify, the examination of X-ray findings and autopsy findings for the evidence of explosive device or projectile, did your remarks include the findings for passengers as well as air crew?

A: Yes, it did.[114]

It seemed clear enough. Most of the lawyers for the interested parties had no questions. But Stuart Goldstein took a stab at eliciting more details.

Q: In the work that you did, are you able to determine if there was a pre-impact fire?

A: In my opinion there was no pre-impact fire.

When Goldstein sought the basis of this opinion, Dr. McMeekin was momentarily and uncharacteristically evasive, mentioning ''the short duration of the flight''. The lawyer persisted.

> Q: But, I mean from the medical evidence, is there any indication of a pre-impact fire or not?
> A: We have seen no reaction to burning in any of the tissues that we have looked at—which would indicate primarily post-mortem.[115]

What he was saying or seemed to be saying was that the pathologists had detected no signs of living bodies' defence mechanisms rushing to mitigate the damage from burning. The death certificates had indicated that all the victims been been killed instantaneously at impact. Examination of tissue showed that the victims must have been dead before the exposure to fire. Thus there could have been ''no pre-impact fire''. *Quod erat demonstrandum.*

 But why not elaborate such an important conclusion in direct testimony? Could there be a chink in the logic? If a fire had broken out in flight, could the aircraft have crashed before the flames reached any of the victims? Could that not lead to instantaneous death without reaction to fire? Smoke and fumes spread faster than flames. Burning or smoldering plastics can quickly release lethal amounts of carbon monoxide, hydrogen cyanide and other toxic fumes. So if an in-flight fire had broken out in the cabin rather than in a cargo hold, some of the occupants might have inhaled combustion products before impact. Hydrogen cyanide—the active ingredient in Zyklon B, the gas used in the Nazi death camps—causes death in a few seconds; carbon monoxide takes somewhat longer. In either case, proof of exposure remains in the victim's corpse.

Post-mortem examination of air crash victims routinely includes toxicological analysis of fluids and tissue. Such analysis is a specialized task conducted separately by a qualified laboratory. Dr. Elcombe's questions to Colonel McMeekin touched only tangentially on this aspect.

> Q: Has any work been conducted at the Armed Forces Institute of Pathology on toxicology?
> A: It has not.

Elcombe dropped the subject.

But just as McMeekin was about to leave the witness stand, Chairman Deschênes returned to the topic of toxicology. "Dr. McMeekin," he said, "having had the occasion recently to conduct an inquiry in a pre-impact fire, I have not become an expert, but maybe I have a couple of questions." He asked if the colonel was aware of the toxicological analysis. But McMeekin was not about to get involved. "Those were done at the Canadian laboratories," he said.

This was not the answer the chairman expected. "You are not aware of the results of those investigations at the Canadian lab?"

"I have heard some discussion but I do not think I could give you a fair discussion."

"Thank you, colonel. I guess the witness is released. . . . We will break for ten minutes," said the chairman.

The total public testimony on the medical evidence relating to 256 violent deaths was over in less than thirty minutes — except for one footnote. It appears that Dr. McMeekin's answers to the chairman didn't put the right words on the record. On resuming after the break, Dr. Elcombe had a clarification.

"Mr. Chairman," he said, "just for the record, could I point out, in response to your question at the end of the last testimony, the toxicology findings for the crew are a part of the human factors write-up for each of the flight deck crew and the cabin attendants."

Elcombe didn't have to ask if the toxicological test included passengers as well as air crew. He knew they did. He also knew that the results would not be presented at the public inquiry.

Hydrogen Cyanide

The inquiry Bernard Deschênes mentioned in the preamble to his questions about toxicology referred to a public hearing conducted a month earlier at Schefferville, Quebec. A twin-engine Beech 18 had crashed and burned on approach to the local airport the previous September; all eight occupants were killed and the aircraft was destroyed by the impact and ensuing fire. During the four-day public hearing, which was set in motion before the Gander accident, the CASB examined a total of thirty witnesses.

In contrast to the single medical witness who testified on the 256 deaths at Gander, eight witnesses had been called to testify on medical factors at the Schefferville inquiry. These included the pathologist who had conducted the autopsies, a second pathologist, two toxicologists and a professor of Forensic Medicine.

The CASB's final report on the Schefferville accident states, "The pathologist who performed the autopsies on all the occupants clearly indicated that death resulted from impact forces, and it was very unlikely that any could have continued to breathe." Suitable samples of blood could be drawn from seven of the eight victims. Elevated levels of hydrogen cyanide were found in six of these samples, including lethal levels in the blood of the pilot.

The investigators could not suggest how a fire might have started in the cabin, and the medical witnesses spurned the notion that the victims might have ingested cyanide or that the laboratory might have mishandled the samples, so the CASB was forced to conclude that the pilot had inhaled fumes before the crash. These fumes could only have been produced by an in-flight fire whose traces had been obscured by the fire that consumed the wreckage.

Participants at the Gander hearing — including Stevenson and Lacroix, who were presiding officers — said they accepted in good faith that the absence of testimony on toxicology corresponded to an absence of significant results. But the impression that the laboratory tests had yielded nothing of significance was far from the truth. Anyone who took the trouble to follow up on Dr. Elcombe's casual remark "for the record" would have found that the results for the crew were not at all clear-cut. Insufficient fluid samples were available to test the first officer's remains. An elevated level of hydrogen cyanide was found in the remains of the flight engineer. Specimens for toxicological analysis could be obtained for three of the five flight attendants; two of the three contained elevated levels of carbon monoxide, one of these also showing an elevated level of hydrogen cyanide. These levels were significant but not lethal. On the other hand, the results for other victims were truly striking: elevated levels of carbon monoxide in more than a third; elevated levels of hydrogen cyanide in more than 80 per cent — in lethal quantities in many cases.

Thus, on the available evidence, the logic that indicated an in-flight fire in the Schefferville accident would indicate an in-flight fire even more convincingly in the case of the Arrow Air DC-8. This inference would have created a sensation at the public hearings. It didn't because it was never stated. The toxicological results had been intentionally withheld.

* * *

Robert Lee, who covers national affairs for the *Ottawa Citizen*, keeps his ears open while commuting by bus to his downtown office. Sometime during the spring of 1988 he overheard a snatch of conversation that would lead him to the story that blew the lid off the controversy brewing in the CASB over the Gander investigation, and eventually to an award-winning series.[116]

Lee's front-page story, "Federal report conceals evidence about Gander crash", exposed the secret of the lethal combustion products in the victims on April 15, 1988 — two years to the day after the close of the CASB's public hearing. By then, the board had circulated a draft report that made no mention of these findings.

On learning of the *Citizen* story, Arrow Air's executive vice-president fired off an indignant letter to the CASB's chairman complaining of the board's conduct.[117] The ensuing controversy set the stage for the abolition of the CASB and termination of office for all the board members.

The story had remained hidden for so long because the investigators had taken elaborate precautions to prevent its disclosure. An early indication of these precautions is found in a curious letter from Peter Boag to Dr. Robin Dodge, acting director of the Civil Aviation Medicine Unit responsible for the toxicological testing.

"As you are aware," Boag wrote on January 28, 1986, "the special circumstances of this accident require that extraordinary measures be taken to ensure appropriate pathological and toxicological examinations were conducted."

Someone unversed in the nuances of bureaucratic prose might well wonder why the head of Canada's national toxicology laboratory needed to be reminded that tests on Canada's worst-ever air crash should be "appropriate". Those accustomed to this kind of letter would suspect a heated prior discussion ending with a demand that a dubious request be submitted in writing. And indeed Boag did reveal a delicate request that couldn't be granted without a record to show responsibility in case of future recriminations:

> The special circumstances of this accident also require that the distribution of toxicology reports pertaining to passengers of the accident flight be restricted.
> Accordingly, I request that only one set of data (report) be prepared for each passenger. The reports, when complete, should be forwarded to me.

Dr. Dodge didn't have to ponder his reply. The next day, he wrote back to Boag, in a letter indicating that the issue had been extensively debated. Covered by this written request from the investigator in charge, he agreed not to distribute copies of the passenger toxicology. His letter also shows that Boag would have liked to suppress the results for both crew and passengers — but it seems there had been a slip-up. "If your request had been received earlier," Dodge wrote, "the reports on the flight crew could also have been directed only to you. However, following the lack of response to my January 9th phone inquiry of how you wished the results handled, the first reports prepared were released, as I had indicated they would be, in the usual manner." That's why the results of toxicology testing on the crew members had to be included in background material for the public hearing.

Apparently the circumstances of the accident were insufficiently special to keep the toxicology lab from retaining a copy of its own results. "The data and lab methodologies will be retained on confidential file at the lab as is normal procedure for all toxicological lab work performed."

On March 5, 1986, Dr. David Elcombe sent the toxicology results on 119 victims to the Armed Forces Institute of Pathology, mentioning that the hydrogen cyanide assays for the passengers were not yet complete. Again the letter refers only obliquely to the issues, the real discussion having taken place off the record.

"As discussed today, we do not intend to present interim passenger pathology or toxicology findings at the public hearings scheduled for sometime in April. No one at the AFIP [Air Force Institute for Pathology] will be called as a witness." These words were written several weeks before the pre-hearing conference held on March 24 "for the purpose of defining the parameters of the public inquiry." It's not clear who decided not "to present interim passenger pathology or toxicology findings", but Elcombe's letter shows that, by this time, the decision to withhold these findings already had been taken.

For the time being, the plan worked. McMeekin's testimony was drowned in the noise of the public hearings. Behind the scenes, however, the toxicology reports were not forgotten. The trail can be picked up in a letter Dr. Elcombe sent Dr. McMeekin on June 20 along with the rest of the toxicological findings:

"Some of the carbon monoxide values and hydrogen cyanide levels are striking," he wrote, "and I look forward to meeting with you and your staff as soon as possible to consider their significance . . . ."

For the time being, though, the evidence that, by the standards applied at Schefferville, would have provided proof positive of a pre-crash fire was simply suppressed. The report eventually submitted to the CASB board in April 1987 didn't even mention the passenger toxicology. The omission was picked up by board member Norman Bobbitt, who had been a presiding officer at Schefferville, but his calls for information got lost in the tempest that led to the approval of the draft report by five of ten board members in the absence of the remaining five—a unanimous decision, as they liked to style it later.

The short shrift given to the medical evidence in the draft report was more than offset by the attention lavished on the investigators' favorite topic, the subject of the grand finale to the public hearing: ice on the wing.

The Ice Was Here, The Ice Was There. . . . [118]

In an otherwise forgettable film, *Three Days of the Condor*, Robert Redford plays the unlikely role of a researcher in the "American Literary Historical Society", a CIA front engaged in combing works of fiction for useful dirty tricks.

In an early scene, the researchers puzzle over a mysterious murder. Evidently, the victim has been shot, but there is no slug in the body and no exit wound. Redford, who hears the discussion in passing, has the answer at once. The murderer pours water into a 38-caliber mould, freezes it and keeps it solid until the crime. Then he shoots the guy with the ice bullet. The cops show up in half an hour. There's just a few drops of water. No bullet. No ballistics."

Redford credits the idea to comic-book detective Dick Tracy. It figures; the thought that a chunk of ice could survive the heat and acceleration in a gun barrel requires the generous suspension of disbelief reserved for the Sunday comics. To me, the CASB investigators' ice-on-the-wing theory called for similar credulity.

No one can doubt the danger of trying to take off in an airplane with ice on its wings. And no one could have been better qualified to explain the danger than Ralph Brumby, an aeronautical engineer from the Douglas Aircraft Company—the only engineer to be heard at the public inquiry. Brumby was a member of the "ice and rain protection committee" at Douglas and the "association of European Airlines" technical committee on de-icing and anti-icing.[119] He was plainly proud of his articles in trade magazines "intended to give awareness of the effect of wing surface roughness [read ice] on airplane flight characteristics", reproduced in the glossy bound Exhibit 18.

The Gander crash reminded Brumby of previous accidents in which "ice or frost was considered to be a contributing factor".[120] He said he had compared flight data recorder traces from four such accidents with the Arrow Air traces: "every one showed aberrations in altitude at the peak altitude that we have seen in this particular flight recorder trace".[121] But the flight data recorder traces of these accidents were not reproduced in the exhibits, and Brumby was not prepared to go beyond generalities. Pressed on specifics, he backpedalled, saying that "there are data in this particular package that were generated that are now, in fact, in part obsolete".

Q: Some of the data is obsolete?

A: Yes, sir.

Q: And you cannot tell us which of the data is obsolete?

A: I have not taken the time to go back through it.

In fact, inferences from the flight recorder data were worse than obsolete. Brumby may not have known it, but he pinpointed the problem when he said, "I have to rely on the Safety Board's analysis of the function of the recorder, and they say that except for a couple of aberrations like that, it seemed to be functioning properly."[122]

Brumby's conclusions about the flight data recorder were based on a mirage. Three years after the public inquiry the American Airline Pilots' Association (ALPA) showed that the output of the Arrow Air DC-8's outdated and malfunctioning flight data recorder was inadequate for meaningful analysis. As Harold Marthinsen, ALPA's director of accident investigation, put it, Brumby based his conclusions on data that were "faulty, or perhaps more to the point, fabricated."[123] Marthinsen, a normally quiet-spoken engineer who had analyzed flight data recorder traces for ALPA for over twenty years, said, "In all these years, we have not seen such abuse of FDR [flight data recorder] data as in this investigation."

At the time Brumby's undocumented comparisons of flight data recorder traces had to be accepted on faith, but his comparisons could have been checked in other respects. He noted, for example, that the notorious Air Florida "Palm 90" flight that crashed into a bridge over the Potomac River in Washington in 1982 "received a great deal of publicity and . . . started an effort of trying to build awareness among the airlines about the effects of [ice] contamination".[124] Ironically enough, the deceased Arrow Air captain was one of the people who had joined in the effort to build such awareness. John Griffin had often discussed the dangers of ice with fellow DC-8 captain Lee Levenson, and the two pilots were well aware that the Palm 90 flight had been brought down by not ice on the wing but by low power on the engines. The Air Florida crew had not turned on the engine anti-ice switch. Ice in the engine inlet, not on the wings, had led to the false instrument reading that tricked the crew into setting the thrust to about 75 per cent of that required.[125] Griffin knew that if the Air Florida crew had applied full throttle, the B-737 could have flown away, ice or no ice.

Brumby's other examples also seem more noteworthy for their differences from the Arrow Air crash than their similarities to it.[126] One of these cited "a DC-8 accident involving Japan Airlines DC-8-62, taking off out of Anchorage." To Brumby the significant factor was that the aircraft "also had been serviced in light freezing drizzle". The National Transportation Safety Board saw it somewhat differently: its official report determined that "the probable cause of this accident was a stall that resulted from the pilot's control inputs." The NTSB also found it significant that the blood-alcohol level in the deceased pilot was over three times the local standard for legal intoxication.[127]

Militant abstainers would undoubtedly find the captain's inebriation the most noteworthy factor in the Japan Airlines crash. In many ways, warnings about wing icing are reminiscent of those about alcohol ingestion. In either case, relatively small amounts can bring dangerous and unpredictable consequences; in either case, there is no lack of evangelical fervor favoring abstinence as the only prudent policy.

The drinking pilot has been a menace to aviation for as long as ice on the wing.* But the social acceptance of alcohol is so ingrained in the Western world that few people talk of outright abolition. In 1984 the National Transportation Safety Board asked the FAA to define minimum blood-alcohol levels for pilots on duty, but the task proved too difficult and the FAA stuck with the "eight hours bottle to throttle" rule.

The failure to devise workable rules to permit pilots to take off with a defined small level of alcohol in their blood mirrored an earlier failure to devise rules to allow takeoff with a small amount of ice. In 1969, the Air Transport Association of America established a "Working Group on Operation with Accumulated Frost, Ice and/or Snow". Operational tests using military aircraft with ice on the wings had proven the feasibility of "operation with limited accumulations of frost, ice or snow".[128] The airlines were eager to use this intelligence to reduce "the out of pocket dollars spent each year for de-icing aircraft".[129] Of course, the technical experts realized that ice on the wings would reduce safety margins during takeoff, but they reasoned that these margins would be reinstated by additional restrictions on weight or runway length. The committee asked manufacturers to determine the restrictions that would have to be accepted.

Douglas Aircraft's cautiously optimistic reaction was conveyed to the Air Transportation Association

---

*Airlines used to contribute inadvertently to alcoholism and broken marriages by putting up crews on layover in the cheapest possible hotels, far from any form of entertainment that could not be obtained in the bedroom or the bar.

in a letter from O. R. Dunn, Director of Aerodynamics. Dunn reported that Douglas had "reviewed the limited relevant data available" and had "arrived at some preliminary estimates" of "the potential performance penalties that would be incurred by aircraft operating with frost on the lifting and control surfaces".[130]

> Based on considerations of the loss in lift and the increase in drag due to an accumulation of frost on the wing and meeting all the performance criteria of the FAR [Federal Air Regulations], the performance penalties for configurations without leading edge devices [such as the DC-8]. . . . if the entire upper surface of the wing is covered with one millimeter [.04 inches] frost deposit . . . the effect would be quite small. . . ."*

On December 5, 1969, Ralph Brumby was on hand at a meeting of the industry group in Washington to discuss the preliminary results transmitted by his boss a few days earlier, but the initiative faltered — not because takeoff with a small amount of ice was unthinkable, but because it proved too difficult to define a workable limit on wing ice.

Similarly, it proved impractical to adopt a limit on blood-alcohol level for pilots. Since there are no social pressures to allow a small level of ice on the wings comparable to those in favor of a small level of alcohol in the bloodstream, the FAA stuck with the "clean wing" rule. And, in so far as ice on the wing was

---

*The sentence continues with a proviso "unless the existing V_2 speeds are less than 1.2 times the stall speeds resulting from having frost on the upper surface of the wings." While it was never stated explicitly in any of the discussions on performance with iced wings, recognition of this proviso may explain the majority report's seemingly inexplicable obsession with "bug speeds".

concerned, Ralph Brumby became an ardent prohibitionist.

The three remaining accidents Brumby cited are also more significant for their differences than their similarities compared to the Arrow Air crash. Consider, for example, the crash of Ozark Airlines Flight 982 at Sioux City, Iowa on December 27, 1968. Brumby notes that the DC-9 "made an approach in light icing conditions, elected not to de-ice and attempted to take off . . . it stalled just after going out of ground effect . . . hit the end of the runway and slid into the trees".[131] He doesn't add that the ground crew warned the captain of ice with sharp points sticking out as much as an inch and extending around the curve of the leading edge for six to eight inches. The investigation also found that the Ozark crew had aggravated the situation with improper thrust settings.*

As Brumby noted, the Ozark DC-9 "slid into the trees". It remained intact and there were no fatalities. Neither the Ozark DC-9 nor any of the other crashes that Brumby cited involved an aircraft with a minuscule amount of ice on the wings and no obvious malfunction. None of the examples duplicated the feature of the Arrow Air crash that most impressed eye-wit-

---

*Brumby's other two examples were: (1.) On November 27, 1978, Trans World Airways Flight 505 crashed while taking off from Newark. The DC-9-10 remained intact after a rejected takeoff and there were no injuries. Failure of both engines after compressor stalls seems to be the most germane factor. (2.) On February 5, 1985, an Airborne Express DC-9-15 crashed while taking off from Philadelphia. The aircraft had been on the ground for forty-four minutes in light freezing *rain* (compare this with less than thirty minutes of very light freezing *drizzle*). The flight data recorder showed that the aircraft rotated to 18 degrees at 6 degrees per second—too much, too fast.

nesses on the site—the total fragmentation of the airframe—and the lack of survivors.

Thus, aside from any consideration of the flight data
recorder traces, the cited accidents give no support to
the thesis that the crash of the Arrow Air DC-8 was
another in a line of tragedies caused by failure to
remove ice from the wings. But that is not to say that
a review of previous crashes can't shed light on the
investigation of the Gander crash.

* * *

A diligent search of the historical record for air disasters where, in Brumby's words; "ice or frost on the
wings was considered to be a contributing factor"
would have turned up another notorious crash of a
chartered flight attempting a winter takeoff after a
refuelling stop in a foreign country.

The accident occurred at Munich on February 6,
1958. The aircraft — a British European Airways
Airspeed Ambassador (called an "Elizabethan" by
BEA)—had been booked by Britain's most celebrated
soccer team, Manchester United. The team and its
entourage were on the way home from Belgrade,
where the players had managed to keep their hopes
for the European Cup alive by battling the Yugosla-
vian team to a draw. The flight had stopped at Munich
to refuel.

Light snow was falling, but the captain decided not
to have the wings de-iced. The crew rejected the first
attempt to take off after noticing power fluctuations
on one engine. They tried a different technique without success, then taxied back to the apron for a conference with the engineers. After some discussion it
was agreed that the problem was caused by the relatively high altitude of Munich airport and could be

cured by advancing the throttles more slowly. The longer takeoff run presented no problem. There was plenty of runway, although the last third of it was covered with a layer of unbroken slush.

By this time the aircraft had been on the ground for an hour and three-quarters. The crew examined the wing from the flight deck and observed only the odd fleck of snow on the part of the upper surface that was visible. The pilots conferred with the ground crew and decided that there was no need to de-ice. But the airplane reached the end of the paved runway before it reached flying speed. As it passed the end, the pilots tried to push the fully open throttles even farther forward, and raised the landing gear. If a nearby house had not ripped off the left wing, they might have made it.

The Elizabethan crashed and burned. Not surprisingly, it didn't disintegrate into fragments as the Arrow Air DC-8 would, and twenty-one of the occupants survived. The twenty-three victims included seven members of the team, their coach, their trainer and a number of well-known journalists.

The weather deteriorated after the accident. Heavy snow fell until the local investigation team, headed by chief investigator Hans Reichel, arrived some six hours later. "The wrecked aircraft was covered with a layer of snow about 8cm [3.2 inches] thick," Reichel wrote in his report. "Under this there was a layer of ice. . . . I found this same condition at all points of the wing, which I examined very thoroughly, with the exception of the part situated above the engine nacelle". He did not comment on the incongruity of the lack of ice near the nacelle. "Apart from the icing," he concluded, "I could find nothing which might have been a cause of the accident."[132]

A team of engineers from BEA inspected the engines and found no malfunction. They could see nothing wrong with Reichel's theory. So on Sunday February 9, three days after the crash, the German authorities sold the wreckage for scrap.

The West German inquiry into the accident opened on April 28 at Munich airport. Like the investigators of the Arrow Air crash, the investigators of the "Manchester United" crash decided on their own which testimony was relevant and how it would be presented. The airport director testified he "felt and saw a rough coarse-grained layer of ice on the surface of the wing. The nose of the wing was free of ice, the engine nacelle and the area behind it were also free of ice. . . . The rough layer of ice on the wing would have been 5mm [one-fifth inch] at the least, if not thicker." No one seemed to note that some 5cm (two inches) of precipitation is needed to produce 5mm of ice. Less than a centimeter of snow had fallen during the hour and three-quarters that the aircraft had spent on the ground; on the other hand, much more than 5cm had fallen in the six-hour interval between the crash and the arrival of the investigators.

Like the investigators' report on the Gander crash, the Munich report blamed the accident on the cause postulated shortly after it happened — the captain's failure to ensure that the wings were free of ice. Like the investigators of the Gander crash, the Munich investigators gave short shrift to all other possibilities and contrary evidence. In both cases, professional pilots were skeptical from the outset. But there was one big difference: the captain of the BEA aircraft, James Thain, survived.

Thain admitted that he had not personally inspected the wings before takeoff, but adamantly denied that ice on the wings had anything to do with the accident. "My opinion," he told Reichel, "is that the aircraft's speed was retarded on the ground, and I think there must have been snow of sufficient depth to retard the speed and not the engines."

If slush or snow on the runway had contributed to the crash, the airport management and the regulatory authority would have to share the blame. Chief investigator Reichel, along with the West German and British authorities, ridiculed the idea.

James Thain was cashiered from BEA, but he fought on to clear his name. With the help of the British Airline Pilots Association he persuaded the British authorities to conduct acceleration tests in slush. The tests conclusively proved that a shallow layer of slush retards acceleration to a previously unsuspected degree. Thain's wife, Ruby Thain, a chemist, provided the solution to the mystery of the lack of ice around the nacelle: antifreeze in the fire extinguisher fluid poured on the nacelle after the crash had prevented subsequent ice formation on adjacent areas.

Ten years after the "Manchester United" crash, James Thain and his supporters finally persuaded the British government to hold an independent inquiry and reconsider all previous evidence One of the witnesses called was Reinhardt Meyer, a pilot and aeronautical engineer who had been among the first on the crash scene. Meyer had provided a written statement, but had not testified at the German inquiry. His testimony came as a shock. Yes, he had examined the wing shortly after the crash, and "there was nothing like frost or frozen deposit: that I know definitely. There was melting snow only."

Had Meyer mentioned this to Reichel? Certainly. Why had he not testified at the German hearing? He had not been called. An examination of his written statement to the German investigators provided another shock: "Meyer's statement was truncated and the most important part of his evidence was concealed."

The report of the British inquiry was published on June 10, 1969. "Our considered view," the authors stated, "is that the cause of the accident was slush on the runway." After eleven years, Thain was exonerated in the eyes of his countrymen and peers, and the ice-on-the-wing theory, which had been so vigorously defended by both German and British investigators finally melted away.

* * *

Back at the CASB's public inquiry, neither Ralph Brumby nor Peter Boag nor Chairman Deschênes was interested in combing the records for cautionary tales about jumping to conclusions.

Of course, any reference to a crash where "ice or frost on the wings was considered to be a contributing factor" is useful only if there was, or probably was, ice or frost on the wings of the Arrow Air DC-8. According to witnesses, there hadn't been any significant precipitation during the hour and a quarter that the Arrow Air DC-8 was on the ground. RCMP Constable Frank Ward had said the snow was so light that he didn't bother turning on the windshield wipers. The helicopter pilots had said the icing conditions started after they began patrolling the crash site. They were not called to the inquiry, but their testimony was not really necessary. For anyone with an open mind, the myth of freezing rain was put to rest by Clarence Bowering, official weather observer at Gander airport at the time of the crash.

Bowering, who retired between the time of the accident and the public inquiry, had spent thirty-five years on the job, thirty-two of them at Gander. It would be hard to imagine a more credible and thorough witness. After he was sworn, he carefully poured himself a glass of water, drew documents from his briefcase and laid them neatly on the table, leaving the briefcase handy on the floor beside his chair. He spoke in measured tones, apologizing for a raspy voice, which he blamed on a throat infection.

Bowering had started his shift some six hours before the accident. Every half-hour, he took and recorded observations — "in accordance," as he put it, "with internationally recommended procedures as established by the World Meteorological Organization."[133]

As he described his observations of an "ice accretion indicator which is made up of a piece of metal similar to what is used in the construction of an aircraft", Bowering reached into his briefcase and pulled out a piece of angled aluminum so the audience could see what he was talking about. He explained that he had examined two of these indicators, one on the roof and another at ground level, every half-hour — noting what had fallen on the surface and then positioning a clean indicator for the next observation. He checked the indicators several times between official observations, but did not replace them. At 5:30 a.m., four minutes before the Arrow Air DC-8 landed, he recorded "light" snow grains and "very light" freezing drizzle, unchanged from previous readings. "I will give you a definition of what is laid down in the manual for very light freezing drizzle," he told the inquiry. 'Very light freezing drizzle' is used to indicate the intensity when scattered drops, flakes, grains, pellets or stones are occurring at a rate which would not wet or cover a surface, regardless of the duration. There was only very small amounts — spotty — and that's why I called it 'very light.'"

Bowering described the evanescent traces as "very, very thin" rough patches, maybe an inch in diameter, covering 10 or 15 per cent of the surface. As he was careful to explain, the precipitation would not have produced a layer of ice — even if it had persisted. In any case, it did not persist.

By the time Bowering took his next reading, at six a.m. — twenty-six minutes after the Arrow Air DC-8 had landed — even the "very light" freezing drizzle had stopped.

> Q: Can you recall what you observed? . . .
> A: There was very little on it then, very little. . . .
> Q: When you say "very little", how would that compare with the amount that was on the indicator at [5:30 a.m.]?
> A: There was just enough to say that there had been a bit of freezing precipitation falling, probably 5 per cent; just a spot here and there.

Since Bowering had changed the indicator just before the Arrow Air flight arrived, this amount—a spot here and there — is certainly more than what would have accumulated on the wing up to that time. Bowering replaced the indicators after his six o'clock observations. Although his next official observation was not due until 6:30, he came around to check at 6:14 and noted that "there was no freezing drizzle adhered to it, just very few snow grains". He issued a special observation noting that all "freezing drizzle had ended and the snow grains had become very light."

Bowering also testified that a total precipitation of two millimeters — eight one-hundredths of an inch —

had fallen over the six-hour period from 3:30 to 9:30 a.m., which included both the arrival and the departure. At worst, a minuscule, unmeasurable amount of freezing precipitation could have been deposited on a small percentage of the DC-8's wing surfaces.

Ross Stevenson, Dave Mussallem and other experienced airline pilots had often observed greater quantities of ice and snow deposited on the wings during the time it takes to taxi from the de-icing stand to the end of the runway. It was absurd to suggest that the aircraft should have been de-iced because of such illusory ''contamination''; it was akin to suggesting that a pilot should dry out for eight hours after eating a liqueur-filled chocolate.

While Deschênes and Boag didn't give a fig for the opinions of Stevenson and Mussallem, Clarence Bowering's testimony destroyed any possibility of blaming the crash on ice deposited by freezing rain. Where, then, could the ice have originated? To those committed to the gospel of ice, the answer was obvious: if the ice wasn't deposited in Gander, it must have already been on the airplane when it landed.

The aircraft had in fact passed through a layer of broken cloud two to three thousand feet thick on approach to Gander. If ice had collected during five minutes of flight in this cloud, and if the crew had not used the DC-8's wing de-ice system, the jet could have landed with some ice on the wing. While no one formulated such a hypothesis during the public inquiry, one snippet of testimony would be cited later in its support. Austin Paul Garrett, one of the two ground workers who refuelled the aircraft, had to go to the cockpit to coordinate with flight engineer Mike Fowler. In response to a series of leading questions from Boag, Garrett said that while in the cockpit he saw some ice on the windshield and spoke about it with Fowler.

> Q: Can you describe what you saw?
> A: Well, when he came in, like . . . ah, he said he picked up a tiny bit of ice on the windshield, and that most of it was melted off. There was just a tiny . . . you know, not very much, left on the corner of one windshield.[134]

A number of questions elicited that the ice was on the lower corner of the right-hand windshield. Boag then put on his best bedside manner:

> Q: And was it thick ice? Could you tell us a little bit about the quantity, or something like that.
> A: No sir. It was just like a glaze. I did not take particular notice of it.
> Q: Was it smooth? Was it rough? Was it clear?
> A: I cannot remember, sir.

Try as he might, Boag could not get any more out of Garrett.

> Q: Could you describe the thickness of the ice at all?
> A: No sir.

Garrett's vague recollection became the only piece of direct evidence that there was ice on the wings during takeoff.

Ross Stevenson explained to me much later that, on many aircraft, pilots can look for ice forming on the windshield wiper pivots as an early warning of freezing conditions.* But there are no windshield wipers

---

*Experienced DC-8 pilots also become aware of picking up ice during the descent by the feel of the airplane and the need to add power to maintain speed. This motivates them to check the wings even more thoroughly after landing.

on the DC-8; the windshield is kept clear by heat from the inside, which doesn't reach the lower corners, so DC-8 pilots watch for ice forming in the corners. If warranted, they turn on the wing de-ice system. "It's like using the windshield wipers in your car," Stevenson said, "When you need 'em, you use 'em."

This explanation clarifies Paul Garrett's conversation with Michael Fowler, which took place after Fowler had conducted his external inspection of the DC-8. Garrett described the conversation more clearly in an interview just after the accident.

"I noticed some ice buildup around the edges of the window and I asked him if they picked up much ice on the way in. He said, 'No, it wasn't too bad, there's a tiny bit left around the window.'"[135]

Documentation filed as an exhibit at the public inquiry portrayed Mike Fowler as "extremely conscientious and thorough in his approach to his professional duties" and reported that his "adherence to standards was noted by many sources".[136] When Fowler said there was a tiny bit of ice left, he could only have meant that the flight crew had been aware of some ice buildup during the approach and had used the de-icing system as required, and that Fowler knew from his pre-flight inspection that no ice remained.

While fuelling the DC-8, Paul Garrett and his partner, Ray Folley, had their heads just under the leading edge of the wing and had to look up to connect and disconnect the fuel hoses. Peter Boag tried asking Garrett about that.

Q: Did you see any ice on the leading edge of the wing?
A: No sir.

183

Q: Is it that you did not see [because] you were not in
a position to see or [because] you did not look?
A: I did not look.

Boag paused dramatically, and in the best Perry
Mason tradition sprang his trap: ''So there could have
been ice there that you did not see?'' ''I do not know.
There could have been.''[137]

The two refuellers had no professional reason to
examine the aircraft for ice. But the four members of
the ground crew who serviced the DC-8 would have
checked, and done the de-icing if necessary on a fee-
for-service basis. No one on the ground crew was
called to the public inquiry, but all four were interviewed by the RCMP.

Pat Fewer, the lead man, had been a ground handler
for thirty-seven years. He explained how the ground
crew reacted to ice: ''Generally, if there is anything
there we bring it to the attention of the crew,'' he said,
adding, *''but there was nothing there.''* As a matter of
fact, Fewer had recommended de-icing another aircraft that had been parked at the airport for some time.
Perhaps ironically, he guided the Arrow Air DC-8 into
the pit and then left it to de-ice the other plane, Eastern
Provincial Airlines Flight 131, which departed while
the DC-8 was being fuelled.

Q: Why did you de-ice the EPA flight?
A: He came at 3:30 and he picked up some ice coming
in. He had some ice on the leading edge of the wing
and a bit on the dome. I brought this to the
attention of the EPA pilot and he told me to go
ahead and do the wing.

Another ramp attendant, Ted West, was responsible
for parking and cleaning the DC-8 as well as for de-
icing it if necessary.

Q: Did you notice ice on the aircraft?
A: No. I never noticed any whatsoever. In my honest opinion I don't think there would be any because there was nothing falling at the time.

A third ground worker, Jack Stuckles, agreed:

Q: Did you happen to notice any frost or ice on the airplane?
A: No, b'y.
Q: Would you notice if there was any frost or ice?
A: Yes, sure. You couldn't help but notice it. There was nothing falling that morning.
Q: Was there any form of precipitation before this?
A: Yes. Early in the morning we had to spray EPA. It was like more or less a haze.

The fourth member of the ground crew, Craig Granter, made it unanimous; he hadn't noticed ice on the airplane either.[138]

Another professional opinion was given by the dispatcher, William Geange, who had nineteen years of experience on the job. Geange, who met the airplane and briefed the crew on the weather, testified at the public inquiry:

Q: Under the conditions that existed that day, would you normally expect a crew to ask for de-icing?
A: No.[139]

So it seemed, at the close of the CASB's public inquiry on April 15, 1986, that no evidence had been presented to suggest any cause for the crash.

In his closing remarks, Chairman Deschênes congratulated himself for holding a good inquiry which was "faster, less expensive, more open and less adversarial than the previous approach". "Determining the contributing factors and causes of this accident will be a most difficult task," he said. "The most important area of further investigative work is the question of the performance of the aircraft, especially as it related to weight and balance, crew handling and the possibility of airframe icing."

The chairman did, however, carefully exclude one part of the investigation from the difficult task ahead. Deschênes' closing remarks, he said, were made not only on behalf of himself, but of the members of the board. This may seem surprising to those familiar with the systematic exclusion of board members from the proceedings. But Norm Bobbitt recalls that Deschênes had mentioned his planned closing remarks. During one of the coffee breaks on the last day of the inquiry the chairman remarked casually that he was going to note for the record his satisfaction with how the question of sabotage had been ruled out.

"I think I have a duty at this stage, in view of recent events, to state the following," he told the inquiry. "Simply stated, and although an extremely thorough wreckage search was carried out, no evidence was found of any explosive action resulting from a criminal act, and no evidence was found of any military explosive devices being carried on board the aircraft aside from limited revolver ammunition".[140]

With the chairman's closing remarks, a curtain of silence fell on the investigation, both for the public and for the CASB board members. The minutes of the board indicate that Peter Boag briefed the board on the progress of the investigation, but those who attended these briefings recall receiving nothing of substance.

The board's official minutes for 1986 contain only one instance where the briefing rates more than a mention. On May 1, 1986, Peter Boag itemized "additional areas of examination" to be carried out by his team. His list, appended to the minutes, is a compendium of generalities such as "review all possible explanations of degraded performance" and "review all pertinent data gathered to date". There is nothing, for example, on the "striking" levels of carbon monoxide and hydrogen cyanide in the remains of the passengers.

On December 5, 1986, "the Chairman briefed the Members regarding his interviews with the Press on the anniversary of the crash. . . . Investigation is ongoing and a draft report is being prepared. Although the Press seems to have ascribed the cause of the crash to the crew's failure to have the plane de-iced before take-off, the CASB is continuing to examine all possible causes including: wing icing; reduced power in one of the plane's four engines; excessive weight; incorrectly calculated reference speed." Stevenson and Bobbitt reminded Deschênes that the public inquiry had ruled out icing and stressed the need for careful examination of all possible causes.

The CASB's annual report for 1986 devotes a single short paragraph to the public inquiry into the Gander crash, the same as for the Schefferville inquiry into eight deaths. The board's total accounting of the Gander inquiry states: "Additional facts that were gathered from the participants at the inquiry have proven useful to investigators in conducting the ongoing investigation." The report also pays much more attention to the CASB's role "in assisting the government of India" with the investigation of the crash of Air India Flight 182 than to its own investigation into Canada's worst-ever air disaster.

* * *

By the time board members learned of the investigators' conclusions in April 1987, I had been named a member of the CASB and I had already begun to suspect that there was something very wrong with the board's operation.

Also by this time, Lieutenant-Colonel Oliver North had been fired from his National Security Council job in the White House basement. The Tower Commission Report had determined that North "was involved in an effort to conceal or withhold important information" about (among other things) chartering civil aircraft for covert transport of arms to Iran and Central America.[141]

North's boss, John Poindexter, had resigned under a cloud. Robert "Bud" McFarlane, Poindexter's predecessor as National Security Adviser, had a continuing part in secret negotiations with representatives of the Ayatollah Khomeini, whom President Reagan had called "a new international version of Murder Incorporated". The day before he was to testify before the Tower Commission, McFarlane attempted suicide with an overdose of Valium. There was talk of impeaching President Reagan.

It was disturbing, of course, to learn that high officials routinely deceived not only each other, but duly constituted authority up to and including the President, as well as foreign governments. But at first there was no reason to connect the monstrous deceptions in the United States with the investigation of an airplane crash in Canada.

It was the curious reluctance of Deschênes and his supporters to account for the downward revision of the victim count from 258 to 256 that stirred my first vague thoughts of a link between the Arrow Air crash and the Iran-Contra secrets. It didn't seem excessively imaginative to surmise that entrepreneurs like Ollie North would spot covert action possibilities in a regular military airlift between the Sinai and the United States.

A look at the Iran-Contra chronology clearly shows that North's frantic efforts to keep the operation together reached fever pitch at the time of the Arrow crash. North was desperate to salvage a floundering deal to secure the release of American hostages in Lebanon — a deal depending on covert transfer of American-made missiles to Iran on privately chartered civil airliners. In a memo to Poindexter on December 4, North proposed a delivery of over three thousand missiles "on or about December 12". The plans pivoted on the clandestine use of B-707 or DC-8 aircraft chartered from little-known private operators based in Miami—airlines that had proved themselves in previous operations of this type, including Arrow Air.

These plans were kept secret from almost everyone in the U.S. government—certainly including the secretaries of Defense and State, possibly including the President. As North put it in his memo, "all parties involved have a great interest in keeping this as quiet as possible".[142]

To fathom the external forces that may have been brought to bear on the investigation of the Gander crash, we need to review the use of chartered civil airliners in the Iran-Contra affair, and in particular the past and planned use of Arrow Air.

BOOK 3

"THAT OLIVER NORTH CRAP"

Part 1

The Iran-Contra Connection

Reporters who converged on Gander to cover the Arrow Air disaster in December 12, 1985, couldn't have known of the crisis unfolding in the White House office of Oliver North, where the cauldron of clandestine activities later called the "Iran-Contra affair" had reached the boiling point.

In October 1984 the U.S. Congress had passed the Boland amendment,[1] cutting off "military or para-military" aid to the anti-government guerrillas or "Contras" of Nicaragua — against the fierce opposition of President Reagan and his right-wing supporters. Since then, Lieutenant-Colonel Oliver North, an aide on the staff of the National Security Council, had organized an elaborate web of covert operations to get around the congressional restrictions. The covert activities, which eventually led to criminal convictions for North and some associates, took advantage of circuitous and possibly illegal sources to keep the Contras supplied with weapons, ammunition and military intelligence.

The Contra supply plan began by soliciting "voluntary donations" from individuals and foreign governments wanting to curry favor with the administration. The funds purchased military supplies on the thriving world arms market. To deliver these supplies to the Contras, North co-opted private air transporters that supported the Central Intelligence Agency's covert operations.

Secretly procured weapons were shipped to various Central American countries, ostensibly for use by local forces. Falsified "end-user certificates", asserting that the weapons would be used by the receiving country, allowed shipments through third countries such as Portugal.[2] Since the weapons were neither procured nor transported in the United States, they would be virtually impossible to trace by a paper trail of hazardous material forms or other documentation.

The key player in the Contra resupply effort was a stocky former air force general, Richard Secord. In the 1960s Secord had helped the CIA plan clandestine air operations in Laos. He had become chief American adviser to the Iranian air force in the 1970s and had helped plan the disastrous mission to rescue fifty-three American hostages in Iran in 1980. General Secord's air force career came to an abrupt end in October 1983, after the Justice department investigated his connections with renegade CIA agent Edwin Wilson and a dubious scheme to profit from the freshly signed Camp David accords by delivering arms to Egypt.[3]

After his inauspicious retirement, Secord capitalized on his contacts in the CIA and the Iranian air force as well as his expertise in supply and logistics. With his partner, Albert Hakim—an Iranian who had made a fortune selling electronic surveillance equipment to the Shah's secret service — Secord set up a network of private arms sales and transportation organizations. Oliver North recruited him to the Contra resupply effort on the recommendation of CIA director William Casey.

Secord became the chief middleman between North, Casey and various footsoldiers in the Contra resupply operation — thus distancing the White

House and the CIA and giving the principals "plausible deniability". To maintain secrecy, Secord set up a maze of foreign firms and Swiss bank accounts to launder the tens of millions of dollars used in the arms deals and in paying off various intermediate profittakers. He was also drawn into the Iran side of the Iran-Contra affair, and was eventually convicted on a criminal charge arising from these efforts.[4]

In late 1984, Secord was preoccupied with sustaining the flow of weapons to the Contras in the face of the congressional restrictions. He recruited a retired U.S. Air Force lieutenant-colonel, Richard Gadd, who specialized in "chartering commercial aircraft for the Pentagon and the CIA in a way that cannot be traced to the U.S. Government." Gadd had reportedly demonstrated his proficiency in "black" airlift methods by transferring helicopters to Barbados before the invasion of Grenada.[5] The records of the congressional committee investigating the Iran-Contra affair give some fascinating insights into his methods.

In one well-documented transaction, Gadd used a Canadian company, Trans World Arms of Montreal, to broker a shipment of "Ammunition, explosives, 40 millimeter shells, and weapons."[6] East-bloc supplies were procured to better conceal the source.[7] Montreal arms merchant Manny Weigensberg had set up Trans World Arms as a sister company sharing a telex number with another of his enterprises, Century International Arms, which received some notoriety and became the subject of an RCMP investigation in 1986, after *Soldier of Fortune* magazine published a photo of a Contra soldier with a case of ammunition bearing the inscription "CIA, Montreal, Canada."[8]

Trans World purchased the weapons from a Romanian source, Romtechnica, who shipped them in an unspecified manner to a Portuguese dealer, Defex.[9] The shipment was picked up from Lisbon by aircraft chartered in the United States. To deliver the "pine-apples" (as he called them) to the Contras, Gadd turned to Southern Air Transport, Inc. (SAT). Like dozens of other small operators in the Miami area, SAT provided scheduled cargo service, flew charters and carried out maintenance under contract.

But SAT had a more colorful history than most. Before 1974, it had been one of a network of "propri-etary" airlines set up by the CIA's Clandestine Serv-ices, also known as the "Department of Dirty Tricks". The airlines were wholly owned and operated by the CIA, but appeared to do business as private compa-nies. Under this commercial cover they provided transportation for the Department of Dirty Tricks' activities, which, according to its charter, included: "propaganda; economic warfare; preventative direct action including sabotage, anti-sabotage, demolition and evacuation measures; subversion against hostile states, including assistance to guerilla and refugee lib-eration groups; support of indigenous anti-Commu-nist elements in the threatened countries of the free world".[10]

The CIA proprietary airlines are best known under the rubric of "Air America", the name of an airline that was incorporated in the state of Delaware (because of lenient incorporation regulations) but which operated mainly in the far east. But "Air Amer-ica . . . is also the generic name used to describe all of the CIA air activities whether under the name of Civil Air Transport, Inter-Mountain, Air Asia or Southern Air Transport." Used in this collective sense, Air America was at one time the largest airline in the world.[11]

At the close of the 1950s, increasing anti-American feelings began to infringe on U.S. interests in Latin America. U.S. foreign policy prevented open intervention, so the CIA decided to set up a civil airline for clandestine operations. The agency also needed a new commercial cover to help it out of a difficulty with the Military Air Transport System (MATS). In 1956 MATS had introduced a new policy that required Civil Aeronautics Board (CAB) certification for all MATS contractors. Existing proprietaries couldn't meet the CAB requirements.[12]

To fill these needs, in October 1960 the CIA bought Southern Air Transport for $300,000. At the time SAT was a small, unprofitable airline with assets of a little over $100,000, but as a fledgling member of the CIA air empire it was able to benefit from favorable government contracts, loans, an infusion of cash and interchange of equipment with other CIA proprietaries. A few quick, lucrative contracts from MATS revitalized the money-losing airline, and within three months SAT became a thriving enterprise with assets of over $2.5 million. Eventually it became Air America's principal operator in the Caribbean and South America.

The CIA proprietaries shared a unique problem. They often generated a profit — an embarrassment because American law demanded that government agencies return profits to the Treasury. But in 1958 the CIA General Counsel decided that "income of proprietaries, including profits, need not be considered miscellaneous receipts to be covered into the Treasury but may be used for proper company or corporate purposes." This ruling opened the door to diverting profits from the tens of millions of dollars generated by the air proprietaries during the early 1960s to other covert operations — and thus sidestepping inconvenient congressional or executive scrutiny.

In early 1985, Richard Gadd asked Southern Air to deliver "about 100,000 pounds of Class-C explosives" from Europe to Central America. Unfortunately, SAT didn't have a jet transport large enough to carry so many "pineapples"—but neighboring Arrow Air was "just available" to deliver shipments in January and March. Arrow made the runs under contract to SAT, using the same type of stretch DC-8-63 that would crash in Gander later that year. The first shipment of 95,000 pounds of ammunition was delivered by way of the Azores and Lisbon on January 22, 1985.[13] To ensure secrecy the flight plan filed in Portugal specified the destination as Guatemala City, but SAT flight logs show that the flight actually landed in Ilopango, El Salvador.[14]

Arrow delivered the second shipment in March. It too consisted of 95,000 pounds of "pineapples", loaded on the DC-8 on thirteen pallets. Southern Air paid Arrow $121,825 for each flight—the first payment going through the Royal Bank of Canada, the second through the First Chicago Bank.[15]

Whatever the legal status of the overall project to resupply the Contras, Arrow Air's role as a subcontracted shipper was not illegal; the deliveries were normal commercial transactions. Counsel to the Iran-Contra Select Committee believed they had found some irregularities in the paperwork, but the prosecutors were not interested in such pettifoggery, and no charges were laid.[16] Nevertheless, those involved in the transaction knew a furtive operation when they saw one.

David Mulligan, vice-president of Southern Air Transport, believed that the cargo "was going to be funneled through U.S. sources in Central America to the Contras" on behalf of the American government. Richard Gadd told congressional investigators that these shipments were intended to "be conducted very discreetly" and that he was instructed that they "should be kept very low key and not be known to the general public".[17]

Endless speculations eventually surfaced about the Contra supply flights. An article in the *New York Times* on January 20, 1987, stated, for example, that "federal drug investigators had uncovered evidence" that "American flight crews covertly ferrying arms to the Nicaraguan rebels were smuggling cocaine and other drugs on their return trips to the United States"[18]. The same article reported that Oliver North had told the FBI to "stop investigating Southern Air Transport, the Miami air freight company involved in the Contra resupply operation and the Iran arms deal"

North's remark links the Contra operation to the other element of the "Iran-Contra" affair. By December 1985, the Contra supply operation had meshed with another clandestine project: selling arms to Iran in hopes of gaining the release of American hostages in Lebanon. This was a little awkward considering President Reagan's oft-repeated policy of no negotiation with the "new international version of Murder Incorporated" — of which he had identified Iran as a charter member — but, behind the scenes, Oliver North and others (perhaps with the tacit approval of the President, perhaps not) had hatched a scheme for an end run around the embargo on arms sales to Iran. Over the opposition of Secretary of State George Shultz and Defense Secretary Caspar Weinberger, Israel was allowed to sell American-supplied weapons to Iran and then restock from the United States.

Initially, the so-called "Iran initiative" got stirred into the Contra kettle through a common need for clandestine logistics. The U.S. and Israeli governments both needed to be able to deny involvement if the sales to Iran leaked, and so the negotiations were to be handled by middlemen without official ties to either.* Private arrangements were made for deliveries through General Secord, Richard Gadd and Southern Air Transport — who were in any event involved in the Contra resupply operation that had recently subcontracted some of the business to Arrow Air.

But Oliver North was no professional spook. In making these arrangements he casually tossed precautions such as "compartmentalization", "need to know" and "plausible deniability" out his basement window. It turned out that the Iran and Contra operations were also linked through dollars, millions of them, skimmed from overpriced arms sales to the wealthy Iranians and funneled to the cash-short Contras. As Pat Buchanan, White House communications director, expressed it, "Colonel North ripped off the Ayatollah and took some $30 million to help the Contras."[19]

Who originated and who approved the diversion of profits from the arms sales to the Contra account? These became the abiding mysteries of the whole Iran-Contra adventure. The answer — quite likely nobody — may well lie in Ollie North's casual use of his favorite arms shippers in both the Iran and Contra capers. Secord, Gadd and Southern Air had legally mixed funds between accounts in their days as a pro-

---

*Robert McFarlane continued to be involved after his resignation as National Security Adviser on November 20, 1985.

prietary. To them, it must have been as natural as diverting the pizza money into the beer account.

Attorney-General Edwin Meese revealed this diversion of funds to the Contras in November 1986. Details of the extraordinary efforts taken by North and the National Security Council to prevent disclosure of the bizarre web of covert activities would emerge gradually over several years. The last spasm to keep the secrets would include the notorious orgy of shredding, and stashing of documents from North's office in the underclothes of the photogenic Fawn Hall.

But back at the time of the Arrow Air crash at Gander, the virtually unknown 43-year-old Marine lieutenant-colonel was desperate to extricate himself and his government from a colossal blunder.

Part 2

Arms For The Ayatollahs

On September 13, 1985, armed with a passport in the name of William P. Goode and a letter from President Reagan, North was dispatched to Wiesbaden, West Germany. The success of his mission hinged on the outcome of a secret flight to Iran by a DC-8 chartered privately through a Florida company. The aircraft was delivering American-manufactured TOW anti-tank missiles, in violation of then-existing U.S. laws. The delivery followed an earlier shipment in August. The secret flights had been arranged by the National Security Council in cooperation with Israel and were intended to secure the release of American hostages in Lebanon.[20]

Depositions for the Iran-Contra committee identified the aircraft operator as "Sur International",[21] a paper company set up in Miami a few months earlier. Two of the three principals were Colombian-American pilots: Herman Durán, a pilot with Southern Air Transport, and Jacobo "Jake" Bolivár, chief of flight operations for Arrow Air. Both Durán and Bolivár were qualified on DC-8s, which were available for lease from Bolivár's employer and Arrow Air's owner, George Batchelor. Arrow Air itself leased its DC-8s from another Batchelor company, International Air Leases, also located at Miami airport.

The DC-8 that delivered the missiles has not been identified, but it has been reported that its registration was temporarily transferred to another shadowy company, International Air Tours of Nigeria, which conducted its operations from a Miami telephone number.

The crew included Captain Herman Durán and another unnamed American of Colombian origin, possibly Jake Bolivár.[22]

The first shipment of ninety-six TOWs took off from Ben-Gurion Airport around midnight, August 20, 1985. When the aircraft landed, the Iranian Revolutionary Guard seized its cargo. No hostages were released. Despite this failure, a second shipment was made on the understanding that the delivery would secure the release of CIA station chief William Buckley.

The same DC-8 and crew took off from Israel with 408 missiles on the night of September 14. The aircraft arrived in Tabriz without incident some five hours later. After it had been unloaded and refuelled it left for Malaga, Spain, but, as luck would have it, the navigation equipment malfunctioned over Turkey. With the Colombian-trained veteran of many covert operations at the controls, the flight vanished from surveillance. It eventually turned up at Ben-Gurion airport on the morning of September 15, where it was quickly hustled to an out-of-the-way hangar.

The circumstances were sufficiently unusual to stir the Israeli media. Technicians told a curious reporter that the airplane belonged to a Miami company called "International Airlines" and that it was being repainted.[23] Its American registration number proved to be false. Israeli censors squelched local stories, but not before they were picked up by the wire services. An Israeli contact roused the National Security Council, and Oliver North showed his ability to influence civil aviation authorities by repressing all further reports of the diverted DC-8.[24] On Tuesday, September 18, the aircraft took off from Israel on a flight plan for Nairobi, Kenya — presumbly for another paint job before heading back to Miami.

On receiving delivery, the Iranians exercised their influence on their Islamic Jihad surrogates to secure the release of one hostage. But it wasn't CIA chief William Buckley—Buckley had died in June "of a pulmonary condition brought on by prolonged interrogation, torture and mistreatment".[25]

Instead, Reverend Benjamin Weir, kidnapped by the Islamic Jihad on May 8, 1984, was released a few hours after the missiles were delivered.* The letter North took to Weir from President Reagan asked for help with information that might be useful in planning a military rescue of the remaining hostages. Reverend Weir brought a message too — it set out his former captors' demands for releasing the other hostages. The terms were not divulged, but Weir did say that if they were not met, the Islamic Jihad would probably execute the remaining hostages.[26]

During the following months, top administration officials — including North's boss, National Security Adviser Robert McFarlane — would be preoccupied with preparations for the summit between President Reagan and Soviet General Secretary Mikhail Gorbachev. With the attention of his superiors thus diverted, North could bestow his unfettered energies on schemes to bring the hostages home.

* * *

On November 24, 1985, a chartered commercial aircraft secretly delivered eighteen American manufactured-HAWK anti-aircraft missiles to Iran. The delivery wound up a two-day comedy of errors orchestrated by Oliver North.

---

*Weir's release was not announced for another four days, in the hopes of securing the release of the remaining hostages.

As with the two previous deliveries of TOW missiles, North's scheme aimed to circumvent the embargo on arms shipments by drawing missiles from Israeli stockpiles and then reimbursing Israel.[27] A total of 80 HAWKs were to be shipped by way of Portugal so as to "hide the fingerprints" of Israeli and American governments. The Iranians paid cash up front, including a healthy markup for various middlemen. Byzantine financial arrangements included a $21 million "bridge loan" from arms merchant Adnan Khashoggi, arranged through a Cayman Islands firm controlled by Toronto businessmen Donald Fraser and Ernest Miller.

By this time North knew that William Buckley was dead, but a contact through Canada reassured him that the remaining five hostages were still alive — albeit in grave danger. North believed that the Jihad was "under immense political and military pressure" to dispose of the hostages because keeping them was becoming too risky. He postulated that "our hostages as well as the French and British could be killed in the near future." But North also knew that the Iranian arsenal was drained almost dry by the war with Iraq. Military material inherited from the Shah was, of course, virtually all made in the U.S.A. Iran needed more American arms, but if the Jihad disposed of the hostages they would have nothing to trade. "It's entirely likely that the only leverage that they have over us (the hostages) may no longer be available in the near future," North wrote in a memo to Poindexter.[28] Little time remained for cutting a deal.

It is not clear whether or not President Reagan approved the plan to fly HAWKs to Iran. Chief of Staff Don Regan testified that Bud McFarlane had informed the President by means of a note on the margin of his briefing at the Geneva Summit. President Reagan told the Tower Commission he couldn't remember.[29] In any event, the plans were closely guarded. McFarlane first mentioned them to Secretary of State George Shultz in Geneva on November 18. Shultz was furious, saying that if he had known, he would have scuppered the deal.[30] But it was too late.

The plan posed a monstrous logistical migraine. Officially, the Ayatollah still denounced both the United States and Israel. The Iranian air force would intercept American or Israeli aircraft trying to enter Iran. The insurers of front-loading Boeing 747s, the only civilian aircraft capable of transporting eighty HAWK missiles (which are about eighteen feet long), frowned on their use for clandestine arms shipments. Given this difficulty, Israel undertook to ship the missiles to Lisbon on a front-loading El Al 747. From there they could be transferred to three smaller chartered aircraft — B-707s or DC-8s that were registered to neither Israel nor the United States.

But there was an unexpected hitch. To arrange facilities to transfer the missiles from the 747 to the smaller aircraft, Portuguese authorities needed to know the nature of the cargo. Portugal supported the American arms embargo to Iran and it wouldn't let the Israeli 747 land.

North dispatched Richard Secord to Lisbon under the alias "Richard Copp" to set things straight — not, however, by revealing his secrets to either Portuguese or American diplomats.* No one thought to tell North's Israeli allies that Secord was using Lisbon as

---

*In *Guts and Glory: The Rise and Fall of Oliver North*, Ben Bradlee, Jr., suggests that Secord took the alias from Dewitt Copp, author of the book he was reading — *A Few Great Captains*.

a base for shipping arms to the Contras — nor were they aware (North himself may not have been aware) that Secord stood to make a personal profit on the deal.

The Israeli plane (belonging to an El Al subsidiary) was scheduled to take off on the afternoon of Friday November 22 — after business hours, to be as inconspicuous as possible. One of El Al's most experienced crews was recruited for the mission.[31] The efficient Israelis had the aircraft fuelled, loaded with the HAWKs and ready to roll at the appointed hour. But there was a delay. A business associate of Secord's had tried to bribe a Portuguese official to expedite the shipment, and the official's superiors had informed the American embassy. The State Department and the CIA had been purposefully kept in the dark. Naturally, the American embassy reconfirmed the United States' policy against shipping arms to Iran.

Secord and North tried another tack. Still loath to confide in the State Department, they turned to the CIA. Not, of course, that they were truthful even with the spooks — they asked the CIA European office to secure landing rights from Portugal for an Israeli aircraft carrying "drilling equipment".

Suspicions aroused, Portugal demanded diplomatic certification. A desperate Oliver North sought help from his boss, Robert McFarlane, who was in Geneva with the President, and McFarlane agreed to call Portugal's foreign minister. Much relieved at this high-level help, Secord signalled Israel that the operation was on. But his difficulties had barely begun: by this time it was past sundown on Friday, and the managing director of El Al was not inclined to risk the wrath of the rabbinate by desecrating the sabbath. The 747 would not be allowed to go.

A series of marathon telephone calls then enlisted the help of Israeli Prime Minister Shimon Peres. Peres was able to get special dispensation from the chief rabbi for conducting a "humanitarian mission", and the Boeing took off at seven p.m.

Unfortunately, when it looked like the mission was going to be scrubbed, Secord's penny-pinching associates had cancelled the DC-8s that had been chartered to take the missiles from Lisbon to Tehran.[32]

Back in Lisbon, the ever-resourceful Secord had another trick up his sleeve; he could divert an aircraft from the Contra resupply mission.

The problem of American markings was not insurmountable. North informed Poindexter that Secord would have the aircraft repainted overnight and put into service no later than noon the next day (Saturday). "So help me," he complained, "I have never seen anything so screwed up in my whole life".[33] And the problems were still not over. McFarlane had less influence with the Portuguese foreign minister than North and Secord had supposed, and he had only succeeded in putting the American government's fingerprints all over the operation. The Israeli 747 still wouldn't be allowed to land in Lisbon.

The 747 had just reached the coast of France when it was ordered to turn back. By this time, there wasn't enough fuel to make it back with the legally required reserves. The flight arrived back in Tel Aviv in the early morning with nearly dry tanks. The charter had expired and the cargo of HAWKs was unloaded.

Back in Lisbon, the aircraft intended to be switched from the Contra to the Iran operation could be restored to its original paint job and registration number. It was free to complete its original clandestine mission before returning to Miami for more mundane duties.

With tens of millions of dollars (certainly) and the lives of the hostages (possibly) at stake, North and Secord were not ready to quit. North contacted Duane "Dewey Maroni" Clarridge, a CIA officer who had worked with him on the Contra operation. Clarridge, in turn, called CIA director Casey, who happened to be in China.*

American law requires the CIA to obtain the President's approval before leaving its intelligence-gathering role to venture into operations. But, with Casey's blessing, the CIA chartered a plane to move the HAWK missiles directly from Israel to Iran. A Boeing 707 owned by St. Lucia Airways (described in CIA documents as "one of our proprietaries") had just delivered a load of fine furniture to Frankfurt in a normal commercial operation. Its CIA manager called the captain on HF radio and diverted the return flight to Israel.** Flight plans were filed by radio with out-of-the-way control towers to minimize "fingerprints".

The 707, flown by a German crew, arrived at Ben-Gurion Airport on Saturday morning, November 23. and Israeli ground personnel immediately shepherded it to a quiet corner. The pilot was surprised to

---

*William Casey died in hospital on May 6, 1987, after an operation to remove a brain tumour, before he could testify at the Iran-Contra hearings.

**St. Lucia was formally based in the Caribbean Island of the same name, but controlled by Airline Consulting in Port Charlotte, Florida. St. Lucia was also used for the CIA's operations in Angola until late 1987, when publicity about its role in the Iran arms deliveries blew its cover and forced it out of business. The description of the captain on the Hawk delivery fits St. Lucia Airways manager Deitrich Reinhardt, a flamboyant West German living in Port Charlotte.

find another Boeing 707 operated by "the company" already there, being loaded with long green-grey plastic-covered containers, each some twenty feet long and three feet square. But this second 707 bore American registration.

Possibilities of repainting it or flying it in formation to hide it from air traffic control were discussed and rejected. Only the non-U.S. registered aircraft would be used.[34] As it could take only eighteen missiles at a time, the plane would make several trips.

Since relations between Israel and Iran precluded a direct flight, the flight plan called for a stop in Turkey to give an acceptable point of origin. But an hour before takeoff the Turkish authorities, suspicious of the "oil drilling equipment" cover story, wanted to certify the cargo. New plans were made to land in Cyprus. As the 707 had been fuelled for the flight to Turkey and would be too heavy to land after the short hop to Cyprus, fuel was taken off. But this was not the kind of operation that uses credit cards; the crew would need cash to top up in Cyprus. It was Saturday evening and the banks were closed. The Israelis chipped in and raised the required amount and the St. Lucia aircraft and its cargo of eighteen HAWKs finally took off at one a.m.

An hour later, North received a frantic call from Israel. The 707 had landed in Nicosia without a valid cargo manifest and Cypriot authorities had detained it — the cargo would be examined in the morning. Again the CIA came to the rescue. Cash changed hands and the 707 took off from Nicosia around six a.m. Sunday morning.

The flight turned out to be as frenetic as the preparations. The "pineapples" for the Contras had become "oil drilling equipment" on various forms required for clearing the flight through Turkish airspace, but St. Lucia Airways had listed the cargo as "medical goods" and the intended destination as Tabriz rather than Tehran. The pilots resolved the ensuing misunderstanding with Istanbul air traffic control by turning off their radios and flying a zigzag route with random altitude changes until they finally crossed into Iran. [35]

The plane landed in Tehran without further incident, but it turned out that the HAWKS loaded in Israel were not the advanced model expected, but older models without the range to shoot down high-altitude Iraqi bombers and Russian reconnaissance aircraft. The Iranians, who had paid $24,720,000 in advance for eventual delivery of eighty missiles, felt double-crossed. No hostages were released. Prime Minister Peres had been compromised for nothing, and the Americans might not replace the eighteen Israeli HAWKs that were now in Iran. The fiasco had further alienated the Iranians and strained relations between the U.S. and Israel. What's more, the Portuguese foreign minister had been drawn in. And the CIA had acted illegally.

Oliver North had a few things to worry about.

To make matters even worse, a personal crisis was brewing at the National Security Council. Bud McFarlane, North's nominal boss, had never got along with the White House insiders, and feuding and nasty rumors had taken their toll. On December 4, 1985, McFarlane, reported to be "exhausted and on the verge of a nervous breakdown," resigned. He was replaced by his deputy, Vice-Admiral John Poindexter. On the same day, Oliver North sent a message to the new boss urging that the Iranians be appeased; otherwise, he claimed, there was "a good chance of condemning some or all [of the hostages] to death in a renewed wave of Islamic Jihad terrorism.'' He followed up with an ominous prediction that continued inaction would ''ignite Iranian fire'' and that ''hostages would be our minimum losses''.[36]

Eight days later, when a tremendous explosion destroyed an Arrow Air DC-8 on its way home from the Middle East and the Islamic Jihad claimed responsibility, American authorities would immediately discount the possibility of terrorist attack. But in the meantime there was a great deal of behind-the-scenes activity in Washington and other parts of the world.

In helping Oliver North solve the logistics problems with the HAWKs, the CIA had violated the ''G. Gordon Liddy rule'' limiting operational support for the White House—a rule imposed in the wake of the scandals associated with the CIA's support of the notorious Watergate ''plumbers''. Congressional oversight committees should have been notified. Many backsides had to be covered.

The CIA's lawyers hit upon a novel solution—retroactive approval from the President. On December 5, one week before the Arrow Air crash, Admiral Poindexter's saw President Reagan for the first time as the new National Security Adviser. He obtained the President's signature on a document that gave unprecedented retroactive approval for ''certain foreign aid and materièl and munitions . . . provided to the government of Iran . . . to facilitate the release of American hostages''. The secret document stated that ''All prior actions by U.S. government officials in furtherance of this effort are hereby ratified. Because of the extreme sensitivity of these operations, I direct the Director of Central Intelligence not to brief the Con-

gress of the United States . . . until such a time as I
may otherwise direct.''

So when the DC-8 crashed in Gander on December
12, Oliver North's warning about the prospects of a
''renewed wave of Islamic Jihad terrorism'' was hidden not only from the public but from the congressional oversight committees and almost everyone in
the United States government.

* * *

In January 1986, President Reagan signed a document
permitting the CIA and ''third parties'' to sell American arms to Iran. In February, SAT delivered a thousand TOW missiles to Israel and then to Iran from
Kelly Air Force Base in Texas.[37] The return flight from
one of these trips brought back seventeen of the eighteen HAWKs the Iranians had rejected, to Israel.*
Other flights may have been rendered more profitable
by ''backhauling'' arms to the Contras.[38]

James H. Bastian—the lawyer who had handled the
SAT account for the CIA and had become sole owner
of SAT after the CIA relinquished formal ownership
—gives us a glimpse of the extraordinary precautions
to keep the flights secret. The pilots flying into Iran
were put on vacation in order ''not to let the rest of
the company know that they were away and have
some excuse for them being away.'' Secrecy was of
the essence, Bastian said, because ''this was supposed
to be a very sensitive area — the Iranian affair — and
that was not in the news at this particular time''.

But Bastian defended SAT's role in the arms deliveries to Iran, suggesting that despite the secrecy,
''except for the uniqueness of providing pilots to go

---

*The Iranians had test-fired one.

into Iran, most of it was business as usual," He failed to see why the congressional committee was focusing on Southern Air. In a revealing aside, he remarked that Arrow Air had also flown "two or three flights" in a similar operation.

Arrow Air's role in carrying arms to Iran seems to have been confirmed by Arif Durrani, an American citizen born in Pakistan and jailed for supplying HAWK missile parts to Iran. In an affidavit filed in his defence Durrani claimed to have witnessed Arrow Air aircraft loading weapons in Lisbon and to have identified Arrow as a carrier of arms to both the Contras and Iran.[39]

* * *

There is another element of Oliver North's behind-the-scenes activity that should be kept in mind — one unrelated to the Iran-Contra affair. North and others in the administration were not relying solely on the missile deliveries to secure the release of the remaining hostages. It will be recalled that North (as William P. Goode) had delivered a letter from the President asking Reverend Benjamin Weir for information to help plan the rescue of the hostages. The plan was for a military rescue mission. Even if he didn't know his own or the other hostages' place of captivity, a newly released hostage would have invaluable information about windows, locks, rotation of guards and so on.

Of course, plans for a military rescue would have to keep in mind the disastrous failure of a previous rescue attempt.

Ronald Reagan himself attributed his overwhelming electoral victory over Jimmy Carter in 1980 to the crisis over the failure to free fifty-two American hostages who had been seized during the takeover of the American embassy in Tehran in 1979.[40] The assault turned into a disastrous fiasco that left eight commandos dead and put classified documents in Iranian hands. Pictures of bound and blindfolded Americans paraded about the embassy compound by hooting Iranian ruffians humiliated all Americans. Criticism and ridicule over the utter débâcle of the rescue attempt, just as the 1980 presidential election campaign was getting under way, was a personal humiliation for President Carter.

Planning for a second rescue mission had begun almost immediately after the diaster of the first. Major-General John Vaught, who had commanded the first mission, remained in charge. His deputy for planning air operations was none other than Major-General Richard Secord.

The failed rescue attempt had relied on a hodge-podge of existing weapons and units. The second attempt, codenamed ''Operation Snow Bird'', would be much more difficult, so it had to be better planned, more comprehensive. A new unit was created for the task, using specially designed equipment. General Secord assembled a fifty-man planning team backed by a force of two thousand support personnel. A fleet of helicopters, to be modified for clandestine counter-terrorist operations in the Middle East, was supplied by the 101st Airborne Division at Fort Campbell, Kentucky. A catalogue of rescue plans was devised to respond to various scenarios depending on the disposition of the hostages and the modus operandi of their guards. Secret negotiations were undertaken with Saudi Arabia, Pakistan and Turkey for permission to pre-position equipment.[41]

Operation Snow Bird never took place. On January 20, 1981 — Ronald Reagan's inauguration day — the fifty-two hostages were released after 444 days of captivity. Amid the rejoicing, Reagan declared that fighting terrorists would be a top priority for his administration. "Let terrorists be aware," he said in his inauguration speech, "that when rules of international behavior are violated, our policy will be one of swift and effective retribution." Toward this end, the special units that had been set up to plan and execute Operation Snow Bird were retained and expanded into a permanent counter-terrorist force, the Special Operations Division of the U.S. Army. A classified unit to provide aviation support for commandos, Task Force 160, was set up in May 1981 at Fort Campbell, Kentucky.

Task Force 160 got its first taste of action three months later. In August 1981, Bashir Gemayel, leader of Lebanon's Christian militia, made a secret trip to Washington to meet President Reagan. The CIA spirited Gemayel from Lebanon, but the task of returning him safely and secretly fell to the Special Operations Division.

Gemayel, his wife, Solange, and his bodyguard were taken from Washington to Cairo in a Gulfstream executive jet. In Cairo, the party boarded one of three specially modified Task Force 160 helicopters, which had been disguised as commercial vehicles and brought from the U.S. in a C-141 transport. The helicopers were refuelled at the coastal city of El Arish and from there, monitored closely by Israeli intelligence, they flew the rest of the journey a few feet over the surface of the Mediterranean to elude Syrian and Libyan radar.

This clandestine airlift took place just as the Multinational Force and Observers was being established to monitor the Israeli-Egyptian peace treaty of August

3, 1981. The presence of over a thousand American personnel in connection with MFO operations in the Sinai created a need for a constant airlift of people and equipment between the United States and the Sinai peninsula.[42] The Special Operations Division realized that the movements associated with MFO operations would provide an excellent cover for a rescue attempt for the hostages in Lebanon.[43]

By early November 1985, satellite photographs and other intelligence had pinpointed the location of five of the remaining six American hostages in a Hezbollah stronghold in the Bekaa Valley. The photos were sufficiently detailed to allow the construction of a scale model showing buildings, streets and guard stations. On November 6, the model was brought from the CIA to the National Security Council to serve as a visual aid in a high-level meeting. North, Poindexter and various officials from the CIA and FBI were discussing options for a military rescue. The results are not known, but one of Oliver North's notebooks obtained by the Iran-Contra committes referred to two operations codenamed T.H.1 and T.H.2, which were apparently alternative plans for rescuing the hostages.[44]

Washington columnist Jack Anderson and other journalists have suggested that a rescue attempt actually took place and failed. No source is given for this information, but it is known that planning for a military rescue mission continued in parallel with other efforts to secure the release of the hostages until the Iran-Contra scandal was exposed in November 1987. Journalist Bob Woodward reports, for example, that Admiral Poindexter sent President Reagan "a top-secret memo about five separate hostage rescue operations" in December 1985. "The first four operations involved other countries secretly assisting the CIA, and the fifth was about Iran.''[45]

Stories connecting the Arrow Air flight on December 12, 1985, with a failed rescue mission have not been substantiated, but it is difficult to doubt that the architects of such plans would not have taken advantage of the MFO operations in the Sinai. Certainly it would have been easy enough to persuade patriotic individuals to help in such a mission of mercy—even if the truth had to be bent along the way.

Part 3
The Dog That Didn't Bark

The blast that killed 248 American Marines at Beirut airport in 1983 registered near the top of the political and emotional Richter scales of the American people. The scandalously lax security measures protecting American peacekeeping forces led to the first serious questioning of President Reagan's foreign policy. The immensely popular president deflected calls for the court martial of senior officers by claiming personal responsibility as command-in-chief. He only escaped personal blame, some analysts believe, because the invasion of Grenada two days after the bombing diverted the public's attention.

So in December 1985, the top ranks of the U.S. administration had reason to be very anxious indeed about any claims that peacekeeping forces had been massacred by terrorists. Allegations of carelessness and ineptitude would be inevitable — and the President could not afford to shoulder the blame again. The political fallout would be disastrous.

Worse yet, if the issue of the Islamic Jihad's terrorism again filled the media, there was a risk that word of the missiles-for-hostages deal might leak out. The administration had reaped a huge public-relations bonanza when Weir had been freed two months earlier—the first return of an American hostage since the triumphant homecoming on President Reagan's inauguration day. The knowledge that the Jihad had extorted American missiles for his release might have given public perception a very different twist.

219

Oliver North served as chairman of the NSC's "Terrorist Incident Working Group" — "the crisis management committee of top counter-terrorist officials from the Pentagon, State Department and National Security Council".[46] One might suppose his warning of a renewed wave of Islamic Jihad terrorism would have cranked up security measures for troops in the Middle East. One might also expect the crash at Gander to give pause to those who knew of North's ominous prediction that inaction would "ignite Irian fire" and that "hostages would be our minimum losses". Yet it was those who had most vehemently warned about the scourge of terrorism who dismissed that very possibility before there could be any reasonable assessment.

There were, of course, many reasons to prevent a thorough investigation, regardless of cause. Consider Arrow Air's part, small as it may have been, in the arms deliveries to the Contras. Oliver North certainly sought to quash an FBI investigation of Southern Air Transport after the loss of a Contra supply flight on October 6, 1986. The surviving crew member, Eugene Hasenfus, claimed to be working for the CIA and papers found on the dead pilot, William Cooper, identified him as an employee of Southern Air Transport, but the American government denied all involvement. When the FBI and the Customs department opened routine investigations of Southern Air Transport, North got his boss, John Poindexter, to call Attorney-General Meese and turn off the heat. North himself worked on FBI Deputy Director Buck Revell.[47]

FBI director William Webster disclosed Meese's interventions on April 8, 1987, during confirmation hearings into his nomination as Director of Central Intelligence. Ironically, the efforts to subvert the investigation of the Hasenfus crash came to light just as the members of the Canadian Aviation Safety Board received the investigators' draft report on the Arrow Air DC-8 crash.

The need to keep the lid on the Contra operation was even more critical in December 1985 than at the time of the Hasenfus flight. The American administration was lobbying fiercely to block the extension of the Boland amendment, due to expire at the end of the fiscal year. The lapse of the Boland amendment would, in the words of Richard Secord, return the CIA's "hunting license" and stop the need for secrecy. Investigators of the Arrow Air crash in Gander could have stumbled unto the arms shipments to the Contras and compromised the lobbying efforts — regardless of the cause of the crash. As Oliver North stated on another occasion, in reference to trafficking with the Iranians, "if it ever becomes known that we have done this, we would be finished in terms of credibility". Even short of public disclosure, any connection between the Jihad and the Gander accident would have compromised North's efforts to quell the opposition of Secretary of State Shultz and Defense Secretary Weinberger.

In short, the were many motives for blocking the investigation — and concerned people had blocked other investigations on no greater grounds. And so — like Sherlock Holmes' famous dog who failed to bark in the night — the watchdogs of terrorism were resolutely silent. Perhaps they were burying bones.

And they would have succeeded, had it not been for the unprecedented power of the Iran-Contra committee to compel testimony and subpoena records. Without such powers, the extraordinary hypocrisy and eagerness to deceive would have foreclosed any speculations about the circumstances of the Gander crash, the conduct of the investigators, and the identities of the puppet-masters who were pulling the strings—if there were any puppet-masters.

Readers of the strange chronicles of the Gander investigation will have to make up their own minds whether there is any significance to "the Oliver North crap", as one investigator termed such speculations.

BOOK 4

A UNANIMOUSLY ADOPTED DRAFT REPORT

BOOK 4

A
UNANIMOUSLY
ADOPTED
DRAFT REPORT

Part 1
Going Once . . .

When I arrived at the Canadian Aviation Safety Board in January 1987, the investigators' report on the Gander crash was not yet ready for board review. Nevertheless, an undercurrent of tension about the conduct of the investigation swirled just below the surface.

The board members had heard very little about the progress of the investigation since the seven-day public inquiry nine months earlier. The minutes of the meeting after the inquiry (May 1-2, 1986) register that "Board Members agreed that the inquiry was a success and congratulated staff for their work" and note that "staff expects that the final report will be published within one year of the accident". But the first anniversary had come and gone. Chairman Deschênes did have some public comments to mark the anniversary. "Although the press seems to have ascribed the cause of the crash to the crew's failure to have the plane de-iced before takeoff," he assured everyone, "the CASB is continuing to examine all possible causes." In addition to ice, he said, the investigators were also looking at reduced power in one engine, overloading and incorrectly calculated takeoff speed. They were also weighing the results of flight simulator tests conducted in Copenhagen and a computer study from the University of Dayton.[1]

Deschênes' remarks alarmed Ross Stevenson, mostly by what was omitted. This was the first substantive indication of where the investigation was headed since the public inquiry had closed eight months earlier. Since then, the press had said almost nothing about the investigation. The press, like Stevenson and the rest of us, was waiting for the investigators to submit a report.

Stevenson had started flying DC-8s when Air Canada introduced them in 1960. When an Air Canada DC-8 crashed while aborting a takeoff from a slick runway at Heathrow Airport in 1962, Stevenson served as the pilot's union representative. He had followed the history of DC-8 accidents ever since. To him, the meteorological evidence at the inquiry had practically ruled out the ice-on-the-wing theory. Nor were Deschênes' remarks reassuring: the items he mentioned were not alternative causes, but contributing factors in addition to ice. It seemed that the investigation had indeed fixed on ice.

Stevenson sent Deschênes a long list of additional areas that needed to be examined. These included:

- —flight control binding*
- —inadvertent ground spoiler deployment
- —deteriorated hydraulic systems
- —inadvertent deployment of a thrust reverser
- —runaway pitch trim compensator

He also asked for the report from the flight simulator tests. Deschênes refused. When Stevenson insisted, the chairman outlined his reasons: "In our task of adopting reports and issuing recommendations . . . ," he wrote,

---

*The possibility that a restricted cable or other mechanical impediment had prevented the pilots from pulling the control column far enough for lift-off was suggested by a pilot who had complained of a slight unnatural "ratcheting" in the Arrow Air DC-8's control column. This and the other items on Sevenson's list had all contributed to previous DC-8 crashes.

"It is very important that we do so without being influenced one way or the other, by anything but the full investigation. When certain parts of the investigation work are viewed in isolation we can begin to form conclusions that would be based on that conclusion alone. . . . I would prefer that the Board receive the complete file with the draft report."

The 100-page draft report finally arrived without fanfare in our inter-office mail a week or so before the meeting of April 22, 1987. It had been prepared with great care. A transmittal letter from the executive director, Ken Johnson, to Chairman Deschênes certified that the report, designated as Draft No. 2, had been approved by various managers. Their approval was registered in a six-tiered signature box. Unlike other draft reports, this one was bound and edited for immediate release.

Before issuing a public report, the CASB was required to send a draft "on a confidential basis" to "any Minister or other person" who in the opinion of the board had a direct interest in the findings. The legislation also required the CASB to give such "interested parties" a reasonable opportunity to make representations before finalizing the report.[2]

Representatives of certain interested parties — the Minister of Transport, Pratt & Whitney, Douglas Aircraft and the United States Army — had kept in close touch with the investigators from the outset. Help from certain representatives is, in fact, indispensable. An investigator needs expert help, for example, to tell if a gizmo in the wreckage is from a model JT3D-7 engine as it should be, or is a mistakenly installed part from a JT3D-6.

As the Dubin Commission showed, the assistance of such interested parties has a darker side. The *Wall*

Street Journal noted, for example, that "a finding that a mistake was made by a manufacturer or an airline can result in enormous liability." Accusations have been made that "these other parties sometimes hinder efforts to determine the cause of the crash" and that instead of providing impartial expertise they "sometimes try to muddy the waters or slant things in certain ways".[3]

William Tench, who was chief inspector of aviation accidents in the United Kingdom from 1974 to 1981, says outside interests "frequently thought it advisable to make an attempt to conceal some fact or occurrence which [they] thought might be damaging to the particular company concerned".[4] Rudy Kaputsin, former chief investigator at the NTSB, makes the same observation; the investigator in charge needs to exercise "heavy duty assertiveness" to resist "pressure from high horse-power individuals" who try to influence every major investigation.[5] But Chairman Deschênes' philosophy left no way for board members to assess how the (necessary) close interaction between vested interests and the investigators might have influenced the report.

Board members could be certain, however, that one large group with an intense "direct interest in the findings" had not influenced the report in any way during the sixteen-month investigation — nor would this group be able to influence the report after reviewing a draft. The CASB did not recognize the next of kin of deceased passengers as interested parties.

Giving the victims' families such status would, of course, have greatly complicated the process of issuing reports. But, as Judge Rolf pointed out in his scathing remarks about the CASB's conduct of the inquiry into the accident at High Prairie, denying the families an opportunity to make representations can cause grave injustice.

As it turned out, the families of the Gander victims would have to wait almost two more years to learn what the investigators had concluded. Their outrage would lead to a congressional inquiry, but not before another two years had passed.

Meanwhile, back in April 1987, the investigators believed their probe of the Gander crash successfully concluded. To them, the draft report was definitive, and review by board members and interested parties was no more than a necessary formality.

* * *

According to the synopsis on the first page, "The Canadian Aviation Safety Board determined that, after leaving ground effect, the aircraft stalled at higher than normal airspeed, as a result of aerodynamic performance degradation caused by ice contamination of the airfoil."

In keeping with a formula set out by the International Civil Aviation Organization (ICAO), the report has two main parts: "Factual Information" to record facts and "Analysis" for interpretations and formulation of hypotheses. Each of these consisted of a number of subsections — on weather, engines, flight recorders and so on. A third section listed "Findings": in this case six "Cause Related" findings asserting the conclusion in the synopsis, and a hodge-podge of twenty-two "Other" findings.

The ICAO format is convenient for systematic indexing and correlation of the thousands of aircraft accidents that occur in different parts of the world. But it's not the best vehicle for clarifying convoluted circumstances or explaining the evidential basis of complex deductions. Despite the impressive heft and seemingly encyclopaedic scope of the Gander report, I found the logic tortuous and difficult to follow.

The "Cause Related" findings pinpointed ice on the wings as the primary cause clearly enough. The "Other" findings seemed to rule out mechanical failures of the type indicated in Ross Stevenson's list. And, by implication, the crew had neglected to check for ice on the wings because they were suffering from "both acute and chronic fatigue." Presumably this fatigue was also responsible for the pilots misplacing markers on the airspeed indicator to trick them into raising the DC-8's nose for takeoff too early, thus rendering the effects of the ice even more insidious. But I couldn't tell if the investigators believed a stall after lift-off was inevitable, or if the tired pilots had failed to take corrective action that could have averted the disaster.

I focused on the parts dealing with the flight data recorder traces, since this aspect was within my own professional expertise and since it formed the basis for the inference of performance as expected with ice on the wings and a stall at higher than normal airspeed.*

For many years, aviation authorities had insisted that new aircraft be equipped with digital flight recorders to store accurate information from dozens of sensors on magnetic tape. But the FDR on the Arrow Air DC-8 was an outmoded type that had been state-of-the-art when the DC-8 received its type certification in 1958. It recorded the aircraft's speed, altitude, heading and vertical acceleration by scratches on a metal tape winding slowly between two spools.

---

*The flight data recorder (FDR) is one of the so-called "black boxes", the other being the cockpit voice recorder (CVR).

. The information from metal-foil FDRs is crude at best, and this tape was much worse than usual. The unit had not been inspected for some time and the spool taking up the tape ran unevenly. The tape's jerky passage made it even more difficult to estimate the time of recorded events. Yet the investigators had made incredibly accurate deductions. They deduced, for example, that the pilot lifted the nose wheel precisely forty-two seconds after the aircraft started to roll, and that after precisely nine more seconds the DC-8 lifted off the runway.

I was immediately suspicious. All my experience suggested that the time of events could not be determined so precisely — not even with modern digital flight recorders. I also knew that even the best-qualified specialists could misinterpret even the best FDR data. The best analysts at the NTSB had, for example, made a timing error of *three and a half seconds* in reading the up-to-date digital flight recorder from the Air New Zealand DC-10 that crashed in the Antarctic in 1979.[6]

Judge Peter Mahon, who conducted a Royal Commission of Inquiry into the Air New Zealand crash, believed the misinterpretation of the FDR data occurred because:

> the transcribers had made the mistake which investigators have often made in times gone by and in different circumstances. . . . Many police inquiries have gone wrong for the same reason. The mistake they made was to first postulate what they had thought had happened, and then treat all information which did not fit their theory as being not correct.[7]

As I examined the investigators' deductions from the comparatively meagre and ambiguous data on the Arrow Air FDR, it seemed that they might have fallen into the same trap. The report asserted that a stall was evident from the FDR data. As a young engineer I had spent many weeks recording and analyzing stalls on wind tunnel models. Yet I could detect no evidence of the stall, which was central to the conclusion that this crash was caused by ice.

The report also alluded to calculations that seemed impossible. It claimed, for example, that the DC-8 had climbed to a peak altitude of seventy feet above the runway. This estimate purportedly came from a "double integration" of the vertical-acceleration data from the FDR — the method used to calculate altitude in inertial navigation systems. But inertial navigation computers use data that are incomparably more accurate than the best available from metal-foil flight data recorders. I could see from a glance at the fuzzy vertical-acceleration trace reproduced in the report that such a computation would be meaningless. The trace was almost flat, with a few small random blips — it reminded me of the electrocardiograms of the hibernating astronauts after the computer HAL went berserk in the film *2001*. The instrument had obviously malfunctioned.

I asked for more accurate plots of the FDR data. They were better, but I still couldn't detect a stall. The altitude and vertical-acceleration traces still seemed meaningless. The heading trace showed that the DC-8 had started a gentle turn to the right soon after lifting off; this was potentially significant. But the most interesting information was on the velocity trace, which showed what appeared to be normal acceleration to a respectable flying speed of 172 knots (about 200 miles an hour), followed by a rapid loss of some 30 knots in about seven seconds before the trace took a few wild swings and stopped.

Using aerodynamic data for a different model DC-8 that happened to be handy, I quickly verified that the acceleration along the runway was close to what would be expected under normal conditions. It was also clear that a weight difference of ten to fifteen thousand pounds and the so-called ground effect were not important factors. The deceleration of 30 knots in seven seconds was close to what could be expected if all four engines suddenly lost power. This, I thought, was surely important.

I carefully read what the investigators had written about the engines. The blades on the front fans had been husked from their shafts like kernel corn from the cob, so all four engines had been spinning at a good clip. But this was only to be expected, even if they had been nominally "stopped". A 200 mph wind howling through the case turns a dead engine into a windmill that spins like a mad dervish. It was evident, though, that the number-four engine—the engine on the extreme right—was turning more slowly than the rest. While the shafts of the other three engines had been twisted in the crash like licorice sticks, the number-four shaft was not deformed.

The investigators assumed that the spinning engines must all be putting out full power. They speculated that branches and shrubs swept into the inlet before impact might have reduced the spin of number four.* It didn't seem to occur to them that low power

---

*The previous crew had reported number-four engine running a bit warmer than the others, and the flight engineer had retarded the power to even out the temperatures. If the flight engineer had done the same when departing from Gander, the power on number four would have been some 2½ per cent lower than on the other three. But this difference would have an imperceptible effect on shaft twist.

on that engine could have something to do with the veer to the right before the crash.

At this stage I didn't study the rest of the report very critically. I overlooked, for example, the significance of the three short paragraphs of "Medical Information" in the "Factual" part. The concluding sentences stated: "All 248 passengers sustained fatal injuries as a result of impact and the post-impact fire. No post-mortem evidence was found that was consistent with the detonation of an explosive device." Of course, this is not factual information but a conclusion. But such lapses were not uncommon in CASB reports.

I also failed to grasp the importance of another "Factual" sub-section, "Fire". "An intense fuel-fed fire developed," said the report. "Substantial portions of the aircraft were consumed in the fire. As a result, it was impossible to account for and examine all the aircraft." This was certainly factual, but the report omitted other significant factual information: the gigantic explosion that raised the mushroom cloud noted by witnesses for miles around, for example. It also said, "All fire damage was attributed to post impact fire." It was indeed a fact that certain individuals "attributed" all the fire damage to the blaze that had broken out after the crash, but the draft report didn't identify those doing the attributing. Nor did it note that experienced investigators, including some on the staff of Tom Hinton, the director of investigations, had other ideas.

To explain this purportedly factual information, the second, analytical part of the report included a sub-section titled "Explosion or Fire", less than two pages long. It too made no mention of the explosion; a reader relying entirely on the draft report would have been left with the impression of a "severe fuel-fed fire" only. The report did raise the subject of an explosion, to be sure—but only as a hypothesis to be dismissed. It allowed the existence of "considerable speculation that the accident occurred as a result of the detonation, either accidental or deliberate, of some explosive device." Such speculation, it explained, "was fuelled by the fact that the military personnel and equipment were aboard the flight and by the increasing world-wide incidence of terrorist activity. Also contributory to this speculation was the point of origin of the flight and the reports by three witnesses of a yellow/orange glow emanating from the lower surface of the air-craft." The report went on to dismiss such specula-tions on the grounds that "Detailed examination of the wreckage . . . revealed no evidence of an explo-sion or pre-impact fire."

At this time I had not yet studied the transcripts of the public hearings, nor was I aware of Irving Pinkel's notions of an in-flight explosion. I accepted the assess-ment on the assumption that the search for evidence of a pre-impact fire had been thorough. But I couldn't accept the statement that added that "the perform-ance of the aircraft was not consistent with a sudden and catastrophic event such as an explosion."

This part of the analysis brought me back to my ruminations on the FDR data. My rough reckonings showed that information from the flight data recorder *was* in fact consistent with "a sudden and catastrophic event". But so far, I had not seen the detailed calcu-lations that I assumed were the basis for the claim. I thought I might have overlooked some subtle factor. Clearly many matters would have to be resolved before the report could be released to interested par-ties. Or so I thought.

The Opposing Teams Line Up

Having immersed myself in a study of the draft report for several days, and being aware that Stevenson and Bobbitt also had grave misgivings about the ice theory, I wasn't sure what to expect when the board assembled to discuss the report on the afternoon of April 22. Investigator-in-charge and nominal author Peter Boag would take the lead in answering the board members' questions. He was flanked by Hinton, Johnson and a gaggle of other staffers, some of whom I had never met.

Deschênes opened the discussion solemnly, reminding us of the importance of the investigation and the eagerness with which the world awaited the CASB's verdict. He complemented Boag and his colleagues for their excellent work, which, in the chairman's opinion, had resulted in an exemplary report that could be approved with a minimum of discussion.

Nevertheless, because of the importance of the investigation, and in order to give us ample time to ask questions, Deschênes said he would allocate an unprecedented amount of time to the review of the sixteen-month investigation: the remainder of the day, the following day and two days the following week. The report could then be sent out for comments from interested parties, which would give them time to respond so that the board could issue its final report before the second anniversary, on December 12, 1987.

As each board member gave an opening reaction, any hopes for a quick *pro forma* discussion evaporated. Six of the ten rejected the conclusion that ice on the wings could have caused the crash — and one was Frank Thurston, retired director of the National Aeronautical Laboratories and Bernard Deschênes' confidant. Later Frank would become a staunch supporter of the ice theory, but that day, he faulted the amount of space devoted to the nefarious effects of ice, proposing that the board would become ''a laughing stock'' if the excess was not moderated. Frank joined the three of us who were aeronautical engineers — Norm Bobbitt, Dave Mussallem, and me — in challenging the technical basis of the ice theory.

But the most vociferous attack came from Ross Stevenson. He announced that the investigators had closed their minds to possible causes other than ice. He reminded Deschênes of his memo about possible mechanical causes, and said the report paid less than lip service to such possibilities. Stevenson was especially critical of what seemed to him a cavalier treatment of important witness evidence. The draft report gave short shrift to three witnesses reporting a yellow/orange glow from the lower surface of the plane.'' One of them was said to have described the reflection of approach lights off the DC-8's belly, and according to the report it was ''extremely likely'' that the other two witnesses were describing the aircraft's landing lights.

Stevenson reminded us that Cecil Mackie had told reporters he could have read a book by the light coming from the plane as it passed over his truck. Stevenson also pointed out that the DC-8 had drifted to the right and was not over the runway lights. The investigators were not able to establish whether the approach lights had been turned on, but even if they were, they would have been set at low intensity — hardly bright enough to read a book by directly, let alone by the reflection from an airplane flying over. There was no way, Stevenson said, that the witnesses could have seen either the reflection of the approach lights or the landing lights — which are not yellow/orange anyway, but white.

The sixth board member who rejected the ice hypothesis was 55-year-old Roger Lacroix. When he was appointed to the CASB in 1984, the former fighter pilot had been a senior official in the Transport department. He had joined the department recently, after retiring from his last military assignment as executive assistant to the Minister of Defence in the previous Liberal government. With his shock of curly white hair and cheerful, outgoing personality, Lacroix contrasted markedly with his fellow French Canadian, Bernard Deschênes, who was, by anybody's description, an introvert and a cold fish.

During the year and a half since the CASB had been formed, Deschênes and Johnson had infuriated Lacroix with a bureaucratic game. Roger had accepted that the tokens of rank — size of office, thickness of carpet and the like—should place him below the chairman. But it galled him no end that Johnson pointedly outdid him in such symbols. Lacroix was well aware of the message in such perks, and equally aware that complaints would label him a puffed-up, self-important General Foggbottom. Consequently, he restrained himself and succeeded in having the worst of both worlds — doing without fitting symbols of his relative rank, and revealing his displeasure just enough to earn veiled jibes.

Frank Thurston, who was greatly proud of his own humility, tauted Roger the most. Thurston's ostensibly light-hearted persiflage particularly irritated Lacroix because Thurston had failed to correct him when he had respectfully addressed the older man as "Doctor". It had taken him a long time to discover that Thurston was not entitled to the honorific, so he deemed Thurston's jibes particularly hypocritical.

Lacroix's frustrations had boiled over at one of the first board meetings I attended. Roger had brought up a procedural point harking back to something before my time and I had let my attention drift. Suddenly I became aware of an animated three-way discussion between Lacroix, Deschênes and Johnson. Lacroix was obviously hot under the collar and the other two were trying to calm him down. As I tuned in, Johnson was explaining in a thoroughly unctuous way that, unfortunately, due to a flaw in the wording of the civil service regulations — a flaw which, by the way, Johnson was striving mightily to correct — board members had to fly economy class while travelling on board business. (The regulations allowed Johnson and senior staffers to fly business class.)

A number of Johnson's staff sat smirking on the sidelines. Lacroix was not about to be trapped into making a scene for their benefit. To vent his frustration, he challenged Deschênes' claim that the legislation gave him, Deschênes, absolute control over where and how board members could travel on board business.

At this time the board had not yet polarized into warring factions, and most members supported Lacroix's contention that procedures should be determined by majority vote. The standoff was defused by a unanimous agreement to secure an independent legal opinion on the issue. As chairman and lawyer, Deschênes claimed the right to frame the question and select the attorney.

The incident passed. But it would return — with an unexpected twist — to haunt our review of the Gander report.

* * *

Bernard Deschênes gave no sign of how he felt about the rejection of the ice theory by a majority of the board. He simply said that the review could now continue with question about the "Factual" part of the report to the investigator in charge and his colleagues.

Peter Boag began answering the flood of questions in a self-confident, expansive manner, fixing his frank, dark eyes on the questioner and gesticulating in a moderate Dale Carnegie manner. As his answers went on and on, I wondered if he had watched Oliver North become an American folk hero by demolishing inane, effete questioners with twenty-minute eulogies on his love for his wife, dog, profession and commander-in-chief.

Boag seemed resigned to idiosyncratic suggestions for improving the grammar, but was adamant that the substance of the report was incontrovertible. He could barely conceal his irritation when Ross Stevenson presumed to suggest an experiment to test whether it was really possible to see reflections from the approach lights on the belly of a low-flying DC-8. "We have all the information we need," the chief investigator said.

As the questioning intensified, Boag became more and more agitated and long-winded. From time to time, Tom Hinton interjected a monologue about the high quality of the investigative process that he had had the foresight to introduce — thus allowing the increasingly exasperated Boag to catch his breath. The game plan didn't seem to be working, and the investigators were hogging the puck behind the net till the end of the period.

The next day was much the same, and after a day and a half of increasingly truculent debate, the meeting adjourned till the following week, leaving a haphazard list of wording changes, questions and complaints about the report. Everyone knew that we were getting nowhere.

Up to this time, I had formed my opinions in isolation. Now I took the opportunity to discuss my doubts about the ice theory with the others who had expressed reservations. I believed that the investigators had overstated the effects of a trace amount of ice, but that this was not really relevant since the FDR showed that something other than ice had caused the crash. I had not thought to question the contention that there was a small amount of "ice contamination" on the wing of the DC-8 in the first place. But Stevenson, Bobbitt, Lacroix and Mussallem had studied this point intensely after the public hearings a year earlier, and soon showed me that a careful examination of weather reports and witness statements not only failed to indicate ice on the wing, but seemed to rule it out. I was particularly impressed that Ross Stevenson, with his vast experience of flying airliners, rejected the ice theory on grounds completely different from those that forced me to the same conclusion.

When the board reconvened on April 28, it was evident that the proponents of the ice theory had regrouped and reconsidered their tactics. Evidently Chairman Deschênes had had a heart-to-heart chat with Frank Thurston, for Thurston opened the discussion by regretting the attitude of confrontation that had developed. The paragraph-by-paragraph discussion of the "Factual" part was getting nowhere. It seemed obvious, Thurston said, that the investigators had over-emphasized the ice theory. His feeling was that most if not all board members would be more comfortable with a finding that the massive destruction of evidence and the poor quality of the FDR data had kept the board from establishing a cause. Thurston then proposed a vote on whether the cause should remain ice or be declared "undetermined".

Up to that time, I had had an apparently congenial relation with Frank Thurston. During the previous few days I had explained my interpretation of the FDR traces to him, and I thought he was as incredulous of the ice theory as I was. I did believe that the investigators had not been able to determine the cause, but Frank's proposal seemed disingenuous. The four of us who had argued against the ice theory insisted that we couldn't reach a conclusion without first establishing the facts, and the discussion dragged on as before. But not for long.

The change in mood occurred when Stevenson referred to "witnesses who saw the airplane on fire." Boag and Hinton interrupted almost in unison, "No witness saw fire," and Norm Bobbitt retorted, "That's a lie." After a moment of stunned silence, Deschênes called for a coffee break. He and Thurston departed for a conference with the investigators.

When we reconvened, Thurston assumed the mantle of the conciliatory elder statesman, suggesting that we excuse the staff and have an *in camera* discussion to put the process back on track.

Everyone let off some steam. Stevenson groused about the staff's "stonewalling", Lacroix unloaded his opinion of Peter Boag, and Thurston countered that "His arrogance is proportional to the pugnacity with which the report is attacked." Ex-speechwriter Bill MacEachern seemed self-conscious about his inability to join in the techno-talk; our role was to show good judgement and common sense and we shouldn't try to be investigators, he chided. Arthur Portelance, the former Member of Parliament and garment industry executive who usually had little to say, gave us a lecture on trusting the people who work for us.

The primal scream therapy seemed to work. In possibly the last unanimous decision taken by the CASB with regard to the Gander investigation, we agreed that it would be futile to continue with a line-by-line review of the report. We would form committees to let those interested in specific areas review relevant material with the most knowledgeable specialists on the staff. One committee would examine the accuracy of technical information. Another would consider alternatives to the icing theory, such as the ones on Stevenson's list. Deschênes proposed, and everyone agreed, that Frank Thurston would chair both committees.

The committees distanced the review of the Gander investigation from the CASB's other business, allowing us to get on with other accident reports. I looked forward to retracing the steps leading to the calculated results with fellow engineers; I had visions of us sitting around a table in shirtsleeves with our calculators in a familiar, fraternal environment—an environment ideally suited to pinpointing flaws in logic.

My hopes were soon dashed. Only one staff member, an insider from the headquarters group, was available to review the calculations. He did have an engineering degree, but he had joined the air force as a pilot immediately after graduation and had never practised his profession. He had not, in fact, carried out or even thoroughly studied the analysis of the FDR data.

We did learn that the data had been read from the metal FDR foil by Bernard Caiger, a respected specialist with the National Research Council of Canada who had since retired. The data had then been interpreted and analyzed at the Douglas Aircraft headquarters in Long Beach, California. No one was available to explain the Douglas calculations, but I received a stack of computer printouts.

Fortunately, the Douglas printouts were largely self-explanatory. The FDR traces agreed with the designers' expectations for the takeoff run but the data from the quirky FDR were not good enough to show if the performance expected for a total weight of 344,000 pounds (as estimated by the investigators) fit better than the 330,000 pounds recorded by the crew. It was evident, however, that the 14,000 pounds difference would not have been a decisive factor.

The printouts also showed what would happen if the aircraft drag — that is, the resistance of the air working against the thrust of the engines — was substantially increased after lift-off while, at the same time, the wing's capacity to generate lift was substantially reduced.* These were, of course, assumptions presumed to correspond to "ice contamination" of the wings.

The Douglas computations were, as far as they went, undoubtedly correct. They showed that if the lift and drag behaved as postulated, the aircraft could stall after lift-off — "could" instead of "would" because the results hinged on additional assumptions

---

*The assumptions were that the lift and drag were normal at the low angle of attack during the ground run, but when the angle was increased for lift-off the drag would rise steeply while the lift would not increase. In more technical language, it was assumed that the coefficient of induced draf substantially increased and the angle of stall decreased by some 5 degrees.

about the crew's actions. The heavier the aircraft, the longer it takes to reach the correct speed; the heavier the aircraft, the higher the nose has to be held in the climb. An airplane will certainly stall if the pilot in control raises the nose for lift-off too early and holds it too high. But since the second pilot is supposed to monitor the takeoff, both crew members would have to ignore their training.

The printouts showed that, with the reduced lift and increased drag, the DC-8 could have stalled if the crew held the nose at the angle normally required for a weight of 344,000 pounds and could have come close to stalling at 330,000 pounds. Again, these results called for a further assumption—that the crew rotated the DC-8's nose for lift-off too early for even 330,000 pounds.

But the many assumptions required to turn the possibilities into probabilities were not spelled out—and led me to a series of arguable questions. Did the unusual lift and drag data used by Douglas adequately represent some amount of ice on the wing? If so, how much? Had there been that much ice (possibly, probably or certainly)? Would the pilots have raised the nose prematurely? Would there have been any warning of impending stall? Would the crew have ignored such a warning (possibly, probably)? If so, why?

Even if we could establish that the assumed effects of ice were realistic and that the crew would have allowed the aircraft to stall, we would have a long way to go to show that a stall was a factor in the crash. The ice theory would not be plausible unless other quantifiable factors — altitude, maximum speed, rate of deceleration, distance covered before impact and so on—matched observations at the crash site or deductions from the FDR.

The CASB investigators had not attempted to link such observations to the Douglas calculations. Instead, they had produced a report of a computer simulation done under contract by two academic consultants at the University of Dayton.[8] One part computed trajectories based on various assumptions about the airplane — altered lift/drag data to represent the effect of ice on the wing (similar, but not exactly the same, as that in the Douglas calculations), an engine not working. Of course, the computer simulations had to include a variety of implicit assumptions about how the pilots would deal with the impending catastrophe.

I believed the increased drag and reduced lift assumed in the calculations was exaggerated in view of experimental studies on the effects of ice. But even if the assumptions were accepted, the computer model showed the aircraft crashing only if one engine was dead as well. To top it off, the combined engine-out/ice-on-the-wing calculations showed the aircraft travelling farther past the end of the runway before it crashed than the 3,000 feet reached by the Arrow Air DC-8.

I hardly looked at a second, more abstract part of the Dayton report, which claimed to work backward from the FDR data to compute the corresponding trajectory—an extremely difficult task under the best of circumstances. The information from the FDR was too sketchy to inspire any confidence in the results; the conclusion depended on a host of unstated assumptions about the actions of the crew, complications due to the right turn and sideward drift of the aircraft, and so on. In so far as the calculations meant anything, they meant that ice on the wings could not, in itself, have caused the crash. But no one on the CASB staff was available to discuss the accuracy, let alone the implications, of the Dayton calculations. Frank Thurston, who chaired the committee supposedly examining the results, endured my increasingly frustrated explanations with benign amusement.

As it turned out, the second part of the consultant's report — the part I had glossed over — would eventually play a crucial part in exposing the fallacy of the ice theory. But not until a lot more water had passed under the bridge.

* * *

While I struggled to understand the mounds of calculations, the second committee was also meeting. There, the atmosphere was more acrimonious. Ross Stevenson couldn't understand what could have driven the investigators to rule out mechanical failure when so much of the aircraft had been destroyed. He was also determined to get to the bottom of what seemed to him perverse treatment of the witness evidence. In the investigators' view, members of the ground crew who hadn't seen ice on the airplane had not been looking; on the other hand, witnesses who reported a fire on the aircraft must have been mistaken.

Stevenson wanted to see the original witness statements. Informed that transcripts had not been made, he insisted on listening to the tapes. He was given boxes of cassettes without a master list. At the time he was commuting from Ottawa to his home in Toronto every weekend, and he took to carrying a portable tape recorder and listening to the cassettes on the airplane and in the airport.

Stevenson was amazed to learn that Loughren and Mackie, the two truckers on the Trans-Canada Highway, had not talked only of a yellow-orange glow. Each man had plainly said he believed the aircraft was on fire as it passed overhead — as the media had reported on the day of the crash. Stevenson found other witnesses, not mentioned in the report, who seemed to corroborate an in-flight fire.

As the committee met during the next two weeks, Stevenson and Lacroix raised question after question and objection after objection with the draft report. They pointed to the witnesses who had said there was no freezing rain while the jet was on the ground, to the evidence from ground handlers that there was no ice on the airframe, to the flight that landed shortly after Arrow Air and didn't need to be de-iced.

Stevenson also ridiculed the contention that the crew had rotated the aircraft (i.e., raised its nose) too early for lift-off — a contention needed to show that the aircraft stalled, given the aerodynamic assumptions used in the calculations.

The investigators had based their deduction of early rotation on the position of three "speed bugs" on an airspeed indicator found in the wreckage. As pilots have no direct way of knowing how far they are along the runway, one looks out to keep the aircraft on the centerline while the other watches the airspeed indicator and calls out speeds. "Vee-1" indicates that the flight is committed to takeoff; 'Vee-r" is the sign for commencing rotation. A myriad of pre-takeoff cross-checks and independent computatins by each of the three crew members ensure that mistakes are not made, the last step being to set the "bugs"—movable pointers that slide on a ring on the outside of the indicator—to mark the critical speeds, which depend on weight and other variables. Stevenson had often seen speed bugs slip spontaneously on their bezel rings, yet the investigators were relying on the position of a bug on an instrument that had been ripped from an aircraft and tossed violently about in a crash. They couldn't be serious.

For his part, Roger Lacroix was most irked by the flight simulator tests done at Copenhagen — the tests that Deschênes hadn't wanted Stevenson to see for fear that he might ''be influenced one way or other'' by something ''less than the full investigation.'' The wrecked remains of the number-four (extreme right) thrust reverser unit looked very different from the other three, much as if the unit was open (i.e., reversing thrust) before impact.Reverse thrust on the right side would, moreover, explain the right turn. So when Lacroix heard that simulator testing was planned, he suggested that tests should include a simulation of the thrust reverser popping open on the number-four engine. But the investigators had not made even a token effort to comply. The Copenhagen simulations were aimed exclusively at the ice hypothesis.

The more I looked at the investigators' report, the less it seemed a technical analysis and the more it seemed a brief prepared for a plaintiff in litigation. During litigation it's not unusual to hear technical experts argue convincingly and at great length that the sun rises in the north and sets in the south, while other experts with equal or greater qualifications counter that, on the contrary, it rises in the south and sets in the north. A well-documented illustration can be found in the multi-million-dollar legal suits following two fatal crashes where wings snapped off Lockheed Electras shortly after the aircraft's introduction into service in 1959.

Hundreds of pages of technical analysis conducted by consulting engineers on behalf of the Allison Engine Division of General Motors proved beyond a shadow of a doubt that the problem originated with a design error in the wing structure. At the same time, hundreds of pages of technical analysis submitted on behalf of Lockheed Aircraft proved beyond a shadow of a doubt that the problem originated with a design error in the engine gearbox. When the mystery was finally solved, both groups of experts were proved wrong.[9]

In the courtroom it is expected that an attorney will stretch all points favoring his client, and minimize the others. Here, however, we were supposed to be looking at an objective report from impartial professionals, presented after a painstaking sixteen-month investigation. The nonsense of it all made me think of the Queen of Hearts saying, "Sentence first — verdict afterwards."

Shaft, Countershaft

The committees met almost daily, but since the board members who supported the draft report saw no need to attend, our exposure of inconsistencies served only to deepen our own convictions. Even Chairman Deschênes didn't attend. His ally and confidant, Frank Thurston, chaired both committees. Although he had not supported the ice theory, it was inconceivable that he would let the committees stray from whatever bounds Deschênes had set. Even so, allowing the committees to function without his close personal supervision was an unusual departure from Deschênes' customary autocratic style.

We soon discovered the source of the chairman's newfound indulgence: he had received the legal opinion on the chairman's powers that Roger Lacroix had demanded some months earlier. On May 13, Deschênes called a special session to tell us he had received an opinion from his former law partner, Robert Décary. Décary's opinion, Deschênes said, "somewhat clarified" the respective roles of the chairman and board members.

As we scanned Décary's detailed 72-page report, Deschênes was even more somber and humorless than usual. Small wonder; the report devastated his claim of absolute authority.[10] In Décary's opinion, the board as a whole "could and should make regulations in which the respective powers and duties of the Chairman and of the Board Members could be set out in the way that pleases the Board." Furthermore, the chairman should remain "totally accountable" to the board as a whole for whatever powers the board might delegate to him. "It must be remembered," Décary wrote, "that the Board is a body corporate that is a master of its own proceedings." Roger Lacroix had been right all along. Deschênes would have to cede to the will of the majority.

"It's not my intention to try to obtain a contrary opinion," Deschênes assured us, "which I could probably get." What he meant, although we didn't realize it at the time, was that it was not his intention to do so openly; he would instead work the back channels with his allies in the bureaucracy. But to buy time for his plans, he had to eat a little crow. He proposed a number of reforms to give board members more say in administration and, more important to me, tacitly allowed that the scope of our review of the Gander investigation could be expanded. The committees would wind up their work by compiling a list of items to be resolved and questions to be answered before further consideration of the report.

The list, "Board Requirements Pertaining to the Gander Investigation", ran twelve pages, fifty items. Deschênes agreed that the "list will be passed to the staff for formal written response and subsequent discussion"[11]. We would start with a fresh look at the revised report in the fall. And so the summer of 1987 passed without further official discussion, but behind the scenes both sides loaded their weapons for the next encounter.

I was as happy for some extra time as anyone. As far as I could see, the arguments of those pressing for quick approval of the report boiled down to their conviction that politically appointed board members were unfit to question the conclusions of "professional investigators". But I could think of only one qualification that placed Hinton and Boag into some kind of high priesthood — the secrets garnered in their qualifying courses on accident investigation. Ross Stevenson and I decided to take advantage of the chairman's newfound respect and learn these secrets. We registered for the aircraft accident investigation course at the University of Southern California — a two-week session touted as the best of several courses available to introduce pilots and engineers to this arcane speciality.

While attending the course, Stevenson and I soaked up all the information we could. We made a point of meeting independent consultants and members of the California chapter of the International Society of Air Safety Investigators. We didn't mention the Gander investigation, but general discussions with experienced investigators convinced us that the technical investigation into the Gander crash fell far short of professional standards.

While the investigators were reconsidering the report in light of our "deficiency list", I reviewed the scientific literature on wing icing and repeated the Douglas computations with a variety of changes in the assumptions. Ross Stevenson continued listening to recordings and reading transcripts of all available witness interviews. The ice theory looked worse and worse.

* * *

Meanwhile — despite his own insistence on the independence of the CASB from all government departments — Deschênes sought advice from Paul Tellier, Clerk of the Privy Council and Secretary to the Cabinet. The quaint title may bring to mind a Dickensian character writing notes on parchment with a quill pen; in fact the Privy Council Office is the nerve center of the government's business, and is often called the Prime Minister's Department.

Paul Tellier was the big enchilada himself, head of Canada's two hundred thousand civil servants. A lawyer from Quebec, like Deschênes, Tellier exemplifies the Canadian civil servant's chameleon-like ability to adopt the color of the politicians currently in power. He rose to high office under the Liberals, and thrived under the Conservatives. Prime Minister Brian Mulroney had picked him for the top job. Paul Tellier had the power, the means and — as we were to find out — the inclination to help Bernard Deschênes.

The plan to deflect the Décary opinion began with a cunningly phrased letter of comfort to Deschênes. In Tellier's opinion, the legislators setting up the CASB had intended to vest all administrative responsibility in the chairman. Unfortunately, an ambiguity in the wording of the law allowed a contrary interpretation — a plausible interpretation, but undoubtedly not the only one. ''We have undertaken to clarify this question,'' Tellier wrote. While waiting for this clarification, he recommended that Deschênes ''continue to administer the Board as before, in conformity with the spirit of the statute establishing the board''.[12]

Deschênes could hardly contain his glee as he called us together and brandished Tellier's letter. He was so happy that he just kept smiling as Bobbitt and Lacroix lambasted him about compromising the board's independence, and pointed out that a formal legal opinion could not be superseded by a civil servant's notions on what the government might do in the future. No matter. The chairman had received all the blessing he needed to run the CASB as he pleased.

The next step to nullify the legal opinion followed in the form of a letter from Transport minister John Crosbie, who wrote Deschênes that his office would be hiring a consultant to ''evaluate the CASB''.[13] The purpose of the evaluation was obscure, but Deschênes was asked to name a ''representative to serve on an Advisory Committee to provide direction to the consultant as required, to receive progress reports and to ensure that all interests are taken into account.''

We found out about this evaluation when we received a copy of the minister's letter along with the answer that had already been sent. The answer, signed in the absence of the chairman by executive director Ken Johnson, said, ''I am pleased, as requested, to nominate Mr. Ken Johnson the Executive Director of the Board to sit on the Advisory Committee you propose to establish.''[14] Johnson would ensure that my interests were "taken into account" just after the Ayatollah Khomeini turned Catholic.

With Johnson's help, the advisory committee quickly selected the Ottawa firm of James F. Hickling Management Consultants Limited to conduct the evaluation. Johnson, as it happened, had recently engaged Hickling to conduct a study of the CASB's data processing needs. He could certainly vouch for the firm's suitability for such a sensitive task.

The consultants got right off the mark, and soon all board members received a long questionnaire. "Do you believe that the CASB operates as an aviation safety organization separate and independent from Transport Canada?" "Does the administrative support provided to the CASB Board meet or exceed your expectations with regard to quality?" And so on. A pair of consultants followed up with personal interviews to allow us to expand on our answers. When they came to see me, there was no mention of the Gander investigation, but I did explain why the administrative support did not meet my expectations. I could not, for example, get any assistance in reviewing scientific and engineering analysis from flight data recorders.

Why did I not believe that the CASB operated as an aviation safety organization separate and independent of Transport Canada? Well, to start with, there was Judge Rolf's report on the High Prairie accident that said board members did not appear to be independent of Transport department. Then there was the failure to investigate the incident involving the department's King Air at the Belcher Islands. I also mentioned a mutually satisfactory understanding between the director of investigations, Tom Hinton, and officials in Transport Canada to restrict the scope of the department's "interested party comments".

The consultants thanked me and left. We wouldn't hear from them again until the next phase of our review of the Gander investigation was in full swing.

Part 2

Going Twice . . .

We received the investigators' response to the "Board Requirements Pertaining to the Gander Report" on October 4, 1987, in the form of a revised draft. No one could say the investigators had disregarded our concerns; each item on our list was covered by a short note ending, "Report amended as indicated," or, more frequently, "No change."

At the same time, we got copies of a report on simulation tests conducted not long before in Los Angeles — mainly to mollify Roger Lacroix, who had insisted he would never approve the report in the absence of simulator studies that looked at possibilities other than ice. The latest studies were a revelation. They clearly showed that various mechanical failures could explain the speed and flight path of the doomed aircraft to the extent that we knew them from the faulty FDR — debunking the claim that nothing other than ice on the wings would fit.

The most interesting case was an extension of the thrust reverser on the number-four engine. When the simulator was programmed to mimic the effects of a malfunctioning thrust reverser just after lift-off, the pilots lost control—whether they had been warned of what was about to happen or not. The maximum airspeed, the veer to the right and the distance covered before the inevitable crash were all similar to the values from the Arrow Air flight data recorder. This might be a breakthrough.

Not so fast. The investigators discounted the new tests. Contrary to first impressions, they explained, careful examination of the wreckage proved that the number-four reverser had been stowed at impact. Any similarities in the flight path must be coincidental.

In fact, the new draft ignored or explained away all evidence at odds with the previous conclusions. It was, in essence identical to the draft we had roundly rejected, but with the most obvious self-contradictions corrected.

An expanded section on "Fire" now admitted that "three eye witnesses reported an orange or yellow glow emanating from the underside of the aircraft." The claim that the maximum altitude reached could be calculated by "double integration" had been quietly dropped. The new draft also acknowledged, albeit indirectly, that factors other than ice might have contributed to the crash, including "inappropriate take-off reference speeds for the weight of the aircraft" as a factor.

On October 14, 1987, the board convened to consider the new draft. Bernard Deschênes was in a combative mood; it was getting close to the second anniversary of the crash, he said, and "The media have started inquiring about the status of the investigation. The eyes of the world are on us." He laid on some sarcastic words about the "laundry list" of questions and criticisms. He might have promised that all items would be carefully considered and all questions answered to everyone's satisfaction, in the aftermath of the humiliating legal opinion repudiating his claims of absolute control, but now the picnic was over.

"We have as many facts as we'll ever get," he said, and insisted that the draft report be sent to the interested parties before the second anniversary. "I intend to run the meeting very tightly," he warned. He then declared that it would be useless to hash over the details of the text. Every board member was to affirm whether or not he agreed with the "cause-related findings".

I protested, arguing that we must first study the investigators' answers to our questions, and Stevenson let loose a long tirade about the unacceptability of the conclusions and the untruthfulness of the report. Most of the others took the stands they had espoused before.

The exception was Frank Thurston. Frank sat through the various outbursts with a benign smile, biding his time. When his turn came, he started out in a rambling, round-about way; he complemented Peter Boag, Tom Hinton and their associates on their prodigious effort at revising the first draft, but said he was still gravely concerned on several grounds. He didn't wholly agree with the cause statement" and felt that "deep stall" should be the "proximate cause" and that "fatigue of the crew should be added as a factor." He added that further debate over "elements of the laundry list" would not change the substance of the report and further delay would be irresponsible. Moreover, the interested parties would undoubtedly make many constructive comments that we could incorporate into a final draft.

It was evident to me that Frank had been conscripted to the chairman's side, but I was thoroughly unprepared for the bombshell he was about to drop. Over the summer, Frank said, his "view on the icing situation" had changed. Initially he had rejected the ice theory. Now, "based on analysis", he embraced it. I was flabbergasted. Ignoring the understanding that gave each member an uninterrupted chance to express his views, I demanded to know the nature of this analysis. Thurston held aloft a copy of a book with which I was very familiar: *Theory of Airfoil Sections*, a compilation of wind tunnel data from the old National Aeronautics and Space Administration, originally published in 1949.[15]

I had used *Theory of Airfoil Sections*, a thick volume including carefully compiled data on the lift of artificially roughened airfoils in wind tunnels, as a research and teaching reference for over twenty years. I knew damn well it contained nothing relevant to claims that ice on the wings had downed the Arrow Air DC-8.

"What has that got to do with anything?" I shouted. The chairman called me out of order and told me to keep quiet until all the other members had had their turn to speak.

Thurston's defection left Bobbitt, Lacroix, Mussallem, Stevenson and me in opposition to the ice theory. The remaining five board members wanted to send the draft to the interested parties.* We all knew, of course, that the regulations governing our meetings stipulated that "where there is an equality of votes, the person chairing the meeting shall be entitled to a second or casting vote."[16]

As soon as everyone had had his say, Thurston tried to move that the report be accepted as the "conditional draft" the law required us to send to interested parties for comment. But Deschênes then made a big mistake, one that irrevocably spoiled the neat plan for dispos-

---

*The interested parties acknowledged by the CASB included: Douglas Aircraft; Pratt & Whitney; Arrow Air; the U.S. Department of Defense; the FAA; the NTSB; the Multinational Force and Observers; Transport Canada; and next of kin of the crew (but not of the passengers).

ing of the report: he accepted my demand for another round of comments. Ross Stevenson rose to the occasion, speaking at length about the many deficiencies in the report — saying openly, for the first time, that he believed there had been a fire on the airplane before the crash. He made an impassioned plea to the chairman not to use the technicality of a second, tie-breaking vote to ram through such an obviously flawed conclusion on such an important accident, over the heart-felt objections of such highly qualified board members. "You pointed out, Mr. Chairman," he said dramatically, "that the world is watching how we handle this investigation."

Something in his diatribe got through. Deschênes seemed shaken and said we could have more discussion before voting on the conclusions. We adjourned for lunch.

The Well Is Dry Gentlemen

When we reconvened after lunch, Hinton, Boag and Johnson formed a glum-looking trio at the end of the table. Six months had passed since they had expected *pro forma* ratification of the report, and here they were, back to square one. Boag was in a foul mood. He seemed to think that removing the specious claims about "double integration" had been a great concession, and stolidly insisted that the FDR gave unequivocal proof that the jet had stalled. I suggested we needed an independent expert opinion. "We are satisfied with the FDR traces," he maintained. "Additional work is not warranted."

Ross Stevenson repeated his lecture about the fatuity of trying to divine meanings from bugs on an airspeed indicator recovered from a violent crash, and Boag matched him with a lengthy discourse "to put the bug speeds into perspective''. Stevenson moved to something else: why had the investigators not reported on the repairs made to the DC-8 after an earlier accident in Casablanca? The board's list had said information was to be obtained. Where was it?

Noting Boag's increasing exasperation, Hinton interjected that the staff would provide that information. But Peter Boag had had enough. He pushed back from the table and said, "Frankly, gentlemen, the well is dry for me. I've done all I can do." He looked ready to walk out. The remark infuriated Roger Lacroix. His face flushed deep red and he started to say something, but words failed him. Hinton put his hand on Boag's arm, the two investigators exchanged glances and Boag stayed.

The questions continued, and Norm Bobbitt raised another issue. The draft report noted that there was no evidence the crew had inhaled toxic combustion products before death. But what about the toxicological evidence in the remains of the passengers? Shouldn't that be included? Boag mumbled something about having to call on Dr. Elcombe, and the discussion drifted to other topics.

Bobbitt and Mussallem joined me in challenging the interpretation of the FDR traces. Boag cited the high qualifications of Bernard Caiger, and Bobbitt grumbled, "In that case, we should be questioning Mr. Caiger." That was a fine idea, and I proposed to make a formal motion to that effect. The chairman didn't want a vote. He exchanged glances with Hinton and Johnson, and said that if Mr. Caiger was willing and available we could ask him to come and answer our questions. We also agreed to ask Dave Elcombe to come.

And, as the minutes record, we unanimously agreed on ways to head off potentially difficult procedural matters. To accommodate members who might be travelling when it came time to vote, ''The Board agreed that when the review is completed, a written motion will be circulated and each Member will have an opportunity to vote on the draft report. The second matter was even more delicate: ''The Board discussed the question of dissenting opinions, and agreed that the procedures for incorporating dissenting opinions on draft reports should be established before the draft report is sent to the interested parties.''[17]

Foolishly, I thought we had made great progress.

* * *

When we reconvened the next morning, Dr. Elcombe was on hand to answer questions on the medical examination of the victims. Neither the public inquiry nor the two draft reports had even mentioned the toxicology of the passengers. Now, six months after the report had first been presented for the board's approval, Elcombe allowed that much work remained to be done on the analysis of the medical findings. There had been a delay because the American autopsy reports did not specify the cause of death. In Elcombe's experience, autopsy reports always gave the cause of death. Very strange.

Had the toxicological examination of the passengers turned up any evidence of combustion products? Before Elcombe had a chance to answer, Tom Hinton piped up to tell us that Dr. Elcombe was still working up those results; they would be provided shortly.

Even at this point I still saw nothing sinister in the evasions and incomplete answers; to me, they just signalled an inept, incomplete investigation that had seized on the wrong conclusion. I joined the other opponents of the ice theory in pointing out the inconsistencies and omissions in the report. Those who supported the ice theory had little to say. The day ended with many questions to be answered by the investigators, and unanimous agreement on various changes to the text.

* * *

The next day, Friday, Ross Stevenson showed up with a tape player and insisted that we listen to parts of the recorded interviews made in the days following the crash. Deschênes decided to humor him.

Stevenson began by playing a portion of the interview with the truck drivers. We heard Cecil Mackie say that he could have read a book by the light that seemed to come from halfway along the airplane.

"Could it have been a fire?" the interviewer asked, and Mackie answered without hesitation, "Yes."

The second trucker, Leonard Loughren, was equally clear. "Almost certainly there was a fire," he said.

Mackie and Loughren had testified at the public inquiry a year and a half earlier, and everyone knew they had seen something. But Stevenson had also brought along tapes of interviews with other witnesses who had reported seeing a fire. One was Judy Parsons, who had told reporters within hours of the crash that she had seen a bright object blow up in the sky shortly before the giant fireball appeared. The investigators had dismissed her account—apparently as a confused recollection muddled by after-the-fact knowledge of the tragedy. But as we listened to Parsons' own words, her account seemed perfectly lucid. And its credibility was enhanced by a number of details. Parsons said she had seen "a bright orange object'', presumably the aircraft, move just over ''the top of the Search and Rescue Building'' before it ''blew up like a balloon in a million pieces.''

The interview had taken place at the spot in the Tilden parking lot where Parsons had been standing. ''The light was *there* and the fireball was *there*,'' she had said, evidently pointing. ''There was a space between them.'' Babbitt demanded whether Judy Parsons had at least been pointing in the right direction. Had anyone ever checked to see if the line of sight from her position in the parking lot to the top of the Search and Rescue building could have plausibly intersected the flight path of the DC-8?

It seemed that even the board members who had firmly supported the ice theory were impressed by the testimony. In any event, they listened attentively. Only Deschênes was distracted. While Stevenson played the tapes, Peter Boag had quietly moved to the head of the table. He leaned close to the chairman, and the two conversed earnestly as Hinton and the other investigators handled questions and comments from the board members.

We were reminded that professional investigators know witness evidence to be notoriously unreliable. Nevertheless, all witness statements had been carefully reviewed. There was no question that the truck drivers had seen reflections or the aircraft's landing lights, and when Mrs. Parsons' indications had been plotted on a map they had pointed in impossible directions. Her statements had not been mentioned because they had absolutely no evidential value. Certainly the interviewer had taken and recorded compass bearings. Certainly these could be made available to the Doubting Thomases on the board.

And so the second week of the second round of the review of the Gander report came to a close, with an agreement to reconvene the following week to receive answers to outstanding questions. First on the list would be a discussion of the FDR results with the expert, who would lay to rest any doubts about the investigators' interpretations.

Bernard Deschênes had confidently expected to wrap up the discussion with a forced vote. His daughter was getting married in France and he had made plans to take several weeks of vacation, although he told us only that he would not be available for the meeting with Mr. Caiger. Deschênes designated Frank Thurston to act as chairman in his place.

Deschênes must have realized that, in his absence, the critics would be in the majority. But Johnson would be hovering in the background as usual. And Bernard Deschênes knew better than most of us could have guessed that Frank Thurston would follow his orders to the letter.

When the Cat Is Away

The prospect of a direct discussion of the flight data recorder results with Bernard Caiger, the National Research Council expert who had actually read the data, boosted my hope that the board might yet be induced to abandon the ice theory and go after the real cause of the Gander crash.

I had met Bernie Caiger a number of times and knew him to be a genuine expert dedicated to his speciality. It was true that he had worked for Frank Thurston for many years and that, now that he was retired, he occasionally served as a consultant to Tom Hinton. But I knew that, given the opportunity, he could prove the investigators' interpretation to be pure baloney.

The Canadian Aviation Safety Board (CASB) was set up in 1984 as recommended by Justice Charles Dubin (left), head of a commission of inquiry into aviation safety launched by Transport Minister Don Mazankowski (right).

Transport Department official Kenneth Johnson (left) became Executive Director and architect of the CASB's policies. In 1988 the Dubin Commission's chief counsel John Sopinka (right), later named to the Supreme Court, reported that the quality of investigations had deteriorated and board members were not given important information.

Entrepreneur George Batchelor set up Arrow Air Inc. in Miami in 1981. Chartered Arrow Air DC-8's shuttled U.S. troops to the Middle East and helped Oliver North's secret schemes in the Iran-Contra affair.

Troops of the 101st Airborne ''Screaming Eagles'' at Fort Campbell Ky. complained of cramped, unpleasant conditions on Arrow Air flights.

Reverend Benjamin Weir was abducted by the Islamic Jihad in 1984 and freed on September 14, 1985 in return for missiles delivered by a DC-8 chartered in Miami. The Iranians felt cheated by the next secret missile delivery on November 24.

Canapress

Canapress

Seconds after lifting off following a refueling stop in Gander Nfld. on Dec. 12, 1985, Arrow Air DC-8 N950JW crashed and disintegrated into unrecognizable fragments. The Islamic Jihad claimed responsibility.

A hangar at Gander airport became a makeshift morgue. The number of victims recorded as 258 on the flight manifest was changed to 256.

Weapons and military ordnance were scattered over the crash site. Shown here are three unexploded mortar shells and some ''nunchakus'' or ''ninja sticks''.

President Ronald Reagan consoled victims' families at a memorial service in Fort Campbell.

Major General John S. Crosby was accused of pressing for early bulldozing of the crash site.

CASB Chairman Bernard Deschênes presided over a public inquiry. Chief Investigator Peter Boag (at witness table) concluded that the crash was caused by a minute amount of ice on the wings. Unknown to most present, Frank Thurston (chairman's right) did not have a valid appointment to the CASB at the time. Board member Roger Lacroix (far right of dais) resigned in frustration.

Investigator Irving Pinkel attributed this hole in the fuselage to an in-flight explosion.

Board member Ross Stevenson (left) with more than 9,000 hours as pilot-in-command of DC-8's insisted that ice on the wings could not have caused the crash. Kenneth Thorneycroft (right), who became chairman when Bernard Deschênes resigned in wake of Sopinka's criticism maintained previous policies.

Benoit Bouchard (left) who replaced John Crosbie (right) as Transport Minister ignored Sopinka's advice and instructed board members to give Thorneycroft their full and unqualified support.

Assistant Deputy Transport Minister Claude LaFrance (left) was forced to resign after the leak of his staff's report rejecting the ice theory. Former Supreme Court justice Willard Estey (right) rejected both the ice and the explosion theories but recommended against further investigation.

Captain Lee Levenson listens as author Les Filotas testifies before the United States' House Judiciary Committee's subcommittee on crime in December 1990. The subcommittee concluded that no American agency had investigated the possibility of sabotage.

I also knew it wouldn't be easy for him to get that opportunity. Hinton would make sure of that. As director of investigations, Hinton was proud of his "exclusive authority". Peter Boag was the official spokesman and Ken Johnson pulled the strings behind the scenes, but in the boardroom Tom Hinton was the supreme field marshal. Whenever Boag floundered, Hinton was right there to help out. "Mr. Chairman," he would interject, "what I think Peter means is that . . . ." or "Mr. Chairman, I think Mr. So-and-So here could answer that question better than Peter." Other staffers would look at Hinton before daring to interject. If signalled permission to speak, they would continually glance his way and defer to the boss at the slightest indication.

In contrast to the combative Hinton, Bernard Caiger was a mild-mannered gentleman. The difference between the two was starkly apparent at a press conference at Caiger's lab the day after the crash. When Caiger, who was introduced as "Canada's top expert in these kinds of recorders", started to answer a question with facts instead of vague generalities, Hinton brusquely cut him off. "We've already explained that under the Act, the cockpit voice recorder is privileged," he snapped. No more questions. No more press conference.

I had found Hinton's none-too-subtle steering of Elcombe's report on the medical evidence particularly galling; we could do without his interference while hearing Bernard Caiger. I consulted the four other opponents of the ice theory. With Deschênes away, we had a majority. We agreed that we could discuss the FDR without help from Tom Hinton or any of his staff.

I went to see Frank Thurston and suggested he notify Ken Johnson that the the staff could devote their valuable time to other work while the board met Mr. Caiger. When Thurston protested, I told him the board could, of course, vote on the issue. Frank got the picture and took my remarks under advisement.

When we showed up for the scheduled meeting on October 21, Johnson and his staff were not in the room. But Thurston advised us that the staff felt very concerned about being excluded, and that ''in the interest of *glasnost*'' we should at least let them ''explain the implications.''

Johnson and Hinton trudged in, looking as though they were about to address the United Nations in a last-ditch effort to avert nuclear war. Johnson distributed a memo registering his complaints about ''the Board's intention to exclude the Director of Investigation and the Investigator-In-Charge''[18] and argued that ''the Board may be breaking the relationship required'' by section such-and-such of the Canadian Aviation Safety Board Act if it persisted with this most ill-considered conceit. Also, he was sad to say, the whole process was unfair. It would ''devastate the morale and pride'' of the staff who had toiled selflessly on the report. Hinton added that Mr. Caiger had analyzed the flight data recorder under contract and was a *de facto* member of the investigation team, which he, Tom Hinton, was dutybound to supervise.

After a brief, predictable discussion we nonetheless voted to excuse all the staff except a secretary to take minutes and Mr. Deschênes' executive assistant.

* * *

The interview with Bernard Caiger turned out to be most congenial. He began by explaining his desperate struggle to extract information from the sub-standard instrument—any useful information at all. Then, with very little prompting, he proceeded to demolish the impression of precision nurtured in the investigators' report.

The report claimed that the accuracy of the speed was "better than plus or minus five knots." Caiger told us he had estimated that the speeds were "probably accurate to plus or minus 5 knots" except for near the peak of 170 knots, where "the five knot error may not be enough."

What about the claim that the aircraft had lifted off fifty-one seconds after brake release? Caiger shook his head sadly. Actually, he said, it was impossible to tell where lift-off occurred just by looking at the FDR traces—in fact, given the poor quality of the altitude trace, the FDR couldn't substantiate that the aircraft had lifted off at all. And of course the precise time of brake release was just a guess. He himself would have put it somewhat after the 33-second mark that had been chosen by the investigators. Somewhere between 35 and 41 seconds, say.

The vertical acceleration trace had been especially difficult. Strangely, only one of every ten marks had been inscribed on the foil. The mechanism was so badly damaged that Caiger was not able to track the problem to its source, but he had made a valiant attempt to read the data. A second, independent attempt by one of his assistants had produced completely different values, so he had had to tell the investigators to disregard the results. "I couldn't produce anything that was reliable on the acceleration," Caiger said, "so I left that out of the picture altogether."

"So then it would not be possible to double-integrate the vertical acceleration to find the altitude?" I asked. The technique was "a perilous business at the best of times," Caiger explained, and it was evident from the outset that it could not be used in the case of the Arrow Air DC-8.

Caiger's testimony demolished the contention that the information from the FDR supported the ice-on-the-wing theory and no other. It also showed the one-sided advocacy of the investigators' report. I felt that from now on every claim, even those seemingly based on incontrovertible fact, would have to be taken with a huge dose of salt. But I kept such thoughts to myself. We all thanked Bernard Caiger for his help, and after he left we congratulated one another for an informative and profitable discussion.

Evidence Review Committee

When the board meeting continued after lunch, Arthur Portelance was the only member present who supported the ice theory beside acting chairman Thurston. Portelance was a pleasant and fair-minded man — balding, plump and bespectacled. His name, meaning "spear carrier", suited him perfectly. He would never have considered voting against his leader in his days in Parliament, regardless of the issue — not because he was a toady but because he believed the infallibility of the leader to be the highest form of truth. He viewed his job on the CASB as an opportunity to support the chairman by offering low-key advice, and felt the rest of us political appointees should take the same approach, leaving technical judgements to the professionals on the staff. By now it was plain that the staff were unwilling, unable or simply not allowed to let the facts decide the conclusion. It seemed equally plain to me that Deschênes had convinced Thurston that the report had to be issued as is. Yet the meeting with Caiger had shown that the facts were available if only we could get around the procedural blocks and diversionary tactics. Deschênes' temporary absence gave us that chance.

I reminded everyone of the wonderful progress we had made in the morning and suggested we use the experience as a model for reviewing other contentious aspects of the technical evidence. I proposed that we proceed as we had earlier, by means of a committee of members working in parallel with other board business.

Portelance thought the idea sensible as long as it wasn't just a delaying tactic. I assured him that the goal would be to produce a compromise report that we could all support, and suggested that the committee's terms of reference include a timetable to prevent delay. On that understanding he seconded my motion. It seemed, momentarily at least, that we had taken a step toward mending the rift between the pro-ice and anti-ice factions.

Thurston was suspicious. He asked if it was my intention that the director of investigations and his staff attend the committee meetings. In my best diplomatic manner, I suggested that the committee could decide on such details as it went along. But diplomacy was not one of Ross Stevenson's strong suits. He interjected that there was no way Hinton would be allowed to attend and obstruct the committee's work. Even Roger Lacroix, still smarting over Boag's "the well is dry" remark, insisted that the investigators had disqualified themselves from further participation.

Thurston dug in his heels. He didn't have the authority, he claimed, to allow a vote in Deschênes' absence. I read him the rule book. He then appealed to my sense of fairness, arguing that such an important issue should not be decided by a minority of the board. I replied that we were just setting up a committee, as we had done before. Moreover, since Arthur Portelance had seconded my motion, an absolute majority of six members was in favor.

Thurston became more and more agitated. He spoke at length and with considerable passion on the unwisdom and unfairness of the motion. "The proposal you are making is absolutely unconscionable," he said. For the first time, I sensed some secret worry behind his maneuvering. I tried to reassure him. The investigators could rest content in the conviction that they had done a good job; what difference could it make what we did in our own boardroom? "I'm not simpleminded, Les," said Frank venomously. "I know what you're trying to do."

What I was trying to do was find the cause of the crash. But there was no use saying so. "Let's have the vote," I replied.

The vote to set up an Evidence Review Committee to study the evidence of the crash at Gander "with the assistance of the staff as requested" passed by six to one, with only Frank Thurston dissenting. All agreed I would be chairman, and Bobbitt, Lacroix and Stevenson were to be members. But it was also understood that all members of the board could attend and participate in the meetings. The members of the Evidence Review Committee would draft terms of reference for ratification at the next meeting.

Lacroix and Stevenson were jubilant. I was cautiously pleased. Thurston was crestfallen. His scorn and anger seemed to have abated. He looked old and sad. "You have destroyed me," he said.

Defunctus

With the Evidence Review Committee controlling its own proceedings, we would finally be able to get the facts we needed to assess the investigators' report— or at least to separate fact from fantasy. I was convinced that we would now be able to reach logical conclusions and, whatever our conclusions, make a case so compelling that they would have to be accepted. The threat of a compellingly argued dissenting opinion would, I thought, help open closed minds.

Still, I was determined to make the Evidence Review Committee as low-key and non-threatening as possible. We would try to write terms of reference that Bernard Deschênes and his supporters could grudgingly accept. I tried to impress on the members of the committee that we would need to accommodate the egos of both Deschênes and the staff.

When the board got together for the next scheduled meeting the following week, Deschênes was still absent. I presented our terms of reference for the committee for ratification, the objective being ''To review evidence gathered during the investigation of the Arrow Air DC-8-63 accident at Gander, Newfoundland on December 12, 1985 with a view to expediting the release of a unanimously supported conditional draft to interested parties.'' The committee would conduct a detailed examination of the physical and documentary evidence and ''interview personnel directly involved in the collection and initial interpretation of this evidence.''

We set out a schedule that would allow the review to be conducted in six weeks. The first item, ''collection and verification of material already sought", consisted of a simple list of documentation that should have been available but that we had not been able to see: "diagram of witness Parson's observations", "original photographs of flight instrument locations", "ice accretion formula"* and so on. In presenting this list, I pointed out that some of the items—like the ice accretion formula — had been requested when we began our review of the first draft some six months earlier. So far we had only received promises that the formula would be retrieved from the files.

The reference to the files gave Norm Bobbitt an idea. The investigators had completed their work and had no immediate need for the files. Why should we have to wait for a clerk to retrieve a file? All the files on the Gander investigation could be moved to an empty office near the boardroom for the convenience of the Evidence Review Committee.

No one objected, but Frank Thurston asked that we confine our work to examination of available documents until Bernard Deschênes returned the following week. We proceeded to clear a number of non-Gander issues on the agenda.

* * *

On November 3, 1987, Bernard Deschênes was back at the office. He had found out about the Evidence Review Committee and had a fit. He labelled the committee "impractical and unfair" and read a prepared statement saying that, while he didn't object to further

---

*This referred to a claim in the first draft that "it was calculated that a thickness of about 0.33 inches of ice would have accreted on the leading edge of the wing while in cloud." The claim had been repeated in the latest draft.

examination of the evidence, "at this juncture the participation of all available members is not only essential, it is a requirement according to the rules of fairness." He waved aside a reminder that the Evidence Review Committee was modelled after the committees he himself had set up the previous spring and insisted that "the exercise shall be carried out by the Board as a whole," meaning that no one but himself could control the meetings. As a final caution, he warned, "if this process bogs down, which I sincerely think it won't, we may have to consider other approaches to accelerate so that we can finalize a draft report by the target date of early December."

Roger Lacroix was livid. He reminded the chairman that the Evidence Review Committee had been established by the majority of the board at a legally constituted meeting according to regulations. Deschênes tried to sidetrack him with a little analogy about how Roger had undoubtedly acted with his junior officers in the air force, which just increased Lacroix's fury. He challenged Deschênes to hold another vote. The chairman refused, claiming he had "the power to direct" changes to the decision. With everybody in a high temper, we adjourned for lunch.

The pro-ice and anti-ice factions went off separately. Lacroix couldn't get over Deschênes' perfidy. Usually he was the perfect gentleman—I had never heard him utter even the mildest form of profanity—but that day, he had been driven beyond endurance. "Up his nose with a rubber hose," he kept repeating.

We managed to calm him down. It was true that Deschênes' action was beyond reason, but there wasn't much we could do without independent legal advice. Boycotting the meeting in the meantime would just give him an opportunity to pass the Gander report. We decided we had to lie low, at least for the rest of the day.

When the meeting resumed, it was evident that the opposition had had its own strategy session. Deschênes called on Tom Hinton to give his views on the review of the evidence, and Hinton spoke like a man who has recently discovered religion—a veil of peace and serenity seemed to have engulfed him. He firmly believed, he said, that his job was "to bring information to the Board. . . . If the Board wants more information, then, by golly, we'll go get it."

He assured us that he was saddened to think that anyone might have believed the investigators had a bias toward the ice theory. "Peter is a very professional, very dedicated, very committed investigator," he reassured us. "Sometimes he's a touch strong about it, but that's how I like 'em." Hinton smiled. "I can guarantee that he did not make up his mind that it was ice," he said with conviction. "There was no preconceived notion."

Hinton suddenly recalled something else. It had been suggested that he himself sometimes had a bit of a stifling effect on those around him. He smiled and shook his head; the notion was just too bizarre. "I really have no intention of doing that and I'll be careful to make sure it won't happen."

Hinton then apologized for failing to deliver the files as the board had requested a week earlier. An inventory was under way for the board's convenience. All the information, including contradictory information, had been carefully preserved, and it would all be delivered into the board's hands.

Roger Lacroix, who had taken particular interest in the ice accretion formula, asked if it was part of the file. Sure, said Hinton, everything relevant was part of the file. The sheer volume was a problem, though, and admittedly the indexing was not as good as it might be.

The chairman seemed well pleased with this new spirit of cooperation that Hinton had trowelled upon us. He noted that the next item on the Evidence Review Committee's agenda was to examine the contention that all engines had been functioning normally when the aircraft plunged into the trees. But it was too late in the day to start on that, he said. He was about to call the meeting to a close, but I wanted to make sure we all knew what was going on.

"You're taking over the Evidence Review Committee?" I asked.

"Yes," said the chairman.

"You're not going to have a vote to rescind the former vote?"

"Well, I would like your consensus to proceed in this manner."

Fat chance of that. Lacroix, Stevenson, Bobbitt and I all denounced the procedure and pleaded with Deschênes to at least give the committee a chance. Deschênes did his best to sound conciliatory; why didn't we turn that around and give his way a chance, he asked. We repeated our speeches with elegant variations.

"Very well, then," Deschênes announced. "It's my ruling that the board will proceed as I have indicated. The committee becomes *defunctus*."

Part 3

Sold

Bernard Deschênes' defunctus ploy shifted the focus of the Gander review from the merits of the evidence to the powers of the chairman. While Roger Lacroix worked on a long "Statement for the Record" outlining Deschênes' contempt of the CASB regulations, I decided that it was time I too put some thoughts on record. I had consistently denounced the quality of the investigation and the analysis, but so far I had not said where I stood on the conclusions.

I wrote Deschenes:

> I now believe that the evidence—all circumstantial— suggesting the aircraft crashed because it tried to take off with ice-contaminated wings is weak and untenable. Circumstantial evidence suggesting there was a fire on board the aircraft before the crash—evidence that is minimized in the draft report—is, in fact, much stronger than that adduced in support of the contaminated wing hypothesis.

> I further believe that the aerodynamic analysis carried out in support of the suggested probable cause is incomplete and faulty. In particular, the conclusion that the wing stalled due to ice is incorrect. . . .

> I urge that the Canadian Aviation Safety Board analyze *all* remaining take-offs recorded on the Flight Data Recorder recovered from the accident aircraft. Such an analysis could lead to vital new evidence. . . .

I was referring to the fact that despite the millions of dollars spent on the investigation, the investigators had not examined the thirty or so previous takeoffs scratched on the metal foil taken from the DC-8's FDR. A specialist could do the work in a few days. I'd bet a bundle that traces for previous, successful takeoffs would show wiggles similar to those interpreted as a fatal stall.

The five of us who opposed the ice theory agreed that we needed impartial advice. We recalled a legal brief that the Canadian Bar Association had submitted to the government while the CASB Act was being drafted. Its author, Professor Rowland Harrison of the University of Ottawa Law School, had accurately foreseen the dangers of the clause about "the exclusive authority of the Director of Investigations."

Norman Bobbitt told us that Harrison was now in private practice with the Ottawa law firm of Stikeman Elliott. Harrison agreed to advise us and Bobbitt started documenting the chairman's follies, pulling together material such as the opinions from lawyer Robert Décary.

We also needed to address Deschênes' appeals to the "spirit" behind the CASB Act. Ever since his legal interpretation had been shot down in flames by Décary, Deschênes had taken refuge in the "ambiguity" of the act. He maintained that this ambiguity had to be resolved in a way faithful to the legacy of the Dubin Commission — and, naturally, he had appointed himself executor of that legacy.

Lacroix recalled that Gary Ouellet, the lawyer who had served as associate counsel on the Dubin Commission, was now an Ottawa consultant; as a matter of fact, his office was in the same building as Rowland Harrison's. Ouellet had drafted sections of Dubin's report. Why didn't we ask what he thought of Deschênes' pontificating?

* * *

When Bobbitt, Lacroix, Stevenson and I met Gary Ouellet, I knew of him only as a lawyer who had worked on the Dubin Commission. I learned later that Ouellet was one of Ottawa's best-connected lobbyists, part of the inner circle who helped Brian Mulroney become leader of the Conservative Party in 1983 and prime minister in 1984. Ouellet is now CEO of an Ottawa lobby group reputed to "pull contracts out of government like rabbits out of hats".[19] But he didn't look or act like a bigshot political insider. He had curly grey hair, a firm handshake and a ready smile, and met us in his shirtsleeves before ushering us into a typical fancy lawyer's boardroom. He was surprisingly well informed about the situation at the CASB.

We described Deschênes' thoughts on the exclusive authority of the director of investigations. Ouellet chuckled. "If you're not satisfied with the director of investigations, why don't you introduce a motion to fire him?" he asked. "Of course you have the authority, you're members of the board," he said, evidently surprised that I should even ask such a thing. By the same token, he added, if the Evidence Review Committee was established by a formal vote it could only be abolished by a formal vote. "Just carry on with your committee," Ouellet advised us.

Might I quote him to counter the chairman's baloney about "the spirit behind the Act"? "Sure," Ouellet said. "If it will help you, write me a letter asking my opinion and I'll send you an answer outlining what I just told you. Then you can use my letter any way you want." I dropped off my letter the next morning and that very afternoon Ouellet's secretary called to ask if I cared to pick up his answer.

"When Mr. Justice Dubin referred to an independent tribunal," Ouellet said in his three-page reply, "he was referring to a tribunal independent of any department of the government, more specifically Transport Canada. . . . It certainly was not intended that subordinates be independent of the employer. . . . The Director of Investigation is the the employee of the Board which means he is the employee of the Board members acting as a group, and it is the Board members acting as a group that hires, fires and otherwise exercises direction over him."

Ouellet also agreed with Robert Décary's opinion that the board as a whole had the authority to direct the chairman. After Rowland Harrison absorbed the load of paper Bobbitt had dumped on him, he became the third top-drawer lawyer with the same opinion.

The way was now clear, I though, to putting the chairman's nonsense behind us and solving the Gander crash.

* * *

Although Bernard Deschênes had rejected the Evidence Review Committee as too exclusive, claiming participation by all members was essential, when we gathered to continue reviewing evidence he was again absent.

The five of us who rejected the ice theory were all participating (Dave Mussallem by conference call from Vancouver) and, as before, Thurston and Portelance held the fort for the pro-ice forces. This time, Deschênes had armed Thurston with an eight-point manifesto. As acting chairman, Thurston was enjoined "to ensure that the proceedings are held in an effective, open and efficient manner." To make sure we got the point about the staff's participation, Deschênes had even provided a sketch of the seating arrangements.

His high-handed manner only made us more aware of due process.

Roger Lacroix, who had worked himself into a righteous lather about Deschênes' perfidy, set the mood. In an emotional voice, he read his thoughts about "the unilateral, dictatorial and illegal action of Chairman Deschênes" at the previous meeting. It was good stuff. We wanted it in the minutes.

Thurston asked if we all understood Deschênes' directives for the conduct of the meeting. I answered by moving that the instructions be ruled unacceptable and that those present establish the procedures, as our regulations allowed. Bobbitt seconded the motion and, waving a copy of the CASB regulations, insisted on a vote. Portelance, who was at heart a reasonable sort, asked if we would at least agree to adjourn while Thurston contacted Deschênes by phone. We saw no reason not to.

While they went off to phone, I wrote out a supplementary motion: "Be it resolved that the Board urges the Chairman to reconsider his decision to nullify the Evidence Review Committee duly constituted by a decision of the Board under Section 20 . . . etc., etc." My purpose was to put our reasons for asking in the record with a page full of "whereases": *whereas* the draft report inappropriately concentrated on ice contamination of the aircraft's wings, and did not adequately examine other possible causes; *whereas* the analysis carried out on behalf of the director of investigations by the manufacturer and the University of Dayton was based on inaccurate and misleading preliminary interpretation; *whereas* the director of investigations could not respond adequately to concerns raised by the board about the draft report; and so on.

To no one's surprise, Thurston reported that Deschênes was not prepared to change his decision. We voted and my two motions passed five to one. Arthur Portelance, who had seconded my original motion to establish the Evidence Review Committee, abstained.

Frank Thurston looked like a six-year-old whose cat wouldn't come down from a tree. But the meeting continued. Having made our point, we could be magnanimous in setting terms. We agreed that Hinton could attend while we questioned two technicians from the laboratory whose analysis of the engines had purportedly proved that all engines were running full blast just before the crash—but we stipulated that we would hear only one witness at a time. Everyone knew this was meant to curb Hinton's interruptions.

We continued by noting that we still had not received the files. Indexing them, Hinton explained, was taking longer than anticipated. I asked that the minutes show that we wanted "the full record of the investigation, including original documents and photographs." To make doubly sure, we listed some items: "the complete medical files on the occurrence, and information on HCN [hydrogen cyanide], as requested earlier"; "original transcripts of all CASB interviews with witnesses" and others.

* * *

Friday, November 13, 1987—an inauspicious day for the defenders of the draft report.

We questioned the two staff members—Bill Taylor, the technician who had supervised the engine teardown, and Don Langdon, the engineer responsible for the analysis—about the engines. Everyone agreed that number four had not been turning, or had been barely turning, when it struck the ground. But what had it been doing a second or so earlier, as the aircraft penetrated the treetops? Could the trees and branches have slowed the engine in such a short time, as the report contended? If so, why not the other three engines? And why was Boag so adamant that number four couldn't have failed? A single engine failure would not, in itself, prevent a safe takeoff, but a failed number four would help explain the right turn and, in combination with ice and a heavy load, the inability to climb.

Tom Hinton soon found his urge to butt in irresistible. At his first interruption, I reminded the acting chairman that witnesses were to be heard one at a time. Hinton persisted, insisting he could shed important light on the point at hand. ''Perhaps Mr. Hinton could make his point in a memo,'' I suggested. Thurston gave me a withering look and, apologizing profusely to Hinton, said he had no choice but to ask there be no interruptions. Hinton kept his cool from there on. (He never did write me about his important insight.)

The CASB's technical staff, it seemed, had more than a few reservations about the conclusion that all engines were at full power. It also appeared that at least some investigators were unhappy with the engine examinations. Taylor and Langdon didn't relish being in the middle of a war between the director of investigations and board members—they mumbled about the pressures of time and the priorities of the investigator in charge—but it was clear they weren't proud of what they had to relate.[20]

Documenting the forensic disassembly of an engine should be similar to documenting a forensic autopsy. We expected detailed photographs of each stage of disassembly and copious, contemporaneous notes. Weeks or even months are often devoted to the examination of engines involved in minor incidents. Yet, contrary to the draft report's inference of a meticulous examination, the conclusions about the Arrow Air engines were based on an undocumented once-over that had taken less than a week for all four engines. A group including observers from Pratt & Whitney, Douglas, the U.S. Army, the FAA and the NTSB had given the engines this once-over without even following a check-list. We had no detailed notes or photographic records, and the CASB investigators had not even bothered to ask for copies of the others' notes.

The engine bleed valves, which snap open when power is cut, provided vital clues to the power at impact. There could be no doubt that the impact with the ground had frozen the number-four bleed valve wide open — twigs inside the valve proved it must have been open while crashing through the trees, and so the power on number four had to be down. But no one had tried to estimate whether the valve had had time to open during the second or so in the trees. No one had tried to establish whether chewing up branches for a second could absorb enough energy to stop an engine's rotation.

Moreover, while it was clear that the other engines were running at impact, there was little evidence to back the claim that they were at full power. For example, the bleed valve on number three was described as "most likely closed" — but no one remembered why.

We also learned that the Pratt & Whitney engineers had cunningly washed their hands of the investigators' conclusions. They carefully noted that their apparent endorsement was "based upon the crash investigation team's observations that engines #1, 2 and 3 were operating at take-off power".[21] In other words, assuming that the CASB investigators were correct, Pratt & Whitney agreed that they were correct.

By the end of the day, the shortcomings of the engine analysis were beyond dispute. Even Frank Thurston seemed genuinely surprised; he did not object to recording in the minutes that the board had concluded that:

—The number four engine was not producing power when the aircraft hit the trees; the report should examine why this engine failed.

—Pratt & Whitney should be asked for assistance in determining whether engines number 1, 2 and 3 were operating normally.

But he couldn't bring himself to support the logical consequence that I put in the form of an other resolution that:

The Board inform the Director of Investigation that it does not accept the thrust of the draft report on the Arrow Air accident at Gander.

By this time Rowland Harrison had given us his considered legal opinion that Deschênes had no right to declare the Evidence Review Committee *defunctus*, and he too had advised us to ignore the ruling and carry on, so my motion also stipulated that the Evidence Review Committee would convene at two p.m. the following Monday. We would start by asking Peter Boag to explain why his team had not attempted to resolve the conflicting evidence about the engines.

The motion passed, as expected, five to two, Thurston and Portelance against—but Frank Thurston said flatly that he would not convey the resolution to the staff. I knew what to do about that. My next motion directed the acting chairman to transmit the board's decision about the Evidence Review Committee to the staff.

The look Thurston gave me was not pleasant, but I knew he would tell the staff they must cooperate with the committee.

* * *

Over the weekend, Deschênes not only learned that we had rejected his rejection of the Evidence Review Committee, but also learned on what basis. Rowland Harrison had sent him a letter stating that, on the basis of such-and-such laws and documents, ''We have concluded that your purported action, at the Board meeting of November 3, in declaring the motion ''De Functus'', was without authority and that, accordingly, your declaration has no effect.'' He had therefore advised his clients to continue with the Evidence Review Committee. This was good lawyer-to-lawyer stuff that Bernard could appreciate. We wondered how he would react. My bet was that he'd read a statement.

Sure enough, on Monday morning Deschênes called a meeting for two o'clock, at which he would make a statement. It was a transparent maneuver to preempt the Evidence Review Committee meeting. I countered by asking a secretary to let everyone know that our committee would convene at 1:30 to discuss our procedures, but would recess at 2:00 to hear the chairman's statement.

Deschênes arrived promptly, Ken Johnson and Tom Hinton close on his heels. I made a show of calling a recess in the committee's deliberations and relinquished the chairman's position, and Deschênes read out a four-page manifesto. He was not pleased with reports he had received of the previous week's shenanigans; "Too much time is being spent on debating the powers of the Chairman," he said. He was saddened by the "entrenched positions of some Board Members." But—wonder of wonders—he had had a change of heart.

Deschênes didn't mention Harrison's letter—which we all knew had been hand-delivered to him during the weekend. He simply announced, "I am prepared to review my directives on a trial basis and to accept to proceed, for the time being, by way of a committee which will report to the full board." Lest anyone think this was total capitulation, however, he listed thirteen elaborate conditions. He was concerned that the review "not go beyond the examination of the evidence obtained by the investigators involved, their investigations and analysis." This seemed to mean that he would allow us to speak to the investigators as long as we didn't try to get new information or interpretations. Furthermore, interviews with anyone not under the direct control of Tom Hinton were "totally unacceptable". Deschênes added that "the director of investigations will continue to be present and to assist the committee in its work."

Then Deschênes moved on to a topic a good deal more to his liking. He informed us that the Minister of Transport had received the report from James F. Hickling Management Consultants — and that they had determined that Deschênes' interpretation of the CASB Act had been right all along. Not only that, but they were of the opinion that "consideration would be given to reducing the size of the CASB" — that is, firing some of the members. Deschênes' expression left no doubt about which ones.

Deschênes dramatically announced his intention to "immediately seek advice and possible action by higher government authorities" on the consultants' recommendations. He had to leave town at once, he said, and would not be back for two days. "In the meantime, please proceed with the Evidence Review Committee."

I asked if Mr. Boag would be attending, and Deschênes' statement that he would seemed to catch Tom Hinton by surprise. Hinton said he hadn't known Boag was required, and then argued that Boag did not want to attend.

By this time Deschênes had had enough grief, without taking lip from the staff. Uncharacteristically, he cut Hinton short and sent him to fetch Boag. Then Deschênes left.

I thanked Hinton but advised him that the committee would not be needing his help. "My strong preference is to stay," he replied. I repeated that we would not be needing him.

"Are you telling me I can't stay?"

"I'm asking you. This is our meeting. You were not invited. Please don't make it more difficult than necessary."

"Am I being instructed to leave?"

"Yes."

Hinton tried to make a statement and I ruled him out of order, but I polled all the members present. Most wanted him to leave at once. Nevertheless, I gave him two minutes.

"If you are going to exclude me from the room, then I think you are inhibiting me from performing my duty," he said. "And I don't think that in doing that

I can allow it to take place. Therefore, I will direct my staff to withdraw."

Hinton and I stared at each other across the room.

"Mr. Hinton will leave and Mr. Boag will stay, please," I said.

Boag stood up, almost smiling. "Mr. Chairman," he said, "I report to the director of investigations as my supervisor. And he has given me his direction that I am to withdraw, so I must."

Roger Lacroix was furious at Hinton's reference to "my staff". He tried to debate the point, but I cut him off. Hinton and Boag left.

We agreed that the investigators' reaction to the committee would have to be considered by the full board. In the meantime, we listed some issues that we would have liked to discuss with the investigator in charge. The matter of the engines headed the list. On what basis had Boag rejected expert opinion that number-four engine was not producing power? Why were various specific clues not followed up? On what basis had Boag told the Pratt & Whitney analysts to assume that engines one, two and three were producing full thrust? Where was all this documented?

There were many other technical points. The analysis of the thrust reversers seemed to be as superficial as that of the engines and the flight data recorder. The report stated that an examination of the tail-skid established that the aircraft had not over-rotated on the runway. Why was there no report and no photos of the tail-skid?

Given the information we already had about the FDR and engines, there could be no doubt that the report had to be changed. Even Frank Thurston had accepted that number four was down. The weakest conclusion the full board could now make was that the range of possible causes was much greater than indicated so far, and much more work was needed to narrow them down.

On that slender note of optimism, I adjourned the first and last meeting of the Evidence Review Committee.

* * *

Up to this time, only a very few people outside the CASB had any inkling of the internal clash over the Gander investigation. But Deschênes and his supporters believed they had a secret weapon: Ken Johnson and his staff had been cultivating reporters covering aviation. The press is always ready to hear another stirring account of how air crash investigators have solved a mystery.

The report from Hickling Management Consultants was also much too good to keep secret, and someone leaked a copy to the *Globe and Mail*. The article in the *Globe* didn't mention that the CASB had been dead-locked five-five on the conduct of the Gander investigation, or indeed anything at all about the investigation. It did, however, offer insight into Deschênes' threat about getting "higher government authorities" to fire some members.

Globe readers learned that a report to the minister said three members of the CASB had "attempted to overrule chairman Bernard Deschênes".[22] The story didn't explain that these attempts consisted of motions at board meetings that the chairman didn't like, but reported that the consultants believed "such moves violate the absolute independence of the agency's director of investigations from any interference." These dissident board members were accused of attempting "to rewrite the director's reports without the director's permission'' and creating dissension "so severe that the agency's effectiveness may be seriously damaged.''

We had not yet learned to pay attention to the press. We speculated on who had leaked the story and wondered which of the five of us were the guilty "three". But we were still more concerned about Hinton and Boag thumbing their noses at the Evidence Review Committee. Despite the bluster and the face-saving conditions, Deschênes seemed unnerved by Harrison's letter; had he not said, "Please proceed with the Evidence Review Committee"?

A court ruling, Rowland Harrison thought, ought to settle the issue of whether the chairman was bound by a majority vote. Harrison had little doubt about how such a ruling would go, but this was not a matter for a private suit. It could be settled quickly, however, if the Deputy Minister of Justice referred the matter to the Federal Court. Harrison made the request on our behalf, noting that the report on the investigation was being impeded by "differing views on what the respective responsibilities of the Board, the Chairman and the Director [of Investigations] should be".[23] The letter elicited a speedy reply from an unexpected quarter — although initially I didn't realize it was a response.

The day after Harrison's letter was hand-delivered to the Justice department, I got a call from an assistant of Transport minister John Crosbie — it was the first time such an honor had been bestowed on me since I'd been appointed to the CASB. Would I and my colleagues who had some problems with the CASB's operations care to meet with the minister next day, say around one p.m?

I explained that Dave Mussallem was in Vancouver and might not be able to make arrangements, but I was sure the rest of us could squeeze in some time for the minister. The four of us would meet Crosbie the next day.

That left us just enough time to deal with the matter of Tom Hinton and his disinclination to allow "his staff" to talk to the Evidence Review Committee.

* * *

Since Bill MacEachern was not present for our regularly scheduled meeting on November 24, I knew I could count on a majority. As soon as Deschênes called us to order, I announced that I had something important to say. I began by reviewing the way Hinton had prevented the investigator in charge from cooperating with the Evidence Review Committee because he felt his own absence would in some unspecified way "inhibit" him from performing unspecified duties.

I credited Hinton with high motives and pure intentions before lambasting his misguided manner, deliberately echoing Justice Dubin's critique of the investigators of the Cranbrook crash. Dubin had not accepted idealism or dedication to safety as sufficient excuse for investigators "to carry out their duties in a manner that they would prefer." Rather, he had decried "the complete insubordination" of those who defied legitimate authority that differed from their personal interpretation of their duty. In recommending the independent tribunal that would become the CASB, Dubin had said that investigators should conduct themselves "pursuant to the directives given . . . by the new tribunal"—not just the chairman, but the tribunal as a whole. In contrast to Deschênes' usual airy generalization about the "spirit" of the Dubin report, I quoted Dubin verbatim, specifying page numbers in his report.

I explained that the Deputy Minister of Justice had been asked to assist in obtaining a definitive ruling on the relationship between the board and the director of investigations, and moved that the board suspend Hinton's designation as director of investigations and the power of investigator given to him under the CASB Act, pending a ruling from the Federal Court on the question of "exclusive authority".

Deschênes hesitated. I reminded him of Section 20 of the CASB regulations, which says that "Any question raised at a meeting of the Board shall be decided by a majority of the votes cast on the question by the members present at the meeting". As if he didn't know.

Deschênes simply said, "The meeting is adjourned," and walked out.

Minister of Funny Lines

Journalists describing the Honorable John C. Crosbie, Minister of Transport, invariably noted his caustic wit and penchant for colorful alliteration, dubbing him "the national quipmaster" or "Minister of Funny Lines". Indeed, many observers attributed Crosbie's loss of the leadership of the Conservative Party to Brian Mulroney to a flippant remark about the French language.

Crosbie had recommended my appointment to the Canadian Aviation Safety Board, but I had met him only once — at a *pro forma* presentation in late 1986. Bobbitt, Lacroix, Stevenson and I were to meet the minister at his office in the Centre Block of the Parliament Buildings. When we got there he had not yet arrived, and the receptionist bid us wait on a long wooden church pew in the high-ceilinged corridor. The four of us sat in a row like students in an English private school waiting for the headmaster.

The minister arrived. He was taller than I remembered, and looked anything but jovial. He wore a rumpled, nondescript grey suit and carried a battered briefcase, and was flanked by two aides in smartly tailored dark blue pinstripes. He looked like the boss, and the aides looked like aides.

After perfunctory handshakes we were ushered into his office. Crosbie sat at the head of the rectangular table; the aides took their places on his left, pulled pads of yellow legal paper from shiny briefcases and scribbled notes throughout the meeting — without a word or look at their boss or any of us. I sat on Crosbie's right, Stevenson and Bobbitt beside me. Lacroix sat facing the minister. Except for the yellow pads, we might have been waiting for lunch in an elegant restaurant.

But there was to be no bonhomie around this table. Nor were we to experience any of Crosbie's celebrated Newfoundland wit. He looked at us evenly, like a headmaster whose patience has been sorely tested. What a terrible thing it was to sue the chairman, he said. It would destroy the public's confidence in the Canadian Aviation Safety Board.

An awful business. Somehow, it fell to me to answer. I said that differing interpretations of the CASB Act had hamstrung the work of our board. Far from wanting to sue the chairman, we were just looking for an authoritative decision so we could get on with the important work of the CASB, including the review of the Gander crash.

Crosbie seemed amazed. The Hickling report described the problem, which was our shenanigans. On the way over, we had agreed not to knock the Hickling report — which, after all, had been commissioned and directed by Crosbie's own office. But the occasion called for a change of plan. I blasted the study, explaining that it had been conducted by former middle-level bureaucrats in the Transport department who had worked under contract for both the Transport department and the CASB, and undoubtedly hoped to do more of the same in the future. In fact, the contract for the study stipulated that the work had to be carried out to the satisfaction of Transport department officials. That by itself, I said, biased the report toward the values and prejudices of the middle managers.

I pointed to a recommendation that investigation of air traffic control incidents — close calls and near collisions — revert to the Transport department. This was clearly self-serving, pandering to those who had signed the contract; it flew in the face of the idea of independent investigations espoused by Justice Dubin, and if implemented it would be a public relations disaster for the Department of Transport and the minister, and worse.

On hearing this tirade, Crosbie seemed to settle down. The annoyed headmaster dropped away and he peppered us with penetrating questions about the operation of the CASB in the no-nonsense style of the shrewd lawyer I suddenly realized he was.

Stevenson kept trying to drag in the Gander investigation. At this time we had not yet started to talk openly of the possibility of sabotage, but at one point I said something mealy-mouthed about the difficulty of getting Chairman Deschênes to focus on certain scenarios other ice on the wing.

"Like sabotage?" the minister asked blandly.

"Yes, sir," I replied.

Crosbie thought about this for a moment, then started talking about our application to the Federal Court. This "legal shit", he said, was not helping anyone. I explained that going to court was just one way out of a horrible impasse. It might be feasible, I mused, to get interpretation of our legislation outside of a courtroom — from someone whose judgement was beyond demur.

"Like who?" asked Crosbie.

"Ideally, someone like Justice Dubin," I offered. "Or, failing that, John Sopinka." None of us had met Sopinka, but since serving as senior counsel to the Dubin Commission he had become one of Canada's most celebrated lawyers. His vigorous defence of Susan Nelles, the Toronto nurse wrongly accused of murdering babies, had made sensational headlines for months. A more recent (but less successful) defence of Sinclair Stevens, Minister of Regional and Industrial Expansion, accused of influence-pedalling, had also generated big headlines. Sopinka couldn't possibly endorse a board so contrary to what the Dublin Commission had envisaged.

Crosbie looked thoughtful and, to my surprise, said the idea might have some merit. He'd think about it. Meanwhile, we should think about the merit, or otherwise, of stirring things up in the courts.

The meeting was over.

* * *

We strolled down the sunny side of Wellington Street for a pow-wow and a coffee at the Château Laurier. It wasn't hard to decide what to do. As soon as we got back to the office, I called Crosbie's chief of staff to tell him how much we had appreciated meeting the minister, and that we were withdrawing our request for a ruling from the Federal Court.

A few days later, Crosbie's office called to say that the minister had engaged Mr. Sopinka to study the CASB.

* * *

On November 25, we found out that Minister Crosbie had also called Chairman Deschênes for a personal tête-a-tête, when Bernard asked us to meet in his office to follow up on our respective chats.

The strain of recent events had taken its toll on Roger Lacroix. Roger, who had been on medication for a variety of ailments, had become more and more despondent about his role on the CASB. He was fed up and in need of a break from all this weirdness, and he told us he just couldn't stomach another useless confrontation. He was going to take his family for a skiing holiday. He had already notified Deschênes that he wouldn't attend the December board meetings.

So Norm Bobbitt, Ross Stevenson and I trooped off to the building across the street to meet Deschênes in the suite of offices he shared with Ken Johnson. I tried to remember whether I'd visited Bernard's office since our first encounter eleven months earlier; if I had, the occasion hadn't been memorable.

Bobbitt opened the pleasantries by telling Deschênes that, sooner or later, the cover-up was bound to unravel. This certainly surprised me, because Norm had never talked about a cover-up; none of us had. To be sure, each of us suspected something more than what met the eye — but so far we had avoided discussing motives, even among ourselves.

I recalled how quickly Deschênes had cut off questions when Arrow Air lawyer Allan Markham uttered the words "cover up" at the public inquiry. Now the same words had no discernible effect; Deschênes looked inscrutable.

We talked obliquely about our respective meetings with Crosbie — sparring uneasily, neither side willing to be frank. We felt we had alerted the government to the problems in the CASB and the Gander investigation. There was not much to say to Deschênes. There was, in fact, not much we could do in the boardroom until we had some feedback. No agenda had been set for the coming meetings, and Deschênes could try to ram the report through. It didn't matter. What was the point of going to meetings if he could avoid votes by walking out, or dismiss results he didn't care for as *defunctus*?

So when the conversation touched on Roger Lacroix's holidays, without having prearranged it, Norman, Ross and I all allowed that we too were ready for a rest. Norman mused that the ski slopes would be much more pleasant than the boardroom. Ross and I talked about wanting to attend another course on investigative techniques. Deschênes sympathized.

We believed we had established the deficiencies of the draft report beyond dispute. Thurston respected Bernard Caiger, so he now understood that the investigators' interpretation of the flight data recorder was pure fantasy, and he had also admitted that number-four engine couldn't have been producing full power. Deschênes would believe him, whatever he thought of us. The chairman wouldn't need out further prodding to know that major changes were essential. If he and his supporters had enough chutzpah they could, of course, fudge up an emasculated version of the ice theory and send the report to interested parties in our absence. If so, some of the interested parties would protest and we'd start the next round with powerful allies. And by then Sopinka should have set Crosbie straight and Deschênes would be forced to play by the rules.

None of this was said aloud, but in short order it was tacitly understood that, like Lacroix, the three of us would skip the December board meetings. We smiled wanly and the meeting was over. It was, as they say in diplomatic circles, an encouraging exchange. As it turned out, it was all based on a hideous misunderstanding — as, in diplomatic circles, encouraging exchanges often are.

By this time Deschênes had to know that we had withdrawn our appeal for a hearing in the Federal Court. He may have thought his hint about reducing the size of the board had provoked the intended effect, or that the minister had read us the riot act and we had capitulated. In any event, we must have seemed like whipped dogs, giving him the green light to put out any report he damn well pleased.

To cheer him all the more, a threat from the courts —one he could not control—had just evaporated. Relatives of the victims had already received compensatory damages, but as of mid-November the courts had yet to decide if claims for punitive damages could also stand. Had such claims come to court, the evidence would have been dissected publicly by the finest minds money could rent. But the district court in Louisville, Kentucky, had just ruled against suits for punitive damages. The wire service report called the decision "a major victory for the Miami-based airline and its insurer." It was also a major victory for the ice theory: the CASB's conclusions would be spared the indignity of second-guessing by lawyers and judges.

* * *

On December 10, the five available members attended a regularly scheduled board meeting. The tape recording they made reveals a jovial Yuletide spirit with Chairman Deschênes in an uncharacteristic cheerful, upbeat mood. And why not? At long last he could preside over the meeting he had expected—or at least dreamed of—when the first draft had been presented the previous April. The board was about to consider a lightly refurbished version of the second draft forwarded in October. Changes were mostly the result of what Peter Boag termed "wordsmithing".

The chairman announced that "some board members had decided not to participate in the review". This was only partially accurate, as no agenda had been set when we told Deschênes we would be otherwise engaged and, during our discussion of the engines, acting chairman Thurston had agreed that we could no longer hope to conclude the review by the second anniversary of the crash.

When the review of the second draft had begun in October, we had agreed that a written motion to accept the draft report would be circulated to allow every member to vote. The board had also agreed to adopt rules for dissenting addendums before circulating the "conditional draft" report to interest parties. But Deschênes now proposed that those arrangements be rescinded and that the report be adopted by voice vote at the end of the day. No one demurred.

The board then proceeded to the draft report. As I listened to the tape some weeks later, I got an inkling of the gulf separating the members then present from those of us who were not there: to them, reviewing the report was a search for compromise language, a battle of adjectives to describe the indescribable. It went without saying that the investigators' work was complete and accurate in all essentials. The board's job was to polish the report here and there, to "make sure we have the best possible product", as one of them put it.

Someone noted, for example, that the word "deleterious" was misspelled. The investigators gratefully accepted the correction of the term "technical log" to the more accurate "journey log". It took a little more discussion to determine whether Canadian limits on pilots' duty hours were more "conservative", "advanced", "stringent", or "restrictive" than those in the United States. More weighty debate centered on the estimate that the Arrow Air DC-8 had weighed 14,000 pounds more than the crew had entered on the load sheet. Yes, Peter Boag allowed, the aircraft could have weighed even more than that. "Peter, would you buy 'at least' 14,000 pounds?" Frank Thurston asked. An easy sale.

The state of the number-four engine cast only a slight shadow when Thurston pointed out that the rotational speed was not "low" but in fact "non-existent". "The bashed-in second stage turbine," he reminded those present, "shows that it was not rotating at impact." But that was no problem—the report need merely note that the "ground impact rpm of the number four engine was *lower* than that of the other three engines." No one could quarrel with that; "It's factual."

When the wordsmithing had been done on everything but the final statement of cause, a thought struck

Director of Investigations Hinton: "It was suggested while we were walking in at lunch time that perhaps the board may want to make a finding with respect to sabotage."

"I think that would be a very good idea," Thurston offered without hesitation. But Bill MacEachern missed his cue. "Wouldn't that be unnecessarily focusing attention on a point for which there is no factual basis?" he asked.

Boag, Hinton and Johnson all weighed in with arguments suggesting that a "negative finding" was necessary to head off controversy, and MacEachern was persuaded. The minutes recorded that "a new finding should be added to make the point that no evidence was found of any military explosive action resulting from a criminal act, nor of any military explosive devices being carried on board the aircraft, aside from limited revolver ammunition. The positive evidence also runs contrary to explosion as a result of a criminal act"[24]. The exact wording was entrusted to Peter Boag. (He decided on "The evidence does not support the occurrence of an explosion either accidental or as a result of sabotage.")

That point settled, there remained only the statement of cause. The second draft had said that the DC-8 crashed because "take-off was attempted with airfoils contaminated with ice and with inappropriate reference speeds for the weight of the aircraft." The latest wording attributed the crash to a "stall with subsequent loss of control as a result of the aerodynamic penalties of airfoil icing." In a way, this change enhanced the report's internal consistency: one "finding" asserted that the aircraft had stalled but no finding said directly that there was ice on the wings. But the chairman wondered if the new wording implied that "only the loss of control is due to airfoil icing, and not the stall." A lively discussion ensued.

Frank Thurston didn't quite agree that the probable cause was a stall. He thought the stall should be termed the "proximate cause". In any event, it was really a "deep stall." Deschênes warned that the term "proximate cause" could lead to legal complications, while Peter Boag was "a little bit uncomfortable with the technicalities of what a deep stall is." He preferred "failure of the flight crew to identify the need for de-icing before takeoff." The tape spun and I thought of Sister Juana Inés de la Cruz, the seventeenth-century Mexican nun who established how many angels could dance on the head of a pin.

When the board discussed whether "inappropriate reference speeds" were necessary or incidental to the crash, the investigator in charge explained that when he had submitted the first draft he had excluded the reference speeds because he believed the aircraft could not have taken off successfully even with the correct speeds; now he was leaning the other way.

In the absence of Ross Stevenson, Boag and Hinton delighted in explaining the techniques of piloting big jets to an appreciative audience. After a lot of good-humored bantering, it was agreed that reference speeds should stay in the statement of cause. Crew fatigue was added for good measure. No one commented on the lack of positive findings on either.

Everybody agreed on the final wording: "the probable cause of this accident was a takeoff attempted with ice contamination of the leading edge and upper surface of the wing. This action resulted in a stall at low altitude from which recovery was not possible. Contributing to the occurrence were flight-crew fatigue and the use of inappropriate takeoff reference speeds.'' No nit-picking logicians were there to ask how the board could be certain these items had contributed to a cause that was still only probable.

"Let the minutes reflect that this is a unanimously adopted draft report,'' the chairman said. Warm congratulations were offered to Peter Boag and other staffers. By five o'clock the meeting is over, the violent deaths of more than 250 human beings accounted for.

* * *

Transport Minister Crosbie's office had set up the meeting with John Sopinka for December 16. Stevenson had some other commitment, but Bobbitt, Lacroix and I met Sopinka in my office.

Sopinka was tall and trim—much the same shape, I imagined, as when he played pro football for the Toronto Argonauts. But he was not at all intimidating. His hair was shorter and greyer than at the Dubin Inquiry ten years earlier and there were a few more crinkles around his eyes; otherwise he looked much the same. We shook hands and he smiled benignly through Daniel Ortega glasses.

After we exchanged pleasantries, Sopinka questioned us about the CASB's operation. It was all very low key. From time to time one of us would attempt to bring up the Gander investigation, but the lawyer put us off. After about half an hour, he thanked us for our help and departed to meet Chairman Deschênes. He had left no clue about the nature of his task or own feelings about the foul weeds that had sprouted from the seeds he had helped scatter with the Dubin Commission.

As we chewed over what we might expect, Roger Lacroix suddenly told us he wouldn't be around to find out. He'd had enough. The CASB had become a farce and he was going to quit. He had already contacted the Privy Council Office and his overture had received immediate and enthusiastic response. Roger Tassé, a former high-octane official at the Justice Department, was to act as intermediary.

With Lacroix gone, only four of us would remain to oppose the five who had so cheerfully endorsed the ice theory and who were now immutably committed to its support. We could only hope that Sopinka would blow the whistle in time to restore due process for the consideration for the response from interested parties.

BOOK 5

FEAR AND LOATHING AT THE CASB

Part 1

Changing of the Guard

In January 1988, I started my second year at the CASB. The board was now utterly polarized over the Gander investigation and, more generally, over the limits of the chairman's authority.

The *Globe and Mail* was continuing its exposure of our internal problems, and to explain the angst unveiled by the leaked Hickling report Bernard Deschênes publicly confessed that "the board had been plagued by internal disagreements".[1] To hear him tell it, these disagreements had nothing to do with the content of reports or the quality of investigations, but arose "over how much the government-appointed directors should interfere with the professional staff's investigations of crashes". (He didn't explain that his concept of intolerable interference included any attempt to question the staff about a completed report.) This line became the theme of a venomous public-relations onslaught.

In the none-too-subtle code, "government appointed" meant "politically appointed", hence politically motivated, while "professional staff" meant competent, dedicated workers unsoiled by crass motives. And the cream of the "professional staff" was, of course, the "professional investigator". Professionals never made mistakes, and their work was probably beyond the understanding, certainly beyond the criticism, of most mortals, let alone the "government appointed". Good board members (if that was not a contradiction in terms) understood this. The message was constantly reiterated, usually in admonitions to restrict ourselves to "judgement and common sense" and not try to "reinvestigate the accident".

It was certainly true that the members of the CASB were political appointees. Five, including Bobbitt, Mussallem and myself, were appointed by the governing Conservative Party. Deschênes, Stevenson, and Lacroix were named by the previous Liberal government, as were Bill MacEachern and Arthur Portelance, who had also served as elected Liberal politicians. Depending on your point of view, Frank Thurston was either the least or the most tainted — he had the distinction of being appointed by the Liberals and then reappointed by the Conservatives.

But the fabric of the CASB hadn't split along a political seam, but between those who doggedly supported the chairman and those who felt compelled to examine the evidence.

Here we were, trying to determine the cause of Canada's worst air crash. Why would half the government-appointed directors and all the professional staff who had any say wilfully disregard the evidence? I started to brood about the dark secrets that might lie behind their intransigence.

At the first board meeting of the new year I tried to smoke Deschênes out. We had already advised him of our determination to continue reviewing the evidence while waiting for interested-party comments. When he took up his old song and dance about what was and was not permissible, I saw a chance to pin him down.

"There's a question I'd like to put to you, Mr. Chairman," I said. "I'm asking you if there's any reason that you know of — even if you feel you can't go into details — is there any reason why our board should not pursue the technical causes of the Gander accident to their roots, in so far as our resources usually allow us?"

"I'm not getting exactly what you want there," Deschênes replied.

"I want to know if there's a reason, some overriding reason, why the Gander accident is a special case and we should not explore the technical causes in as great detail as we possibly can."

Deschênes gave a rambling reply to the effect that he was trying to reach a consensus.

"You did not answer my question, Mr. Chairman."

Deschênes still didn't understand. "If there is a need, if a specific board member wishes to have something, I certainly will consider it," he said.

Norm Bobbitt saw my exasperation and weighed in, insisting on an answer. The chairman kept bobbing and weaving but Bobbitt persisted. "No, Mr. Chairman, you were asked a very specific question and you did not give an answer." "Let me be more explicit," I finally said. "Is there any reason — of national security, or of anything else — that would stop us from looking into this accident as far as we can?"

Deschênes hesitated, then said, "From a national security point of view, I . . . ," trailing off.

"I'm just trying to give you some examples of possibly higher reasons than aviation safety."

"No," he admitted, "I don't know where national security would apply."

"Okay" I said. "so it isn't national security. Is that your answer?"

"Yes. But I'm not admitting that we haven't gone as deep as we could or should."

This was less than a ringing declaration of our duty to get to the bottom of the crash. But weren't going to get any further that day. There would be other days. I let myself get sidetracked.

The Hearts & Minds of the Interested Parties

The review of interested-party comments offered our last chance to keep the door from closing on the Gander investigation. Those of us who opposed the ice theory wanted to push that door open and let the sun shine on the evidence, while the pro-ice members were determined to nail it firmly shut. They had already made a good start by setting the first of February as a deadline for receiving comments—less than the usual sixty days, and certainly not much time to analyze the report on one of the ten worst aviation disasters in history. The Christmas and New Year's holidays further reduced the time available for meaningful review.

The ice theory imputed neither direct nor indirect blame to government agencies or manufacturers; in fact, it absolved them. They, at least, had no reason to take issue with it. Snippets of the report might, however, contradict corporate dogma or conflict with the desired corporate image. The representatives who had worked with the CASB investigators during the past two years should have ensured that such flaws were eliminated before the draft report was even presented to the board, but the home office would want to make sure.

Still, there was little danger that petty demurrals from these interested parties would derail Deschênes' schedule. The CASB had a long-term relationship with government agencies and the big companies through the investigation of past and future accidents. Behind-the-scenes accommodations would assure that the formal responses met the deadline and brought no surprises.

The official response from the Multinational Force and Observers consisted, for example, of a short one-page letter from the Secretary-General. There were no substantive comments, but the MFO did see fit to "commend the board for an exhaustive study into the causation of this tragedy." To supplement this official response, the MFO's legal counsel "informally" conveyed a longer letter to the CASB's counsel "at your suggestion during our telephone conversation".[2] This more detailed response took issue with items important to the MFO. The draft attributed a small oversight in paperwork to an "MFO representative" — the MFO's response pointed out that the individual in fact, represented the U.S. Army, not the MFO, and suggested that the draft report correct "any possible implication that the MFO may have had a role in such departure documentation and supervision."

Douglas Aircraft conveyed its list of suggestions directly to Peter Boag — the board members had no need to trouble with trivial wording changes, inaccuracies or typos. This back-door system to ensure bland, complimentary official responses saved a lot of trouble.

But this unofficial system did not extend to all interested parties — to the operator of the airline who was responsible for selecting and training the crew, for example. Arrow Air had taken a severe hammering at hearings held by the Armed Services Committee of the United States Senate in 1986. These hearings, conducted in the aftermath of the Gander crash, looked broadly at the practice of military charters rather than at the crash itself. Witnesses charged that Arrow actively urged crews to economize by cutting corners on maintenance and discretionary services such as de-icing.[3] If the crew's hectic schedule had made them too tired to notice or care about ice, the company might be faulted for not scheduling enough rest time. All kinds of hints and innuendos had, moreover, arisen about the condition of the sixteen-year-old DC-8. Arrow Air and its attorneys could be expected to review the draft with some care.

While it was not Bernard Deschênes' habit to keep us posted on contacts with interested parties (or anyone else), he let slip that Arrow had asked to make an oral presentation to the board, an option specified in the legislation. He didn't think this was a good idea. His account of the request was vague, but he did mention that the call had been made on Arrow Air's behalf by Walter McLeish, the Ottawa aviation consultant who had served as Transport Canada's senior aviation official at the time of the Dubin Commission.

I called McLeish for details. It turned out that Arrow was eager to make an oral presentation, but that Deschênes had nixed the idea on the ground that all the other interested parties would have to have the same opportunity.

Well, of course all other interested parties had to have the same opportunity. I told McLeish that if a formal request for an oral hearing came to the board, I for one would strongly support it, and several other members would probably do the same.

When Arrow's telex arrived, the chairman, Thurston and MacEachern argued against an oral hearing but the others in their camp thought it wouldn't look good to turn Arrow down. In the end only Mac-

Eachern held out, and an oral representation was arranged for February 29. Meanwhile, the ploys to keep the investigation going — and the counterploys to shut it down — accelerated.

It seemed prudent to us to check if other interested parties knew that Canadian law gave them the right to seek an oral hearing. A round of calls brought us a major break. In the haste to wrap up the interested-party stage, the professional staff had made an incredible oversight — they had neglected to send Theresa Griffin a copy of the conditional draft. For the party with the greatest "direct interest in the findings" was not the operator of the airline, but the captain of the aircraft — or, more precisely, the captain's estate.

The CASB had done its best to avoid finding fault or assigning blame, but pinpointing the causal factors in a crash unavoidably leads to blaming someone. In the case of the Gander crash, the ice theory pointed an unwavering finger at Captain John Griffin.

By all reports Griffin was an exceptional airman. A former air force officer with an undergraduate degree in physics and a Master's degree in Finance, he had flown 707s for Pan American Airlines for ten years before joining Arrow in 1982. He became Arrow's chief pilot and vice-president and manager of Flight Operations, but had recently returned to line operations. He was forty-five years old at the time of his death, and left behind a widow, two children and, as we were to find out, many loyal friends. To imply that his negligence had been directly responsible for the deaths of over 250 people entrusted in his care not only cast a slur on his memory, but left his estate vulnerable to legal suits.

A copy of the report was mailed to Mrs. Griffin on March 16, 1988, more than three months after copies had been sent to the other interested parties. Deschênes gave her until April 15 to reply. By then, however, much more bilgewater would have swirled under the bridge.

Odd Man Out

After his initial surprise at our refusal to play dead, Bernard Deschênes had become strangely subdued and almost compliant.

On January 26 some half-dozen steel cabinets had been delivered to an empty office near the boardroom and Hinton and Johnson had assured us that they, contained every scrap of information gathered during the Gander investigation. Deschênes had also agreed to hire an independent expert to re-examine all four engines with a view to determining the power output prior to impact. Incredibly, he accepted my renewed urging for an independent read-out of the previous takeoffs recorded on the FDR, and he even asked if I could recommend someone for the job; and I proposed Ray Davis, a retired flight recorder specialist from the British government's Accident Investigation Branch. I thought Deschênes had finally come to his senses.

Along with the files, we finally received the results of the toxicological analysis conducted on the remains of the passengers some two years earlier. There was no index or summary, but a glance through the computer printouts showed lethal carbon monoxide and hydrogen cyanide levels in victim after victim.

Something was very wrong here. Anyone could see that this data was sensational. Experts had argued for days over the relatively simple toxicological findings in the Schefferville crash, yet the results on the soldiers on the Arrow Air DC-8 didn't even rate a mention at the public hearings or the report.

What the hell could this mean? Could there be an innocent explanation? Fortunately, the chairman allowed us to question Dr. David Elcombe, the physician responsible for the medical aspects. He certainly gave no indication that he had anything to hide or to be embarrassed about. Nor did he seem cowed by the presence of Tom Hinton, Peter Boag and his nominal boss, Ken Johnson.

Elcombe said he was still puzzling over this very interesting case. The analysis, he told us, remained "ongoing" and "far from complete." He seemed unconcerned that his colleagues had wrapped up the investigation and that the report had already gone out to the interested parties.

Yes, Elcombe had been in Dover, Delaware, to observe the autopsies. Yes, he was aware that the report said the autopsies had been conducted under "the control and supervision of the CASB." Yes indeed, if the autopsies had been conducted under the control and supervision of someone at the CASB, that someone would be Dr. Elcombe. But on reflection Elcombe felt that the words "agreement and liaison" would be better than "control and supervision".

We should be aware, Elcombe told us, that if the Americans hadn't done the autopsies they would have had to declare the soldiers lost in action, and that would have brought Lord knows what kinds of complications. He didn't elaborate, but I'd heard that Vietnamese peasants had switched from salvaging scrap iron to scavenging bones they could pass off as missing American soldiers — surely Elcombe didn't think the unemployment problem in Newfoundland was *that* bad?

As Elcombe spun his tale, Tom Hinton, back to his old tricks, jumped in and told everyone that the CASB and Newfoundland's deputy Minister of Justice had been "most heavily involved in the decision to conduct the autopsies in the United States." Bernard Deschênes also became momentarily reanimated, telling us he had personally discussed the arrangements with the provincial Deputy minister. But he wouldn't say if he had also discussed the matter with other senior officials or politicians.

Eventually, we got to the question of hydrogen cyanide in the remains of the passengers. "The toxicology results began to roll in around March 1986," Elcombe reported — that is, before the start of the public inquiry. After some sparring about the completeness of the analysis, we elicited that at least two passengers apparently decapitated at the moment of impact had none the less somehow managed to absorb lethal levels of hydrogen cyanide or carbon monoxide.

Frank Thurston was better informed than the rest of us. It seemed he had the job of finding a closet for this skeleton. "On your list of decaps there were only two decaps with high CO," he said to Elcombe. "How many non-decaps had high CO? What percentage?" (The secretary who transcribed the tape was puzzled here, noting in parentheses, "does not have answer" and "other small questions here." I couldn't decipher the exchange either — not at the time, and not when I listened to the tape later.)

Then Frank got up a head of steam. "Could we look at the gross statistics of the pickup as a percentage of the decaps plus non-decaps, because I think we would have to deduce that if the decaps were statistically divorceable from the non-decaps this could only happen if the pickup of CO happened after it crashed and not before. Do you think there is an identifiable difference between the populations, decaps and non-decaps?''

Elcombe seemed to understand this gibberish. ''It seems to me the decaps don't have soot,'' he replied. ''They don't have CO. The other victims do, though not all the victims.''

Thurston was well satisfied. ''Then it has to be after impact,'' he said. ''There is no other explanation.''

''That's my feeling,'' answered Dr. Elcombe.

I was flabbergasted. The report stated that ''the post-mortem examination of the flight crew and cabin crew and all passengers of the aircraft did not reveal any evidence of pre-impact explosion or pre-impact contaminating event such as fire''. It didn't say *conclusive* evidence, or even *convincing* evidence. It said *any* evidence.

''Do you think that the decapitated victims — the ones that had the high levels of toxic gases — can be considered, just by themselves, to provide *at least some* evidence that's consistent with pre-impact mortality?'' I asked Elcombe.

''I think that when we begin to deal with individual upon individual, one at a time,'' he replied without hesitation, ''we can presuppose just about anything. And that doesn't get us anywhere. I think I have to look at the group findings. And these two people fit within a group of certain injury patterns. Now, I asked that question to pathologists and others, and their conclusion was: No, this represents post-mortem injury.''

''But my question really is,'' I said, ''do you not think that one exception is sufficient to disprove a generality?''

319

No, he didn't. "In pathology," he explained, "there is a kind of thinking that is called the odd man out. Does the odd man out illustrate anything that could be this? But so far, nobody seems to have fastened that on there."

I still couldn't understand what on earth he was talking about. I tried another tack. Did the American authorities document the negative finding? That is, did they describe what they looked for and how hard? Seizing the opportunity to leave the topic of toxicology, Elcombe said that there was no evidence of pentetrating metal fragments before death. That wasn't what I had asked, I said. Was the conclusion as to the cause of death documented?

Elcombe repeated what he had told us earlier: inexplicably, the autopsy reports had not specified cause of death. Was there then a consolidated report on the absence of evidence of injuries due to explosion or fire before death? Well, as a matter of fact, the U.S. Army always produced an "after action report", but Elcombe had not seen one on the autopsies of the Gander victims. He undertook to get one.

I was out of steam, but Stevenson had another approach. How did the proportion of victims with high carbon-monoxide and hydrogen-cyanide levels compare with observations in other accidents where it was known that there had or had not been an in-flight fire? Stevenson rhymed off some examples: Cincinnati, Tenerife, San Diego.

Elcombe didn't know, but Hinton weighed right in with a long discourse to the effect that comparisons would be meaningless. His tone was pleasant, without the combative edge it often displayed. But I couldn't understand how Hinton, Deschênes and the others could be so nonchalant. Their mask of competence, such as it was, had been shattered. The most innocent conclusion from what we had just heard was that, after more than two years, a major part of the investigation was incomplete. Sensational information that "came rolling in around March 1986" had been withheld from the board members, the public inquiry and the interested parties.

Yet, Deschênes and his supporters seemed at ease now that the medical evidence had been put to rights. They took care of some other loose ends just as casually. Hinton informed us that, unfortunately, he couldn't find an expert to re-examine the engines; he had a list of names he had approached. What about the additional analysis of the flight data recorder? Hinton produced a letter from Ray Davis, the British flight recorder expert, to Ken Johnson. Amazing as it might seem, Davis had declined the work, which might cost the CASB "perhaps even as high as $50,000." "I do not consider that the work would justify the cost," Davis wrote.[4]

A consultant declining a $50,000 contract because the prospective client might be wasting his money? This was something new in my experience. I'd never met Davis, but I could call him. He turned out to be a pleasant chap, not at all averse to a chat with someone who shared his fascination with FDRs. It seemed there had been a misunderstanding. Somehow, after a flurry of calls from CASB staff members, Davis had been left with the notion that someone in Canada was dissatisfied with Bernard Caiger's work on the Arrow Air flight data recorder. Davis had known and worked with Bernie Caiger for many years; he had the greatest of respect for him. He would not get involved in any effort to discredit him—not even for $50,000.

321

Something was certainly going on. Why, at this late date, were they trying to create a paper trail to document valiant but unsuccessful efforts to comply with the board members' requests?

Mr. Sopinka Reports

On February 16, 1988, the *Globe and Mail* published its third and most detailed article on the CASB's internal problems.

When a reporter called a me few days earlier, I felt we could no longer leave the public record to Deschênes and those who had leaked the Hickling report. For the first time since joining the CASB, I talked to the press. I spoke guardedly about the need to improve the scientific and engineering analysis in the CASB's investigations. I didn't talk of the conflict among board members or about any investigations. Nevertheless, the story revealed, more or less accurately, that some of us had sought legal advice on the board's procedures and had sought to suspend the director of investigations. And, for the first time, it linked the problems in the CASB to the Gander investigation.

"Dissent has been a factor in the two years the board has taken to write a report on the Arrow Air crash," the *Globe* reported. "Several dissatisfied board members have refused to accept the investigators' conclusions." The article also reported our opponents' explanation for the board's ills: "dissident board members may be motivated by personality conflicts" — an all-purpose mantra they would continue to rely on up to the end.

Around the same time, Roger Lacroix told us that his resignation was definite. The Privy Council Office had proved remarkably accommodating. We tried to talk him into staying at least long enough to take part in the Arrow Air oral presentation, but he had had enough. "It's not worth ruining my health over this," he said. On February 19, he resigned.

On February 23 the chairman let slip that John Sopinka had submitted his report to the Minister of Transport a few weeks earlier. I asked if he had seen it. "I've glanced through it," Deschênes replied. Could we see it? The minister hadn't given him permission to show the report to the board members. So Norm Bobbitt called Crosbie's office and had copies sent over for the rest of us.

Sopinka presented his conclusions in the guise of "advice in respect of the proposed Act relating to the Multi-Modal Transportation Accident Investigation board" — a format that got the minister off the hook for looking into the internal affairs of a supposedly independent agency. But there was no mistaking the basic issue, and it was suddenly clear why Bernard Deschênes had been looking so hangdog of late. Sopinka had thoroughly debunked the chairman's theories and stewardship.

There was no doubt, Sopinka wrote, that Dubin intended the chairman to exercise "such powers and duties as are delegated to him by the Board" and that "the claim to exclusive power of the Chairman over the staff was unfounded. Deschênes' "rationale for separation of the investigators from the Board" was "an erroneous concept". Sopinka was not as blunt as Gary Ouellet, but there could be no doubt of his low opinion of "accident investigators who view themselves as independent of the Board members".

In Sopinka's opinion, this misguided view of independence was all the more dangerous because the "investigators are dependent on Transport Canada for career progression inasmuch as the CASB is not large enough. . . . '' They ''may therefore be seen as keeping one eye on their own advancement and therefore tending to soften their criticism of Transport Canada.''

Sopinka didn't seem worried about the supposed personality conflicts. He noted that ill winds had been howling long before the worst of this supposedly bad lot had arrived on the scene. ''The view in the industry, '' he reported, ''is that the competence of the investigators and the quality of their investigations has deteriorated in recent years'' — that is, since the CASB had been formed. The problem had become critical because ''the CASB Act as it has been interpreted in practice and by the Hickling Report has fragmented this agency to such an extent that it has crippled its effectiveness.''

It was hard not to feel elated at the seemingly complete endorsement of our position. Bobbitt called Crosbie's assistant to thank him for sending over the copies, and the assistant was friendly but guarded. Nonetheless, he conveyed the impression that we had been vindicated and that Crosbie would heed Sopinka's advice.

But anyone who thinks this was the last word just doesn't understand the power and persistence of the Ottawa bureaucracy.

Some Interested Parties Speak Out

The official response from the FAA administrator had little to say about the report on the Gander investigation. His single-page letter ended with a generous bouquet: ''I would like to commend the CASB ''for the depth and detail of its investigation and the thoroughness of the preparation of this document.'' The chairman of the NTSB was equally gracious and concise: "The National Transportation Safety board staff has reviewed the draft report and has no comments to suggest. They find the report thorough and complete, and the conclusions logically supported by the text."

By contrast; Jon Batchelor, chairman of the board of Arrow Air, fired in a lengthy rejoinder with no accolades. "We must express our concern that some of the Findings and Conclusions are either not supported by the evidence, or are inconsistent with the 'Factual Information and Analysis' set forth in the draft report," he wrote, in a letter prefacing an inch-thick binder of comments and criticisms. Arrow repudiated the conclusion in much the same terms as we did. Ice on the wings couldn't have caused the crash. There was no evidence that the crew was fatigued or used incorrect takeoff reference speeds. The most pungent comments were found in a nine-page attachment outlining the reactions of "Mr. Humphrey Dawson, an aviation law specialist with the law firm Stafford Clark & Co. of London [England]."

As might be expected, Dawson took issue with various nuances deemed to reflect unfavorably on Arrow's corporate practices. But the lawyer's indignation went beyond the direct interests of his client; he found "the reasoning and reporting on matters such as icing muddled to the extreme", and recognized, as we did, that despite the labyrinthine cataloguing of the evils of ice, the report didn't even try to determine how much ice might have been on the wing and how it could have got there. "I find the report very inadequate and unprofessional in this regard," Dawson wrote.

The lawyer's sensibilities were also offended by the strained logic concocted by the conditional draft's wordsmiths. The conclusion cited crew fatigue as certainly contributing to a cause that was only probable. The report presented neither direct nor indirect evidence of crew fatigue. On the contrary, ground personnel at Gander had attested that "crew members were alert and in excellent spirits." In fact, the report provided a textbook case of a circular argument: the normally competent crew could have missed the ice only if they were fatigued; hence they must have been fatigued, which explained why they had missed the ice.

We looked forward to the oral presentation.

* * *

Jon Batchelor, son of Arrow founder George Batchelor, led the delegation. He was accompanied by Humphrey Dawson; Richard Skully, formerly Arrow's vice-president of Operations; and Captain John Coe, a senior DC-8 captain.

The members of the Arrow delegation took turns punching holes in the report. Captain Coe denounced the ice theory from the same pulpit as Ross Stevenson: experienced DC-8 captains recognize ice building on the aircraft during approach by the need for additional power. It was not uncommon, Coe explained (as Stevenson had), for ice to form on the corner of the DC-8's windshield; "If there was ice on the windshield I would have someone check the wing." Yes indeed, Coe could vouch from personal experience that the wing could be free of ice after landing even if a bit of ice had collected in the corners of the windshields during descent.

Dick Skully, who had served in senior positions with the FAA for over thirty years, had some interesting intelligence. His contacts in the FAA were apparently too rusty to secure a copy of the agency's formal response, but he had obtained a copy of the "comments made by the Miami FAA office that were submitted through channels." Skully knew head office would have edited these comments. Nevertheless, he wanted to share the unabridged version. He read portions indicating that, unlike the FAA administrator, "professional pilots and experienced investigators within the FAA" rejected the ice theory:

> The board has concluded that ice was present on the leading edge of the wing when it landed at Gander. In our opinion there is evidence available . . . which makes this conclusion arguable. There is further evidence presented both in the subject report and in the public testimony given at Ottawa that makes the conclusion of additional ice accumulation during the ground time even more suspect. . . . The aerodynamic analysis of what occurred during the landing and subsequent take-off . . . is where we feel the board may have made significant errors. . . . It seems at this office that there is considerable ground to question the board's scenario of ice build up.

His insider experience at the FAA assured Dick Skully that the FAA's official response would incorporate the substance of this "unabridged version" from the regional office. Skully quoted the regional officials freely, taking it for granted that their report was freely available—on request or, at worst, under the Freedom of Information Act.

Deschênes showed scant interest in this detailed internal report that belied the FAA administrator's bland generalities. The CASB, he said, would secure a copy directly. It never did. In fact, it would take a congressional hearing to pry a copy from the FAA—

and by then, two and a half years later, few people would care.

In making its presentation, the Arrow Air team's strategy was to point out errors without speculating on alternative causes. But at one point, after Captain Coe had finished a long exposition of why ice couldn't have been the cause, one of the board members who supported the ice theory challenged him to give another explanation.

Coe recalled an incident where a compressor stall caused an inboard engine to lose power on takeoff. The DC-8 yawed violently — "I thought we had hit something," he said. "If it had been an outboard engine, the yaw would have been even worse. And of course we only weighed about 250,000 pounds, so we didn't lose any airspeed. If the airplane had been heavier. . . . " He left the thought hanging.

Then he suggested something else. "Beyond that, I've had some thoughts as to some sort of sabotage," he said slowly. "Because the facts and the findings don't really support anything else that might have caused that airplane to fall out of the sky like that."

Coe seemed to want to leave it at that, but everyone was waiting for him to say more. "Sabotage has been, in my mind, a very real possibility for a number of reasons," he finally said. He looked around at his colleagues and added, "But I don't think this is the place to go into them." Then he talked some more about compressor stalls.

I didn't think this genie should be stuffed back into the bottle so easily. When it was my turn to question Coe, I took him back to the subject. "Captain Coe, earlier you were asked to speculate on what you thought might have caused this accident, and you mentioned that in the back of your mind you had thought about sabotage. You then said that this was not the place to go into the reasons. I would suggest, in view of what was said by way of introduction by the chairman and by Mr. Batchelor, that this is exactly the place to go into whatever information or thoughts you may have that could help us make this report as complete and accurate as we can possibly make it."

Coe didn't seem comfortable with this line. He said he had "no cold hard facts" but had conjectured about sabotage "on being aware of the world situation at that time" and on "witness reports of fire from the underside of the aircraft."

Now that alternative causes were up for discussion, Dick Skully recalled something that had been bothering him for a long time: the "incredible arsenal" found on the Transamerica military charter shortly after the Gander crash. This was the first time I'd heard anything about it. But Skully could no longer remember the details. The documentation had been passed to CASB investigators two years earlier, he assured us.

The next day I asked for the material from Arrow on the Transamerica incident. There was nothing in the files. I called Jon Batchelor and he immediately forwarded copies of documents that lawyer Allan Markham had tried to introduce at the "pre-hearing conference" in March 1986, and I read about the discovery of hoards of dangerous explosives in the soldiers' personal baggage. Yet Deschênes had ruled evidence of the slap flares, detonator cords and other explosives smuggled aboard another DC-8 military charter irrelevant to the Arrow Air investigation.

Apparently the topic was too irrelevant for him even to keep the documents on file.*

At the end of the day, Deschênes thanked Arrow Air for the helpful representation. The board, he said, would carefully consider the points raised. If there were extensive modifications, the board would consider sending out another draft. "We are all in this together," the chairman said reassuringly, "each one contributing his share to the process."

We already knew who would be together with us for the next contribution to the process. Miami lawyer Carl Hoffman had also asked to make an oral representation, on behalf of Flight Engineer Mike Fowler. We would hear Hoffman on March 22.

* * *

To help make his case, Carl Hoffman also brought along Captain Charles Basham, who had spent ten years in command of DC-8s at Eastern Airlines. It should have surprised no one that Basham's testimony corresponded to what we had already heard from Ross Stevenson and John Coe: ice on the wings could not have caused the crash.

Hoffman had also consulted an aerodynamicist, who had examined the flight data recorder traces. I was gratified to hear that Hoffman's consultant hadn't detected a stall any more than I had.

Hoffman reviewed Mike Fowler's extensive experience. Fowler was not a young pilot putting in time as second officer on his way to a ticket to the left seat; he was a career flight engineer with twenty years of

_____

*As far as I know, neither the board members who supported the ice theory nor the investigative staff ever showed the slightest interest in the Transamerica incidents.

experience in the United States Air Force, all over the world and in all kinds of weather. There was no way, Hoffman insisted, that Fowler would not have checked the aircraft thoroughly for ice, especially since it was Fowler himself who had chatted with refueller Austin Paul Garrett about "the little bit of ice" on the lower corner of the right windshield.

Hoffman was certain that ice hadn't caused the crash, but he was vague about his own theories — perhaps something to do with binding of the elevator, possibly complicated by some ice on the wings. Again, the chairman thanked the interested party for the representation and promised that the board would give the points raised serious consideration.

In truth Bernard Deschênes was in no position to make promises about what the CASB would or would not do. Almost two weeks earlier, he had submitted a letter of resignation to the Privy Council Clerk, Paul Tellier. He had said he was resigning at the end of the month because he had accomplished all his goals.

Out with the Old

April 1, 1988, fell on Good Friday, and federal offices were closed. We didn't find out about the chairman's little April Fools' joke until Tuesday. But Bernard Deschênes' surprise announcement of his resignation misfired; the *Globe and Mail* scooped him on April 5 with the next instalment on the strife in the CASB. Reporter Ross Howard had obtained a copy of John Sopinka's tale of woe, and his story was not the kind of epitaph Deschênes would have wished.

"Toronto lawyer John Sopinka, an expert in regulatory tribunals who served as counsel to a 1983 (sic) royal inquiry into aviation safety" had condemned Deschênes' "assumption of virtually exclusive power to run the board and block certain investigations," the *Globe* reported. "Well-qualified members" of the board "with extensive academic and professional experience in aviation" had been "limited in their role by the rulings of the chairman."

By almost incredible coincidence, Prime Minister Mulroney had just shuffled his Cabinet, and former Immigration minister Benoît Bouchard had replaced John Crosbie as Transport minister. Political staffs were in turmoil. No one was available for comment.

There wasn't much time to reflect on (or revel in) the *Globe* story. The chairman wanted to meet the board members in his office. Executive director Ken Johnson was standing by as Deschênes handed out copies of an announcement of his resignation, dated March 31 but marked "To be released April 5, 1988." Not only had Bernard resigned, but his successor was already installed. "Mr. Ken Thorneycroft, former Inspector General, Transportation Safety to the Minister of Transport has been appointed new chairman of the board." Deschênes also handed us copies of what he called a "self-explanatory" letter of resignation, date-stamped three weeks earlier and mentioning even earlier meetings with the Deputy Clerk.

Ironically, in view of the story in the *Globe*, Deschênes declared himself well satisfied. In his own eyes he had attained the goals he had set for himself when he accepted a seven-year appointment four years earlier, and having presided over "development of an agency with high operational and ethical standards" he was content that it was "appropriate at this time that another person should have the opportunity to head the present board." It was always hard to tell what Deschênes was thinking, but it was easy to see that Ken Johnson was feeling no pain. It fell to him to tell us about the new chairman.

Johnson could barely contain his enthusiasm. Ken Thorneycroft was a super human being and a terrific administrator, Johnson had known and esteemed him for many years, as a matter of fact it had been Thorneycroft who taught Johnson to fly many years ago. . . . Memories momentarily overwhelmed the executive director. He told us that Thorneycroft had taken him in hand "when I could hardly tie my shoes" and taught him all he knew.

Deschênes said he figured he'd go back to his law practice. On my way out, we shook hands and I wished him well. "We'll be thinking of you as we continue with the Gander review," I sid.

For the first time since I had met him, Bernard Deschênes broke into a broad smile.

In with the New

The chairman of the troubled Canadian Aviation Safety board has retired and been replaced by a senior bureaucrat from the federal Department of Transport.
Globe and Mail, April 6, 1988

When the Canadian Aviation Safety board was set up in 1984, Don Mazankowski—the designated sniper on the opposition benches—blasted "the blatant patronage appointment" that ensconced Bernard Deschênes at the head of the new agency. But as time passed, bygones became bygones, and although Mazankowski was now Deputy Prime Minister in the Conservative government, Bernard often displayed his hope and expectation of gaining the chairmanship of a grander safety agency, on the lines of the NTSB.

Deschênes' unexpected ouster had the earmarks of a palace coup—the coincident leak of Sopinka's criticism, a new minister who couldn't have studied the dossier, a backdated resignation letter, a postdated press release. . . . But we had no way to distinguish design from coincidence or to guess which puppet-master was pulling which strings.

What was certain was that Ken Thorneycroft had been para-dropped into the CASB without warning. Benoît Bouchard, the minister who made the formal announcement, had been on the job only one day when the appointment went into effect—an appointment that turned a ''senior bureaucrat from the Department of Transport'' into the head of a tribunal created specifically to be independent of government departments, particularly Transport.

When that tribunal, the CASB, was being set up in 1981, interested observers had kept close tabs on every step. They raised mighty hell when Jean-Luc Pepin, the Liberal Transport minister of the day, had the temerity to put his senior aviation official, Walter McLeish, on a study group. ''Putting a fox in charge of the chicken coop,'' the opposition Conservatives charged.

Yet Thorneycroft's ascension evoked no alarm at the Canadian Air Transport Association. Gordon Sinclair, the association's president, thought instead of Holy Scripture; the new chairman would need ''the wisdom of Solomon, the strength of Samson and the patience of Job to run the board,'' he told Canadian Press.[5] Bouchard's press release lauded Thorneycroft's ''technical skills and proven leadership qualities''.[6]

Back in 1981, Gordon Sinclair had himself been ''a senior bureaucrat from the federal Department of

Transport.'' Later he succeeded Walter McLeish in the department's top aviation job. He stayed around long enough to welcome Ken Thorneycroft to Transport before moving on to take the helm of the air carriers' lobby group. As they say, it's a small world in the fast lane.

Notwithstanding the kudos, it was almost impossible to find out what exactly our new chairman had been doing at the Transport department. The press release was vague: "Ever since his appointment in 1984 as Inspector General of Transportation Safety, Mr. Thorneycroft had made substantial contributions to Transport Canada's reviews on safety policy, practices and administration''. In fact, Ken Thorneycroft had been the first-ever Inspector-General of Transportation Safety, and no one I asked could tell me what the job involved, much less the nature of his substantial contributions. But the new chairman's earlier career was clear enough. Before joining Transport, Thorneycroft had served thirty-seven years with the Canadian Armed Forces — starting as a fighter pilot and retiring as a lieutenant-general and deputy commander of NORAD.

Thorneycroft's path from high-ranking military officer to high-ranking Transport official was well beaten. The Canadian military-industrial complex is less developed than that in the United States, and the revolving door for retiring senior officers rarely opens on the executive suites of major corporations. The preferred method of stretching military pensions passes through the civil service. So when Thorneycroft left Transport, he took with him connections to a number of old comrades-in-arms — not ephemeral political connections like Bernard Deschênes', but solid connections where it counted, in the permanent civil service.

Since Roger Lacrois was also a retired air force general who had done a stint at the Transport department, I tried calling him to get the scoop on the new chairman. "Is he going to be like Deschênes?"

"No. Thorneycroft has a sense of humor."

"Come on, Roger," I pressed. "I've got to know what to expect from this guy."

"Well, don't expect him to rock the boat," Lacroix finally said. And that's all I had to mull over as the other board members and I trooped over to pay our respects to our newly installed leader.

My first impression was of all-embracing greyness: grey suit, grey tie, grey complexion, bushy grey eyebrows, receding grey hair combed straight back. I wouldn't have thought him a former jet jockey. Tall and cleanshaven, he looked slightly stoop-shouldered and weary, though close observation revealed an erect military bearing.

The new chairman greeted us pleasantly. He spoke almost entirely in clichés. It was nice to meet us, he said. He wanted to exchange notes with each board member, one on one, but first he needed a little time to get his feet wet. I needed to exchange a few of those notes right now. How, if I might be so bold, had he got appointed?

Thorneycroft showed no sign that he resented such boldness. His good friend Ramsey Withers, Deputy Minister of Transport, had mentioned that the job had come open and "asked if I wanted to throw my hat into the ring." Thorneycroft had thereupon "dialogued with the folk in the Privy Council Office" and here he was.

Had the dialoguing yielded any clues on dealing with the difficulties that had been so unfortunately ventilated in the press, I asked. Not really, but it was understood that the government was planning to replace the CASB with a new streamlined, air-conditioned, multi-modal organization, and Thorneycroft expected to head it. He said he was aware of Sopinka's report but didn't want to comment.

Unfortunately, the new chairman didn't have a lot of time to spare on his first day for dialoguing with board members. He had to conclave with Ken Johnson. Ken had arranged a meeting with all the staff, and the chairman and the executive director had to get their heads together. No need for the board members; the board meeting would take place the next week, as scheduled. Carry on. Troops dismissed.

Well, that was just great. Just got here, and already under Johnson's thumb. Still and all, I thought, there might be some hope. Thorneycroft obviously knew what Sopinka had said. Both the oral representations had savaged the ice theory even more than our internal review. And Thorneycroft had played no part in approving the conditional draft; he didn't have to save face. In any event, he had already had a long and distinguished military career. Surely he wouldn't flush that down the tube.

Part 2

The New Leader

On April 12, 1988, the CASB's new chairman presided over his first meeting of the board. Fortunately, the Gander investigation couldn't raise hackles; it was temporarily on the back burner while the secretarial staff tried to arrange a date for hearing a representation on behalf of Theresa Griffin.

Our meeting began on a cordial note. The chairman hoped "that our work can be fun as well as productive." He was suitably piqued on learning that board members had received some documentation only the previous afternoon—the pertinent agenda items were deferred and the chairman instructed the staff to try to do better next time. We were impressed.

Tom Hinton was on his all-time best behavior. Any point raised by a board member got a complete answer or a promise of an answer forthwith. Only once was there a hint of the old Hinton, when Stevenson complained about how the weather was summarized in one report. "There are thirty pages of weather information here," Hinton snapped, banging a manila folder on the table. Thorneycroft smoothed things over and everything was quickly back on track.

During the coffee break I strolled to the files and examined the one Hinton had brandished. There were eleven pages of weather data. It hit me that Hinton wasn't just exaggerating; he didn't think of numbers the way I did. Thirty pages? A dozen? What was the difference? Either way, it was too much to put in the damn report. That was what he was trying to tell the insufferable Stevenson.

I began to realize why I had so much trouble communicating with Hinton, Boag and others who thought the way they did. At heart they were poets or something. When George Bush mentioned "a thousand points of light" in accepting the Republican nomination, I knew it was just verbiage—I still wondered why he didn't say a hundred points of light, or ten thousand, or a million. These guys extended this kind of thinking to the investigation of air crashes. Was it a third of an inch of ice or a quarter-inch? Who cared? Either way, it was too damn much. Numbers were just more adjectives for wordsmithing. . . .

Meanwhile, Chairman Thorneycroft's first meeting was continuing smoothly—the rest of that day and all of the next. We cleared a large backlog of reports. A job well done—I went home with an empty briefcase.

But I hadn't left the CASB completely behind. For the first (but by no means the last) time, a reporter called about the Gander investigation. Robert Lee of the *Ottawa Citizen* asked, was it true that half the passengers on the Arrow Air flight had inhaled hydrogen cyanide?

Well, the board was still struggling with that particular investigation, and I wasn't in a position to discuss the details. That was fine, he said. Could we discuss the effects of hydrogen cyanide in general terms? I explained that Dr. Elcombe was our authority, but Lee countered that he had a deadline. How about some information now? I'd do my best, I said.

Lee wasn't playing dumb to lead me on; he really wanted to understand. I explained that smoke from burning plastics contained hydrogen cyanide and other deadly fumes and that medical examination could show if victims had inhaled such fumes before death. Sometimes it was obvious that a victim who had inhaled fumes had been killed instantaneously in the crash—when his head was crushed or severed by the impact, for example. In such a case, it was reasonable to conclude that there had been a fire while the aircraft was still in flight.

"Well, what about Gander?" Lee asked again, and I reminded him that the investigation was not complete.

"If I told you I'm going to stick my neck out and go with a story that half the passengers had high levels of hydrogen cyanide, would I get my head chopped off?"

"If reporters got their heads chopped off for being wrong, there wouldn't be many of you guys left."

Lee was pleasant but persistent. Had I voted for the conditional draft? That was no secret—I sure hadn't. I hadn't even been there for the vote. Would I have voted for it if I had been there? Not too bloody likely. When we were done I hadn't said anything specific about Gander, but I'd confirmed that he had a hot story.

The hot story made a hot headline for April 15—"Federal report conceals evidence about Gander crash." "The aviation board's draft report says ice . . . was the probable cause. . . . But nearly half the board members contend that the icing theory leaves too much evidence—such as the cyanide—unexplained." Lee had it right. He wouldn't get his head chopped off. But others might — the article alluded to board members "who spoke on the condition they not be identified." The story was picked up by the wires and television. Questions were asked in the House of Commons.

Later, looking through CASB files, I found an interesting insight on how a new Minister of Transport handles a hot issue. David Bell, a senior official in the Transport department, had prepared "a suggested pressline for questions in the House on the Gander crash report." It read, in its entirety:

Issue:	The Gander crash report.
Response:	The investigation of the air crash is ongoing. A final report has not been produced. Until the report is released, it would be inappropriate to comment.
Issue:	Judicial Inquiry.
Response:	Until a final report is available it is impossible for anyone to assess if further action is necessary.
Issue:	Conflict among Board members.
Response:	A new chairman, Ken Thorneycroft, has recently been apponted and is working to ensure the Board continues to fulfill its mandate.[7]

In the House, Bouchard stuck to the script.[8] The new CASB chairman was "aware of the story" but had no comment. He didn't really need this grief just two weeks after taking on the job. Bernard Deschênes poked his head out of retirement long enough to defend the report: "The investigation was one of the most thorough carried out, I would say, worldwide. All possibilities were, I think, examined.[9]"

Ross Stevenson and I had planned to spend the next week at the FAA Training Center in Oklahoma City. We left on the 16th, Saturday. But we didn't stay. A fax from Thorneycroft called us back for an urgent meeting "in view of the leaking of the draft report" on Gander and "allegations that members have discussed confidential matters with the media."

The situation was heating up. Members of the opposition had a copy of the draft report. The minister wouldn't budge from the pressline prepared by officials, so the press hounded Thorneycroft. Arrow Air reacted with incredulity and outrage at the CASB's "contempt of accepted international standards in producing the draft report that is not in accord with the evidence" and "concealing the truth" at the meeting two weeks earlier.[10]

Arrow's lawyer, Humphrey Dawson, demanded answers to a long list of questions. When had the toxicological tests been done and when had the results been made available to the CASB and the United States Army? Why was there no reference to the toxicology in the draft report? "Please supply a full copy of blood and toxicology testing reports." "Does the CASB accept that cyanide poisoning is unlikely to arise from an aircraft stalling because of the excessive build-up of ice?"

The emergency meeting of the board convened on April 21. Frank Thurston had the flu. Dave Mussallem couldn't get to Ottawa and the chairman refused to let him participate by conference call. No staff were present and minutes were not kept; no tape recording was made.

It seemed that the Arrow Air charges were not the topmost concern. "Extremely serious charges have been made against some board members," the chairman said. The article in the *Citizen* alleges that board members gave out confidential information on the condition that they not be identified." The evening edition had also reported that Mrs. Griffin would make a presentation to the board on May 12. That date had just been set; the implication was obvious. Even worse, an opposition Member of Parliament had obtained a copy of the draft report. Thorneycroft intoned all this as if quoting from the Book of Revelations. Did anyone know anything about these matters?

I related my conversation with Robert Lee and agreed to follow up with a memo. I had said nothing off the record and I hadn't asked not to be identified. I didn't add that I was certain Lee had talked to Roger Lacroix and some of the investigators before he talked to me. But I'd taken the trouble to do a bit of research after getting Thorneycroft's summons. Photocopy records showed that at least seventy copies of the draft report had been printed the previous December. I pointed out that many copies had been circulated, among our own staff and, presumably, by the interested parties. I could personally verify that the contents of the draft report were well known within the FAA. Former staffers and former board members were familiar with it. The date of the meeting with Mrs. Griffin was, moreover, not a secret; Mr. Lee could have learned about it from one of our staff or, indeed, from Mrs. Griffin herself.

My remarks enraged former newspaper man Bill MacEachern. He couldn't abide the thought of "highly confidential" information getting to the press. He had held his tongue up to now, MacEachern told us (I recorded that in my notebook, annotating it "Not true"). He could hold it no longer.

Events of the past week had suffocated all his good will, he riled at "those board members who broke their oaths of office." "The attempt to deflect blame on the staff is odious," he said. "There has been a deliberate campaign to undermine us from within." This was "another round in a vendetta" against upright citizens. "It burns my ass that the leaked information is inaccurate and given in a grossly misleading context."

When MacEachern finally wound down, a number of other members began attacking board members who stooped to talk to the press. Stevenson took my side with a slashing counterattack; it was a good thing some information came out in the press, he said. For the first year and a half after the crash, he had learned more about the investigation from the newspapers than he had in the boardroom. He gave it his best but there was no way he could match MacEachern's bombast.

So what were we to do? The chairman said he had already sent a note to the minister—I wondered if he had forgotten he was no longer in Transport, but said nothing. He had also called the staff together and reminded them of the need for confidentiality. They were working on a draft response to Arrow Air, which we could consider tomorrow. In the meantime, Thorneycroft had a little litmus test for the board members' loyalty: we should all sign a letter to the *Citizen* denouncing Robert Lee's story.

Norm Bobbitt, who had been very quiet up to then, had a thought. The leak might be regrettable, he said, but surely the overriding issue was the truth of the information. The credibility of the CASB wasn't going to be restored by writing letters to the editor. Bobbitt could think of three options. We could try to resolve things internally, we could reopen the public inquiry and openly explore the additional evidence, or we could ask Cabinet to invoke its option of appointing a judge to hold an inquiry. In Bobbitt's opinion it was too late for the first option, but either one of the other two would show the public and Arrow Air that we had nothing to hide.

The letter to the editor was sidetracked. I prodded the chairman with the question I had put to Bernard Deschênes: was there any preemptive factor — be it of national security or foreign policy or whatever — that restricted us from going full blast after all the causes and contributing circumstances of the crash?

Thorneycroft didn't waffle as Deschênes had. "No," he said. "Nothing should get in the way of the truth." He was not at all averse to reopening the public inquiry. I was surprised and pleased.

It suddenly seemed that we had found a theme we could all warm to. Who could object to the truth? A consensus appeared to be developing around the idea of more public hearings. Even MacEachern, who was still bristling, supported the idea — as long as it was understood, he said, that it was to clear the air, and not because the original investigation had been bungled.

It was getting late, and we adjourned.

* * *

Our "shirtsleeves" meeting resumed the next morning, still without minutes. The chairman distributed copies of a draft response to the questions from Arrow Air's lawyer. The staff had answered the questions in a style I had come to know only too well: disclose the barest essentials; avoid blatant lies. Questions were interpreted narrowly, at times disingenuously, and answered with irrelevant or out-of-context protestations. The crux of the reply was that "it is not the

Board's practice to include detailed toxicological findings" in its reports.

But it proved impossible to dodge all Humphrey Dawson's artful questions. Dates of issue confirmed that the results of the toxicological analyses had been "forwarded to the board's investigators" before the public hearings — and provided to board members after the conditional draft was already released. We also learned that "Colonel McMeekin had discussed the outcome of the toxicological testing with the CASB." (I made myself a note to look for a record of that discussion in the files. I never found it.)

Dawson didn't ask vaguely about "the CASB" but specifically about the chairman and board members. The suggested response said: "The chairman of the CASB normally does not receive copies of laboratory results although, of course, such material is available to the chairman, board members or staff as required at any time."

I circled this sentence. The evasion seemed so transparent as to be laughable. Who cared what the chairman did "normally"? We were talking specifics here. When did he get these results and whom did he discuss them with? And "as required" by whom? I put a second, smaller circle around the "as required"; the phrase perfectly illustrated the doctrine of exclusive authority, which would let the director of investigations decide who "required" what. Had Deschênes or any of his apologists ever read Milton's thoughts on hypocrisy — which was not bloody likely — they might have recognized this as a perfect illustration of "the utility of interpretations".

There was lots more of the same in the six-page letter. I filled the margins with notes.

We started to discuss the reply. Somewhere along the line, Thorneycroft had already acquired the religion. He was pleased with the staff's response and wanted to send it out that day. Bill MacEachern's only complaint was that some answers might be too revealing: they "may have to be fudged", he mused, with commendable candor.

The board split along exactly the same lines and with the same intensity as before. It was as if the new chairman had been the old chairman's faithful understudy during the past year. Deschênes' departure had changed nothing.

But I really couldn't believe they'd have the chutzpah to send this baloney to Dawson. How could anyone think he'd be taken in? Imagine his reaction, I asked, to the promise to send "an additional factual report which will contain all the factual information in the conditional draft plus all the releasable factual evidence gathered by the CASB investigators." I did an impersonation of Dawson encountering this phrase: was this "releasable *factual* evidence" supposed to counter the "releasable *fictional* evidence" already sent? Who determined what constituted "*unreleasable* factual evidence"? How did you combine the "releasable *fictional*" and the "*unreleasable* factual" to reach your conclusion?

Thorneycroft was not amused. He insisted that the reply must go out.

Bobbitt and Stevenson were with me. There was no way we would be party to such a letter. But Thorneycroft still wouldn't allow Mussallem to participate by phone, so it was four against three; they could send the letter, but references to "the board" thinking this or believing that would be a sham. If they were bound and determined to send such a letter, it must have a disclaimer.

"Can you suggest any words?" Thorneycroft asked. I suggested; "I am sending this letter on behalf of the majority of members attending a special meeting of the CASB board. I would like to make it clear that some members strongly dispute its contents."

Thorneycroft accepted the wording. The letter was sent by fax, with our disclaimer but no other significant changes. I'd never get another concession so easily.

* * *

On Monday morning, April 25, we were back at full strength—Thurston recovered, Mussallem allowed to participate by conference call. The topic was Gander, and specifically whether the CASB should reopen the public inquiry.

I had consulted Roger Lacroix during the weekend. He had kept in touch and knew all about the Arrow Air letters and the oral representation to be made on behalf of Theresa Griffin. He didn't exactly regret his decision to leave the board, but he was keenly interested in the Gander case, and said there was ample precedent to allow him to take part in the continuing discussions. Members of the AARB, the CASB's predecessor, had come to CASB meetings for over two years. Former AARB members Max Friedl and Ron Baker had, for example, joined a meeting as recently as February 24, 1987, to help wrap up the report on an accident that had occurred three years earlier.* Friedl and Baker had taken active roles throughout the

---

*A PWA 737 had caught fire on the ground in Calgary after a failed compressor disk forced a rejected takeoff, in March 1984.

consideration of interested-party comments and the approval of the final text.

Roger recalled yet another precedent. He was hazy on the details, but said that a year after the CASB had begun operations there had been some kind of hitch with Frank Thurston's reappointment. The original appointment had terminated during the summer of 1985, and paperwork to do with Thurston's reappointment had been delayed—summer holidays or something. Eventually everything had been straightened out, but in the meantime Frank had continued to participate.

I tried to think of a way to make Roger's participation palatable to the rest of the board. I proposed that both he and Deschênes be invited to join the continuing deliberations, putting my proposal in the form of a motion: "Whereas former members Deschênes and Lacroix had spent a great deal of time delving into the evidence related to the causes and contributing factors of the crash and in so doing acquired invaluable insight . . . etc., etc."

I listed various rationales, mentioning the valuable contribution made by Thurston in the interval between his initial and subsequent appointments. I didn't consider this a strong point, but it elicited a strange reaction from Thurston. He huffily explained that he had had a contract to act as a board member, to bridge the gap. His intensity surprised me. What was this about a contract? I pressed for details. Thurston brushed me off, but I could tell I'd hit a nerve. I filed a mental note to check out Frank's appointment.

There was some half-hearted sparring about whether or not Deschênes or Lacroix would even be interested, but—by now—our discussions had fallen into a predictable pattern — terminal schizophrenia.

Bobbitt, Stevenson and Mussallem were with me, the others against. Thorneycroft revealed surprising bitterness, seeing Roger Lacroix behind the leaks to the press. I readily admitted I'd been talking to Roger, and Thorneycroft advised me to stop. Roger might be in serious trouble, he said, for violating the "no pissing clause" in the contract he'd signed with the government. When I asked, Thorneycroft backtracked, claiming not to know about Roger's contractual arrangements.

When all was said, there was no way Roger Lacroix was going to participate. We got back to the question of new public hearings. But there had been a change of mood over the weekend—Thorneycroft and all the members who had supported the ice theory now solidly opposed further public hearings.

Frank Thurston, who had not been there for the first part of the discussion, resumed his favorite role of elder statesman. "The report won't serve as it stands," he said gravely, "but a public inquiry wouldn't work because it can't deal with the intricacies of the technical issues." It would become "supersaturated with lawyers" and thus "totally useless". As Thurston went on his tone became more strident, and he accused those urging a new public inquiry of wanting a witch hunt.

MacEachern picked up the theme and elevated the stridency level to the stratospheric standards he had set the last time. The only charges of cover-up, he said, were coming from within this board, and he resented the attempt to hold him hostage by having a public inquiry to disprove these farcical theories of a "Turkish-Mongolian conspiracy". He was definitely against.

Thorneycroft didn't enter the fray except for voting. Bobbitt, Mussallem, Stevenson and I were for reopening the public inquiry, the other four were against. The chairman cast his vote against. We would not be having new public hearings into the Gander crash.

There was one more thing.

Thorneycroft had arranged a press conference for the next day. Arthur Portelance would assist him in case there were any questions in French. The rest of us were not invited, thank you. Carry on.

Full and Unqualified Support

Another front-page story in the *Ottawa Citizen*, set the stage for Thorneycroft's press conference. Robert Lee had interviewed Arrow Air vice president Harry Weisberg and had obtained copies of the exchange of letters. He had also interviewed airline captain Lee Levenson, a close friend of John Griffin, who was helping Mrs. Griffin prepare an oral submission.

From the evening news I learned Thorneycroft's position at the press conference. "Let me state categorically, there was no cover-up, and there won't be any cover-up."

The morning *Globe* gave more details. Thorneycroft "denied an *Ottawa Citizen* report of coverup in the board's investigation of the crash and said he was bound by law not to discuss anything in preliminary reports of the investigation." Sensation-seekers would just have to wait for the final report, which the chairman promised would reveal all and explain all. He did add that the leaked information was "out of context and, to a large degree, incorrect." And he praised the investigators who had prepared the report, suggesting that their "magnificent reputation throughout the world" spoke for itself.

Pressed to explain why the draft report had not disclosed the hydrogen-cyanide results, Thorneycroft seemed to forget his own strictures against discussing preliminary reports. The results had not been included, he said, because there had been no evidence that they contributed to the cause of the crash.[11] His nonchalance about the evidence was a bad omen.

So was a letter that Benoît Bouchard sent to each member of the CASB. "As the newly appointed Minister of Transport, I am writing to you to explain my position concerning representations made to my predecessor on internal difficulties," Bouchard wrote. Presumably he was talking about the reports from Hickling and Sopinka, but he didn't spell it out. "Mr. K. Thorneycroft, the new Chairman and Chief Executive Officer of the Canadian Aviation Safety Board, has the full support of the Government to direct the operations of the Board to fulfill its mandate. As a Board Member you are expected to give your full and unqualified support to Mr. Thorneycroft in the interest of all Canadians who rely on the Canadian Aviation Safety Board to play its role in maintaining air safety in Canada."

It was a curious and unusual letter — the bureaucratic equivalent of a burning cross on the front lawn. It could be (and soon would be) construed as blatant political interference into the affairs of a supposedly independent agency. But Bouchard certainly hadn't got a sudden irresistible urge to intimidate me just on his own. He hadn't held the Transport portfolio long enough to learn the way to his private bathroom, let alone to be briefed on the Sopinka report. Thorneycroft had told me himself that he had received no instructions on how "to direct the operations of the Board."

Somebody had pulled some mighty powerful strings. But to what end? Who was threatening me with what? Who could have tricked Bouchard into sending this potentially embarrassing letter?

It was not too difficult to unravel the semantic machinations. The words "and chief executive officer" added weight to the key phrase "you are expected to give your full and unqualified support". The switch from the first-person "I am writing," to the impersonal "you are expected" was not just rhetorical happenstance; "I expect" or "the government expects" could be construed as improper interference, but in a pinch "you are expected" could be explained away — why, of course, it was the Canadian public who expected, there was no interference there. Similarly, the conclusion to "to direct the operations of the Board to fulfill its mandate" remained implied but unstated: "as he sees fit."

In short, Bouchard's letter displayed a low-grade Machiavellian cunning similar to the answers the CASB had just sent to Arrow Air.

The physical appearance of Bouchard's letter betrayed haste and sloppiness quite out of keeping with the office of a senior minister. Any secretary knows how to stuff an envelope, but Bouchard's letter had been folded by a rank amateur — either that or a big shot not used to assembling his own mail. The letter hadn't even been proofread; the last sentence mentioned "fulfillment of your imnportant [*sic*] collective responsibility."

I compared the signature on my copy with the one on Stevenson's. Nope — the letters had not been signed by machine. An aide must have grabbed the copies from a typist and dispatched them after hastily obtaining the Minister's signature. What was going on?

The operation didn't display enough political smarts to be the work of senior aides, and it was far too sloppy to have passed through the deputy minister's office under normal circumstances. But just days earlier, Deputy Transport Minister Ramsey Withers had been replaced by Glen Shortliffe, a newcomer from the Privy Council Office. The resulting confusion — rearranging office furniture, moving files, the changeover of aides and secretaries — would have provided an ideal environment for a bit of free enterprise from the middle levels of the bureaucracy. The forces that had toppled Bernard Deschênes — whoever they were — were trying to complete their work—whatever it was.

Regardless of what lay behind Bouchard's letter, we knew what to do with it — ignore it. Assigning incorrect causes to crashes was certainly not "in the interest of all Canadians". Sopinka had provided the map for fulfilling the board's mandate, and if Thorneycroft wasn't aware of it, we could help him.

But the bozos who had foozled Bouchard's letter had taken one more hilarious pratfall. Copies had been rushed to the names on an outdated list of CASB members, and the copy addressed to Roger Lacroix was duly forwarded. Roger was not only a former CASB member but also a former executive assistant to a Liberal Cabinet Minister; nothing prevented him from discussing the minister's curious missive with whomever he liked. Copies of the letter surfaced in Parliament and led to embarrassing questions. An opposition member said the letter made a mockery of the independence of the CASB, while government spokesmen weakly insisted that it was intended only to "encourage cooperation".

Bernard Deschênes, erstwhile champion of the CASB's independence, defended the minister's right to send the letter. I too absolved Bouchard—because I knew he'd been had.*

If Things Are Rotten in Denmark

The representation on behalf of Mrs. Theresa Griffin was scheduled for May 12. Then the Board would have a second face-to-face with Arrow Air a week later, in consequence of a barrage of complaints and questions about Thorneycroft's excuses for the suppressed toxicological evidence. We had innumerable meetings about these and about other accidents.

As promised, the "releasable factual evidence" was sent to the interested parties. Someone versed in the art of historical engineering had cleaned up bits that might be misunderstood: for example, the photo caption originally reading "Close-up view of three 81 millimeter mortar rounds" had been expanded, in the copies sent out to read "inert, practice-type mortar rounds".[12]

Thorneycroft kept harping on the leaks and the unspeakable treachery of consorting with the press. He really seemed to feel that this was the crux of the CASB's difficulties. "If things are rotten in Denmark,

---

*By the time the media got wind of Bouchard's letter, much bigger news about leaked ministerial documents was on everyone's mind. On April 26, Global TV reporter Doug Small read excerpts from a summary of the forthcoming budget on the six o'clock news. The opposition demanded the finance minister's resignation but instead, criminal charges were brought against Small and others. The charges were dismissed when RCMP Staff Sergeant Richard Jordan revealed that he had been taken off the case after refusing to lay unfounded charges "intended to please elected officials". I'd have reason to remember the Doug Small case a year later, when I met Jordan in the course of his duties.

to coin all I know from Shakespeare," he said at one point, "then you don't wash your dirty linen in public." He was not trying to be funny.

Thorneycroft's other major theme was that "no more contribution to aviation safety can come from the report." All the board members who supported the ice theory seized on this as the ultimate answer to their critics: the report served to deliver a "safety message" about the dangers of ice on the wing and anything that diverted this message must be suppressed.

On May 12, the CASB convened a special meeting to hear the representation on behalf of Captain Griffin's estate. Theresa Griffin had assembled a formidable team of volunteers to review the draft report — no lawyers, no hired guns, just her husband's friends and associates.

Her chief representative, DC-8 pilot and longtime friend Lee Levenson, was joined by another colleague, Dick Moore. With over 10,000 hours of DC-8 time, Moore belonged to the same select fraternity of airline captains as Ross Stevenson. The two pilots were joined by Jerry Rusinowitz, who had put in over 5,000 hours as a flight engineer on DC-8s and countless hours explaining the DC-8's innards to aspiring aircrew.

Levenson, a former director of the American Air Line Pilots' Association, had also obtained technical assistance and moral support from the American and Canadian Air Line Pilots Associations. ALPA's director of Accident Investigation, Harold Marthinsen, and Keith Hagy, an engineer on Marthinsen's staff who had helped with the review, accompanied the team. CALPA representatives Ray Bicknell and Martin Martineau completed the group.

Naturally, the Griffin representatives roundly rejected the ice theory. They had gone through the same reasoning as we had, but there were other, more personal reasons. Everyone who had known John Griffin described him as a meticulous pilot. A mechanic in Cologne recalled that he "was a very correct person . . . .You know, he would never take an airplane that was not perfect."[13] Griffin's friends recalled that he wouldn't drive to the airport without first conducting a walk-around inspection of his car. His children had to study the instruction booklets before they were allowed to use household appliances. The CASB's own assessment described him as "a competent and conscientious airman" and "an effective professional".[14]

Dick Moore had brought along recent photos of a DC-8 in the bay where John Griffin had parked the Arrow Air jet on December 12, 1985. He showed us photos of the top of the wing taken from the stairs at the front door. This was what Griffin would have seen, he told us. If there had been the slightest indication of ice, Griffin would have called for de-icing. While he was in Gander, Moore had also taken the trouble to speak to the ground handlers who had serviced the DC-8. He distributed copies of their signed affidavits. "I am an operator of the de-icing equipment," Ted West had written, "If de-icing was needed I would have noticed it."

Harold Marthinsen a small, stocky man with thinning white hair and wire-rimmed spectacles, who looked more like the friendly grocer in a Norman Rockwell painting than a crackerjack aeronautical engineer — had examined the metal-foil flight data recorder tapes from just about every major American airliner accident since the 1950s. He and Keith Hagy had not been able to detect signs of a stall on the Arrow Air traces, any more than the rest of us had. Marthinsen said he couldn't understand why the thirty or so other takeoffs on the tape had not been examined, and volunteered to do the work himself at ALPA's headquarters in Herndon, Virginia.

The Griffin team attributed the crash to some unexplained mechanical failure — certainly involving a massive power loss on the right side, quite possibly involving an uncommanded deployment of the number-four thrust reverser.

The conclusions were not surprising. What did surprise me was the detail with which Levenson and his team had pored over the available documentation. In a bravura five-hour presentation, they catalogued a long list of inconsistencies and errors in the report and the supporting documents. Many details were subtle technical points overlooked by previous critics. Rusinowitz recognized, for example, that a diagram of the plumbing for the wing fire-bottles had been printed as a mirror image. He also outlined a number of seemingly plausible malfunctions that could have caused spontaneous deployment of a thrust reverser. The investigators' claim that there was no evidence of precrash malfunctions appeared more and more facile — whether or not you could follow all the technical details.

Levenson said that the Griffin team needed five or six weeks to submit their written comments. Thorneycroft thanked them for their valuable presentation and promised to provide whatever information was needed, and to make arrangements for Marthinsen to inspect the FDR tape. The board would consider the written submission carefully.

The meeting adjourned on a cordial note.

* * *

On Tuesday, May 17, it struck me that Thorneycroft was being strangely close-lipped about arrangements for the meeting with Arrow Air the following Thursday. I decided to find out what was going on. Time to call Jon Batchelor to thank him for sending the information on the Transamerica fiascos.

Jon was not available but his father, George, came on the line and was very affable. Jon had been pleased to help out with the Transamerica stuff, he said, but unfortunately, he wouldn't be able to come this time. But Harry Weisberg and Humphrey Dawson would be along. I remarked that the Transamerica incidents had been really hair-raising, and George Batchelor recalled other incidents involving live explosives on military charters. On one occasion his daughter-in-law, a former flight attendant, had had to pick up live hand grenades that some young yo-yo had dropped in the aisle.

The next evening, Weisberg called me at home. He and Dawson had just arrived in Ottawa; how about a chat before the meeting? Breakfast tomorrow? No problem. I met the pair in the cafeteria of the Holiday Inn on Dalhousie Street. I freely admitted my conviction that ice on the wings had had nothing to do with the cause of the crash. (Thorneycroft had, after all, shown no qualms about publicly stating his view of the irrelevance of hydrogen cyanide.) I said I shared Arrow's views about the "muddled thinking" in the draft report, but I pointed out that my interests might not coincide with Arrow's. I wanted to find out what had happened—regardless of the commercial or other implications. I mentioned Michael Mooney's book on the Hindenburg, and said I didn't think the world should wait twenty-five years before discovering the truth about Gander.

We chatted for a while about the philosophy of accident investigations and the aviation business in general. Then I left for the office. My first chore when I got there was a visit to the chairman. I briefed Thorneycroft on my tête-à-tête with the Arrow representatives and he listened impassively.

Thorneycroft opened the meeting in the boardroom with his favorite theme. "Premature release of the draft report has caused a lot of anguish and interrupted a process designed to keep things confidential until the final report is out on the street." Arrow Air agreed on the evil of leaks. "All publicity is bad publicity," Weisberg said. But, unlike Thorneycroft, the company didn't see the leaks as the *causa sine qua non* of all problem. Arrow simply couldn't "understand why a major item like the toxicology report was not mentioned in the report." All the information should have been put into the hands of the interested parties. The disclosures, Weisberg said, cast doubts on the fairness of the CASB's public inquiry.

Dawson added that the report was unacceptable as it stood, and that he could think of only three possible explanations for recent events:

— the investigators hadn't considered the hydrogen-cyanide findings to be relevant;
— they had known the results were relevant but had chosen not to investigate;
— there had been a major attempt to conceal the truth.

He dismissed the first possibility; the fancy footwork at the public hearings showed an awareness of the sensitivity of the toxicology results. The second possibility, suggested by the numerous errors and inconsistencies in the draft report, would flow from a preconceived idea of the cause.

At this stage, Dawson suggested ominously, a rational person could not avoid considering the third possibility. Could national security apply? Was there a political view that the deaths must be attributed to natural causes?

Thorneycroft said he couldn't see that there was a major attempt to conceal the truth, and didn't feel comfortable even looking at the scenario. He left a more substantive answer to Thurston, but it turned out to be the same thing in more words.

The idea of a major attempt to conceal the truth "cannot stand up," Thurston assured us. "It is a fact that political interference is, by law, zero. It's built into the Act." To Thurston's knowledge "there has never been the slightest attempt by anybody in the security business to influence our work in any way at all."

I waited for Frank to comment on attempts by someone in some other business, but he didn't pursue the line. Instead, he gave us his own analysis of the problem at hand.

"It may look from a distance," he sighed, "as if our processes are somewhat ramshackle." No one disagreed. "The thing that went wrong with this procedure is the leakage to the press." The press, it seemed, just didn't understand that from the moment Adam bit the apple, man was doomed to imperfection. "It's almost impossible to find any report in the world's literature," he said sadly, "that you can't find doubts about." It would take "a major research program" to satisfy the skeptics, but "total investigation could delay a report for ever." Frank sighed again. "There's very little in terms of safety to be gained from further work," he concluded. He then passed the ball to Arrow Air. "I would like to elicit from you gentlemen, how much greater depth of investigation would you press for?"

In person, Humphrey Dawson was not the pit bull suggested by his letters—not quite a lapdog either but a Welsh terrier, say. Thorneycroft's answer to his letter, he said, "raises more questions than it answers." He politely requested all the pathology and toxicology data, which must be reviewed by an independent pathologist. He also asked for the information on the armaments recovered at the crash site and the list of what was thought to be on board. He repeated that all information should be released to all interested parties. Perhaps the outstanding questions could be answered in a two- or three-day technical review with them.

Thorneycroft promised to send all the information. He denied that the investigators had come to a premature conclusion about ice. He explained that the board had promised to consider further representation from Mrs. Griffin before making a decision on how to proceed.

The Critique

So, as we neared the end of May 1988, the CASB's report on the Gander investigation was just about where it had been a year earlier. The chairman and half the board members were hell-bent on putting out a report that blamed the crash on ice, but external pressures were setting back their plans. Ken Thorneycroft had promised to provide missing information and make appropriate revisions, as had Bernard Deschênes before him. Could we hope that these revisions would address the issues?

The prospects seemed dim. The board members who supported the ice theory clung to their discredited conclusions like a drunk to his bottle, and showed no interest in such things as the number-four engine or the thrust reversers. To some extent, this was our fault; we had not assembled the case against the ice theory in a coherent form. Arguments we advanced at board meetings were not detailed in the minutes. Our memos to the chairmen had not even been put in the Gander file. There were no transcripts of the meetings with the Griffin or Arrow representatives. On the other hand, the conditional draft—with its numerous references to the dangers of ice, de-icing procedures, alleged computations and specious assumptions — was always there for ready reference.

I reflected on this imbalance during the meeting with Dawson and Weisberg on Thursday. Our next meeting was scheduled for Tuesday, May 24. I decided to compile our objections to the ice theory into a single, detailed review. We had been hashing over the arguments for more than a year now; it would be just a matter of consolidating my notes and adding comments.

Bobbitt, Mussallem and Stevenson all agreed it was time to put the anti-ice arguments on the record, and I said I'd crank out a report over the weekend in all our names.

I wrote in a flat, academic style, with careful citations and cross-references to let the reader check data and challenge the assumptions and logic. There was no need for narrow interpretations or fancy debating tricks. I could afford to bend over backward to acknowledge ambiguities and weak spots. Our case was cast-iron solid.

Ross Stevenson's extensive review of the witness evidence was a valuable source of direct quotes, but I used codes to identify witnesses who had not testified in public. Without the key to the witness names, nothing would be revealed that was not already available to anyone with a copy of the draft report. Of course, by now this included the media and opposition politicians. Compiled into a single report — equivalent to about fifty single-spaced pages — the case against the ice theory was devastating. It would have been hard to make a much better case against a mid-air collision with an Unidentified Flying Object.

I finished typing in the early hours of Tuesday morning. I called the report "Critique of the Ice Contamination Hypothesis Presented in Conditional Draft No. 1." The secretaries had copies for the chairman and the board members by ten o'clock. I had no idea how they'd take it, but the critique should help the interested parties — particularly Lee Levenson, who was preparing the written submission on behalf of Mrs. Griffin. I'd let my fellow board members mull it over before proposing further distribution.

At the same time, they could mull over some other pertinent news. John Sopinka had been appointed to the Supreme Court — directly, without a berth on the bench of a lesser court. The move was not unprecedented, but it was unusual nevertheless. I asked a secretary to make each board member a copy of a sound bite from one of the news clips on Deschênes' resignation the month before, in which Sopinka had said, "When they [the appointed board members] came to evaluate a written report by the accident investigators, they were not given the factual underpinnings that they needed to determine whether or not the investigators had got at the true causes of the accident."[15]

Right after our meeting I took a copy of our critique to Roger Lacroix. He was impressed, and suggested sending copies to the interested parties right away. But I was planning to make a formal motion to that effect in the boardroom. Let those who objected to the exchange of information do it on the record. Still, I confess I had some fear that Thorneycroft would find a way to keep our critique under wraps.

Roger then pointed out that some of his former air force colleagues worked in the section of the Transport department that reviewed material from the CASB. Ultimately they would have to prepare the department's formal response to the report on the Gander investigation. They had had legitimate access to the draft report all along, but were constantly bombarded by the pro-ice position. They should know the strength of the case against the ice theory.

That sounded like an good idea; it would be wise to get a reaction to the critique from someone who wasn't known to be predisposed in its favor. I called Jim Stewart, head of the Aviation Safety Analysis section in the Transport department and the department's official point of contact with the CASB. Stewart's groups was responsible for initiating the department's interested-party comments to the board's reports. One of his staff had served as the Transport department representative on the Gander investigation. His staff had reviewed the draft report and would ultimately review the final one.

Jim Stewart was president of the Ottawa chapter of the International Society of Air Safety Investigators and I knew him moderately well. He was surprised to hear from me, but agreed to give a quick reaction to our comments. Tom Hinton often sent stuff to Stewart that the board members hadn't even seen, and I didn't mind turning the tables. I dropped off a copy of the critique at Stewart's office on the way to work the next morning.

Good as his word, Jim Stewart phoned me with his reactions the next day. They were polite generalities. The critique made a very interesting case, he said, but he didn't go into details. I realized that I might have put him into an awkward spot.

I had no idea of the spot I was in myself.

On May 26, just two days after I distributed the critique to the board members, our analysis was the subject of a front-page story in the *Ottawa Citizen*. I hadn't planned on this, even though I can't say I was surprised or sorry. I hadn't told Roger what he should or shouldn't do with the copy I'd given him, and I never asked him what he had done.

Thorneycroft summoned me to ask what I knew about the leak. When I told him I'd given copies to Lacroix and Stewart, he looked grim but didn't say much. His official reaction arrived a few days later, in the form of a letter expressing disappointment about my giving "a confidential working document" to "persons outside the Board." As the chairman saw it, my "disregard for the basic requirements of procedural fairness . . . compromises a statutory confidential process approved by Parliament." Not only that but it constituted a "display of attitudinal bias" that had "adversely affected the Arrow Air investigation." I was, moreover, guilty of "extremely poor judgement at best and an illegal act at worst."

It was quite a blast. I wondered about the "illegal act" bit. Rowland Harrison advised me to take the letter seriously; the term "attitudinal bias" carried some rather significant legal repercussions. Thorneycroft's letter looked like the tip of the lever to pry me out of the CASB. Rowland suggested I write back rejecting the allegations and asking for a retraction. I did.

I didn't get a retraction. But my response seemed to turn off that particular plan. And there was more to come.

The next missive from the chairman tried another tack. "As you are aware, the government introduced a new security policy applicable to all federal departments and agencies," he wrote. "Privy Council Office was contacted regarding your respective status on this subject." I was requested to fill in forms to update my security clearance by "the limitation date" of June 30, 1988 — two weeks hence.

I was not aware of any new security policy, but I certainly knew how J. Robert Oppenheimer had been drummed out of the U.S. Atomic Energy Commission. I'd also heard something about Senator Joseph McCarthy's creative use of security clearance investigations in the 1950s. Talk about recycled rubbish.

I returned Thorneycroft's letter with a handwritten note suggesting that there must be a mistake because my security clearance had been updated when I was appointed to the CASB. That was the last I heard of the security clearance gambit. But it was clear that Thorneycroft — or whoever was pulling his strings — had declared war.

Part 3

Long, Hot Summer

As the Ottawa summer dragged on, the cold war between the pro-ice and anti-ice factions turned scorching hot.

The wire services and other media circulated Robert Lee's stories from the *Ottawa Citizen*. Revelations about the hydrogen cyanide, Arrow Air's discontent, Bouchard's letter, and the "Critique of the Ice Contamination Hypothesis" all prompted questions in the House of Commons. Sensing a political hot potato, the opposition launched its own investigation of the investigation. Marc Laframboise, an energetic young researcher with the Liberal caucus, flooded Canadian and American agencies with Freedom of Information requests. He had little luck with his own government, but he learned that American agencies had been more curious about the Gander crash than they had let on. Liberal Member of Parliament George Baker, whose riding included the town of Gander, told the House of Commons that a request through the Freedom of Information Act had revealed that "the FBI was sufficiently convinced of the possibility of an on-board explosion that it launched a world wide investigation."[16]

"Is the government aware," Baker demanded, "that the FBI conducted this worldwide investigation . . . ? Is the government further aware that 239 of the 289 pages of evidence ensuing from this investigation were withheld, considered too dangerous to the national security of the United States to be released?'' For the material that the FBI released to Laframboise was heavily censored — words, sentences, paragraphs, and complete pages had been crudely obliterated. Whatever information had been protected, the blacked-out pages would prove a potent image on television. But that would come later.

The Canadian government maintained a posture of self-righteous non-involvement. ''We have a lawful and legitimate process that takes it out of the political realm and puts it into the hands of professional investigators. . . . '' The solution would come through ''the next major initiative of the Government, . . . a Transportation Accident Investigation Board for air, rail and marine'' with ''the appropriate powers and the appropriate division of responsibility between the board and the professional investigators''.[17]

Spokesmen for the government and the CASB's official representatives were singing from the same hymnbook, the phrase ''professional investigators'' a favorite refrain. The CASB's public relations office declined to answer questions about the Gander investigation on the grounds that the law prevented discussion of the confidential report, while PR wizards scoured the land for friendly reporters to do gee-whiz stories about the neat equipment and international standing of the ''professional investigators''.

As the publicity increased, reporters started calling us more and more frequently. Curiously, Canada's self-styled national newspaper, the *Globe and Mail*, didn't follow up its scoops on the Hickling report and Sopinka's study. The *Globe* didn't even run wire service reports on the questions in Parliament. Between the time of Deschênes' resignation and the publication of the Gander report eight months later, I got only one call from the *Globe*. It came right after the stories on our critique from Paul Koring, who had recently taken over reporting on military and aviation matters.

At that time I was hewing to the party line of not discussing details of the investigation with reporters. That was no problem, because Koring was only interested in one obscure point of aerodynamic theory — where would ice first form on an airplane flying through cloud?

Everyone who has driven through winter storms has seen ice collecting on the flat back of a rearview mirror. Our critique suggested that ice formation favored flat surfaces perpendicular to the air stream where the flow separates from the contour. The point I was trying to make was that ice might form on the corner of the windshield before it formed on the wing — might or might not, depending on the circumstances. I was trying to emphasize the difference between the generally streamlined shape of the airplane and the parts where streamlining is compromised — the windshield is kept flat for good visibility, landing-gear doors hang out during descent and so on. I was also drawing attention to the testimony of ground handlers like Ted West, who said that when they see ice on the wings they also see it on other parts of the airframe.

I had undoubtedly oversimplified. But the issue was a secondary consideration, an aside to the main point. Yet Koring was insistent. He had heard of experiments that compared ice formation on two streamlined objects of similar shape but different size. The smaller object collected proportionally more ice.*

I tried to explain that the issue of ice buildup on similar streamlined shapes was irrelevant to the comparison between streamlined and blunt shapes—and that, in any event, the whole question was of marginal significance to our critique. But I wasn't successful. Koring didn't write a story at the time, and I forgot about the call—for awhile

Counteroffensive

June 21, 1988. The chairman had called a special meeting to present the investigators' response to our critique, which they had had for about a month. Hinton, Boag and a group of their colleagues were in the boardroom at the appointed hour, along with two strangers. Thorneycroft explained that our guests were consultants who had been retained to assist the investigators. One was an aeronautical engineer from the Department of Defence, the other a scientist from the National Research Council. The investigators' rejoinder, a document somewhat thicker than our critique, was distributed. A glance showed lengthy appendices by the consultants.

For the first time since I had joined the CASB, I was truly and royally browned off. This was just too damn much.

---

*Suppose half an inch of ice has built up on the leading edge of a wing near the fuselage, where the wing is ten feet wide. If the wing tapers to five feet near the tip, we would expect ice near the tip to be more than a quarter of an inch thick—that is, proportionally thicker than at the root. Whether the ice near the tip exceeds half an inch—that is, whether it is actually thicker than at the root —depends on a variety of complex factors.

Thorneycroft had consistently blocked all our efforts to obtain outside assistance on contentious technical issues — the engines, the flight data recorder and so on. The CASB's "professional investigators" hadn't been available to help with any analysis we suggested. Now we learned that they retained outside experts to help rebut members of their own board. Thorneycroft had berated me for giving copies of our critique to Lacroix and Stewart, who had had legitimate access to the draft report. He had insinuated that I had committed "an illegal act" — forced me to retain legal counsel to defend myself. At the same time, he condoned the distribution of my paper to these hired guns.

I restrained myself until the chairman finished a syrupy welcome to our "guests". Then I told him I had some important remarks for the ears of board members only, and wanted everyone else out of the room. Thorneycroft balked, but saw I meant business. He asked everyone but board members to clear the room, and investigators, secretaries, guests and hangers-on trooped out.

I reminded Thorneycroft of his reprimands about giving a "confidential working document" to persons outside the board, asked if he was going to give a similar reprimand to whoever had given copies of the same document to these outsiders. He retorted that it was quite proper for the investigators to seek assistance with their review. He refused to say whether he had approved the hiring of the consultants, and wouldn't say whether he knew the investigators had given out the document when he berated me.

I demanded he withdraw the accusations in his letter. He stolidly refused.

He was sullen and defiant, but also rattled and inarticulate. It seemed to me that he was defending a strategy — the refusal to extend the investigation, the "illegal act" letter, hiring these consultants — that he didn't understand himself. There was no point going on.

It had taken a month and the aid of outsiders to get this response. Did he expect us to discuss the material that had just been handed out before we even had a chance to look at it? This, at least, Thorneycroft could deal with. He agreed to postpone the discussion until the following week.

In the meantime we could study the investigators' response. It leaned heavily on the "independent assistance" of the two consultants — like a last-ditch submission that tries to dazzle the court by pulling expert witnesses out of a hat to save a case that was going down the tubes.

I tried to look at the rebuttal from a neutral perspective. It was organized under the same headings used in the critique, so a casual reader would see a nominal answer to all the issues we had raised. There was nothing terribly wrong with the consultants' contributions, except that they largely missed the point. The scientist from the National Research Council gave, for example, a dissertation showing that, on the basis of a page of mind-numbing assumptions ("a medium volume diameter of cloud droplets of 20 micrometers", etc., etc.), the flight through clouds on approach to Gander "could have produced an ice thickness of about 7 mm (0.3 in)." The upshot was that we hadn't conclusively proved that it was impossible for some ice to form on the wing during descent.

As I read, I also realized what Paul Koring had been driving at when he called a few days earlier. They thought they had a ''gotcha'' with the ice on the windscreen. Based on the research Koring had cited, the ''conclusion that ice on the windscreen was not evidence of ice on the leading edge'' was ''based on an incorrect foundation and is insupportable'', the investigators said. ''When ice forms on the windscreen of an aircraft there will almost certainly be ice forming on the leading edge of the wing.''

Well, maybe they were right. Captain John Coe had testified that, on occasion, he had noted ice on the windscreen during descent when inspection after landing showed none on the wings. Perhaps the ice melted from the wings during the approach. But the issue was a side show, mooted by the next obvious question. When ice ''almost certainly'' formed on the leading edge, did it also ''almost certainly'' form elsewhere on the airplane — on parts conspicuous to the ground handlers, say? And of course the whole question of ice formation in the clouds was immaterial if Captain Griffin had turned on the aircraft's wing de-icing system during the approach.

Well, a brief for the defence didn't have to hand points to the prosecution.

On second thought, I realized that the investigators' response was not really a brief to the court. The call from Paul Koring had betrayed the true audience. The response was intended to neutralize the effect of our critique on the media.

I also noticed, in passing, that the key to the witness identities I had carefully segregated from our report had been made available to the consultants, who had freely identified the witnesses. So much for confidentiality.

Under the circumstances, a rebuttal of this depressing rebuttal seemed hardly worth the effort. I soon got a clear sign that it was not. As the struggle between the pro-ice and anti-ice forces became known among the CASB's own staff, I received an occasional telephone call offering encouragement on behalf of past or present staff members. While such calls were usually uplifting, they didn't often contain helpful information. But one call to me at home relayed a useful tip: if I wanted to know what the CASB staff really thought of our critique, I should have a look at file number such and such.

I made a routine request for the file. The most recent addition was a report from the chief of the Systems Engineering Group in the CASB's laboratory, titled "Comments on the Critique of the Ice Contamination Hypothesis" and dated June 13, a week before we had got the official response.[18]

While this report still strove mightily to find nits to pick, many entries against significant items in our critique were simply annotated "Agreed". Most significantly, the report concluded that "An engine failure in addition to ice contamination is considered necessary to explain the crash of the aircraft." In other words, the best opinion within the CASB itself had concluded that ice on the wings could not have caused the crash.

It was no longer possible to avoid the conclusion that the discussion of the causes of the crash was a sham. Any opinion contrary to the official dogma of the ice theory would be suppressed, even one coming from the ranks of the "professional investigators". The professed respect for confidentiality and the contempt for leaks to the press were as empty as the professed interest in the correct cause.

But Thorneycroft still had to act out one final chapter in the charade. Lee Levenson wanted to make another oral presentation when he delivered the written submission to the conditional draft on behalf of Mrs. Theresa Griffin. Thorneycroft had promised to weigh this submission before issuing a final report; moreover, the board had heard a second oral presentation from Arrow Air, and it couldn't refuse the same treatment to the pilot's widow.

* * *

On July 13, 1988, the CASB assembled to hear the second oral presentation on behalf of Theresa Griffin. The four board members who contested the ice theory were there, along with chairman Thorneycroft and Arthur Portelance.

The atmosphere was charged with a weird electricity. The four of us on the anti-ice side welcomed the occasion and the participants, seeing the last faint hope to open up the investigation. But the relationship between the chairman and Mrs. Griffin's advisers had soured over the past two months.

Lee Levenson believed the CASB had reneged on the promise to provide information. He was particularly bothered by roadblocks preventing the examination of the flight data recorder tape. The board had voted to let Harold Marthinsen examine the tape at the American Air Line Pilots Association facility on condition that a CASB investigator accompany the tape and observe while it was examined, but—for reasons that were never explained in the boardroom — Marthinsen was later told the tape couldn't leave Canada. If he wanted to see it he would have to come to Ottawa, where he wouldn't be able to use his own equipment. After discussions with Hinton and Thorneycroft, Marthinsen had examined the tape in

Ottawa, only two weeks earlier. His analysis was still incomplete.

Thorneycroft had told us that Levenson wanted more time to prepare, but not that the decision about the FDR had been altered, so I had agreed to ask Mrs. Griffin to continue as scheduled, on the understanding that we would accept supplementary material when it became available.

The written submission we had received the day before was definitely testy. The presentation, Levenson complained, had been "hampered by our inability to obtain important information", and he contrasted the runaround with the courtesy accorded to Douglas and Pratt & Whitney. "We are greatly concerned that the Board's investigators, in their efforts to better understand the DC-8 systems, may have been guided by non-neutral interested parties," he wrote. The submission charged that the conditional draft "projects probability into factuality by inference and innuendo and was not consistent with a conclusion that is based upon professionally constructed assumptions."

There was something else that really bothered Mrs. Griffin. "Captain Griffin's notebook was recovered from the wreckage," her submission noted. "His family was never notified that this article of his personal effects had been recovered. We believe that quoting excerpts from Captain Griffin's personal notes exhibits *extremely poor taste* and is an unfortunate effort to denigrate his professional reputation."*

Ironically, the board had unanimously agreed to remove these irrelevant passages when we had first reviewed the draft in the spring of 1987 — but such

---

*Emphasis in original.

changes had been cancelled *en bloc* by the board members who had approved the conditional draft.

Thorneycroft opened the meeting by reading a prepared statement. The Griffin submission, he said, "makes what I think are pejorative comments. It questions the performance of the staff and the board members. It charges bias and it strongly implies dishonest practices." He insisted that "the investigation was done in a professionally, internationally recognized manner following ICAO methodology. . . . All reasonable scenarios were examined based on the evidence available. . . . As you know, professional investigators don't jump to conclusions. And they have no reason to take a biased stand." He asked Mrs. Griffin to withdraw these statements.

Levenson replied that there was "no intent to imply dishonest behavior," but declared that they would stand by their opinion that the draft demonstrated "a bias towards a theory that ice contamination caused this accident." It was not an auspicious beginning.

The Griffin submission was some 120 pages long, plus lengthy extracts from manuals and service bulletins. It went through the usual problems with the ice theory and the performance calculations, and also gave a detailed critique of the superficial analysis of the mechanical systems. Levenson and his team had thought a lot about the engine fire extinguisher bottles. One had almost certainly been fired on purpose, but Thorneycroft — who knew a lot more about this than I did — said the location of this bottle couldn't be determined because "the fire bottles were removed by agencies concerned about the possibility of explosives."

Since their last presentation, the Griffin team had moved beyond what *didn't* cause the crash. A review of problems with DC-8 thrust reversers suggested that a single failure in the thrust reverser system could put the engine into reverse thrust at takeoff power with the thrust levers in the forward position. This, they suggested, was the most probable cause of the crash. The deficiencies in the DC-8 reverse thrust system should be corrected, whether one accepted the theory or not.

After Levenson and Jerry Rusinowitz had finished the technical comments, Mrs. Griffin, who had observed this and the previous meeting quietly, said she had a statement. She spoke in a calm but determined voice: "I am outraged that the staff recovered and referred to my husband's log without ever acknowledging to me that they had it. I think that they kept it with malicious intent to assist them in proving their preconceived theory that icing was the cause of this accident." She said she intended to make the CASB's ineptitude a public issue.

Within the usual scope of his activities, Thorneycroft was a master of social niceties. If someone retired, got married, broke a leg, died, or had a baby, he was right in there with just the expression Miss Manners would recommend. But Mrs. Griffin's remarks were harder to deal with—so he simply ignored them. He thanked her for the presentation and assured her it would receive detailed consideration. Then he went to close the proceedings. But I couldn't leave the widow's statement just hanging there. "Mr. Chairman, Mr. Chairman," I called out, so loudly that he couldn't ignore me.

"I see one of my colleagues has a comment," he acknowledged, none too pleased.

I said that I would like to supplement the chairman's opening comments with a personal view, and told Mrs. Griffin that I for one regretted some of the material in the conditional draft. As for the mutual accusations of pejorative remarks, "They may be regrettable but they're not parallel. The draft report impugns the memory of the dead crew who can no longer defend themselves. The submission calls into question the professional work of a group of people who are not named individually. Each of these people is alive and capable of rebutting anything that may be unfair or incorrect. I regret anything that may have impugned the memory of your husband, Mrs. Griffin, and that of the other crew members. And I assure you that I take everything that was presented here without any offence at all."

It wasn't great eloquence, but it was the best I could do on the spur of the moment. Mrs. Griffin, Levenson and the others nodded appreciatively. Thorneycroft and his supporters glared with undisguised contempt.

* * *

The next day, members of the opposition Liberal Party arranged a press conference for Mrs. Griffin and her advisers. They denounced the ice theory and attributed the crash to an uncommanded deployment of the number-four thrust reverser. The Canadian Air Line Pilots Association and the Liberal Party supported Mrs. Griffin's call for a judicial inquiry.

When questioned by the media, Bobbitt, Stevenson and I backed this call. The *Ottawa Citizen* reported that "the directors have never before criticized the investigation"[19], while Thorneycroft said it was "extremely unfortunate" that we had "broken the board's solidarity". He maintained that a judicial inquiry was not necessary, and he was backed by Gordon Sinclair, his old pal from the Transport department, who was now president of the Air Transport Association of Canada. Benoît Bouchard continued to reject any calls for a judicial inquiry: "We will have a final report," he said, "and at that time the minister will consider the recommendations."

In the meantime Bouchard had introduced legislation to replace the CASB with a five-member multimodal board. The legislation would die on the order paper when Parliament adjourned, but Thorneycroft couldn't know that, and he was banking on it to solve his problems. He became even more isolated, and made even less pretence of heeding the board as we continued going through the motions of meetings to deal with a procession of other accidents.

* * *

With the oral submissions out of the way, it could be argued that the CASB had satisfied its obligation to the interested parties. It was time for Boag, Hinton and their colleagues to revise their report in light of the representations. Their task was much what it had been a year before, but working under a new chairman was an advantage. Thorneycroft was as committed to their cause as Deschênes, but less likely to impose idiosyncratic requirements — and he clearly enjoyed the support of top bureaucrats. On the other hand, a new set of critics had joined those within the board. Worst of all, the investigation had come to the attention of the media.

The strategy for the final report included a push to stake out the high ground with the media. Critics and potential critics were to be neutered with references to the confidentiality of the investigation. At the same time, a campaign was launched to polish the image of the CASB staff and to again build up the myth of the "professional investigator". Reporters were invited to the laboratory to marvel at the high-powered microscopes and banks of computers. No opportunity was missed to preach the exclusive authority of the director of investigations. Every scrap of good-news potential was exploited: A French-speaking female investigator provided good grist for a human interest story and, not quite incidentally, spread the gospel.

The media blitz gave us an inkling of Thorneycroft's intentions. He was bent on issuing the final report before the third anniversary of the Gander crash; he had no intention of circulating another draft.

The media also provided our only clues about his plans for dousing the criticisms. Apparently the investigations were confidential only to critics; Peter Boag told Canadian Press that "it appears that the plane's number four engine, the outboard far right engine, may have been operating at 60 percent of maximum thrust at the time of the crash".[20] Was Boag conceding that ice couldn't in itself explain the crash? Heretofore, he had never given the slightest sign of willingness to eat crow, but perhaps he was now being forced to nibble a little.

* * *

Then out of the blue, we heard that "American servicemen, Gander firemen, and as many as 60 members of the Public Service Alliance of Canada who responded to the crash of an Arrow Air jet three years ago are starting to develop serious medical problems."[21] The rescue workers were reporting "liver problems, nausea and dizziness which doctors believe may be linked to toxic fumes released from the burning wreckage." The union representing many of the workers demanded an independent medical examination of everyone who had been at the crash site.

These reports, vague as they were, gave an immediate urgency to finding out exactly what had been in the mysterious crates loaded in Cairo. Suppose they had contained dangerous chemicals that were affecting the health of the people who had mucked about in the wreckage — surely those in the know would pass along the information. But what if the only ones who knew had perished in the crash?

Dave Owen, the first CASB investigator on the site, told me he didn't think there had been any hazardous material on the aircraft and wasn't worried about his own health. Nevertheless, I wrote to Thorneycroft urging medical tests for any of the CASB staff who had been on the site. Outside the CASB, the government moved quickly to defuse the issue, announcing that it would appoint a medical team to study the complaints. Dr. Rosemary Marchant, a specialist in occupational health at Dalhousie University, was appointed to do a study, which would begin sometime in November and would continue for an indeterminate period.*

By then the air crash that killed General Zia of Pakistan in August 1988 had been attributed to deliberate sabotage. The *New York Times* reported that "a chemical agent may have been used to cause incapacitation of the flight crew", quoting Pakistani investigators as mentioning "high levels of phosphorous and antimony" and "traces of pentaerythrito tetranitrate".[22] After I drew this to the attention of Dr. Marchant, an official of the Health department eventually wrote to thank me for my interest.

_____

*The report, which was released in June 1989, attributed the symptoms to traumatic stress.

While the burst of health problems deflated the campaign to flood the media with stories about the prowess of "professional investigators", the government's willingness to launch a study ensured that the medical problems would not become an issue in the CASB report. The net effect was to restore the status quo.

"The Journal"

The campaign to enlist journalists had an unanticipated downside: some of them weren't satisfied with bromides and visuals. Kevin Tibbles, a reporter for the news program "The Journal", was one.

"The Journal" appears on the CBC national network six nights a week, after the ten o'clock news. Tibbles wanted to do a story on the Gander crash, and Roger Lacroix urged our side to cooperate. But, wasn't the report-in-progress supposed to be confidential? Was this a good time to talk to the press? How much could we say? Wouldn't we get hammered? After all, the other side had a whole public affairs department to spin the story of devoted professionals impeded by a few renegade political appointees.

Tibbles had a good come-on: Thorneycroft had already agreed to cooperate. We didn't want to give free reign to the other side, did we?

Thorneycroft had publicly declared that he didn't believe hydrogen cyanide was a factor, so we should be able to say that we didn't believe ice was a factor. John Sopinka had said publicly that the appointed board members had not been given the facts they needed to determine whether or not the investigators had got at the true causes, and we could certainly go along with that. We'd do it.

Roger Lacroix, Ross Stevenson and I met Tibbles and his crew at the Ramada Hotel on August 31. He told us he had interviewed Thorneycroft the day before, but he wouldn't say how it had gone.

As we waited for the crew to set up, I chatted with the sound technician, a slightly built man called Richard. He seemed more interested in psychology than electronics. "Thorneycroft wouldn't sit down. He insisted on standing for the whole interview with his hands on his desk behind him," he volunteered "as if he was trying to cover it up. Positively Freudian." Richard grinned and shook his head. "Positively Freudian," he repeated. It was a good sign.

Ross, Roger and I sat in a semi-circle around Tibbles and talked freely within the limits we had agreed to. It was the first of many interviews to come.

* * *

On September 13, Ken Thorneycroft circulated a memo to all board members.

> As you have certainly noticed, there is currently a new wave of media discussion of the Board's internal activities with respect to the investigation of the Arrow Air accident. Furthermore, my sources indicate that a reporter from one of the major news wire services will be interviewing certain Board Members. . . . public discussion at this stage is not serving aviation safety . . . any discussion of the investigation analysis or potential findings should take place only within the Board's process until we have completed our review. . . . If you cannot abide by these guidelines, I request that you advise me immediately.

The "new wave of media discussion" bothering the chairman was the extension of his "good news" campaign to include contrary opinions, and his sources were at least partially right—a reporter from UPI was coming to my home for an interview that evening. As for the "Journal" interview of both Thorneycroft and us, it was to be aired as a full-edition documentary a week later. But Thorneycroft already knew that.

The "Journal" documentary turned out to be a good summary of the impasse—at the CASB in general, and with respect to the Gander investigation in particular. As expected, Thorneycroft and Boag deflected awkward questions on the grounds that the investigation was incomplete. And Thorneycroft did indeed plant himself between the camera and the secrets on his desk, covering up as much of what was behind him as possible. The symbolism would have escaped me without the helpful psychology lesson from the sound technician. Thorneycroft denied that information had been withheld from board members. "Mr. John Sopinka is entitled to his opinion," he said.

Roger, Ross and I disowned the ice theory without speculating on what might have caused the crash, but "The Journal" set the stage for such speculation with two eye-witness interviews; Ted West said he would have felt ice on the airplane had there been any, and Cecil Mackie again reported that the glow from the DC-8 had seemed "like a steady flame". Moreover, West and Mackie illustrated a point we had been trying to make from the outset—ambiguities in witness statements could be settled by talking to the witnesses.

Kevin Tibbles had also interviewed Lee Levenson and Dick Moore, who had analyzed the draft report for Mrs. Griffin. The two pilots were convinced that the number-four thrust reverser had somehow popped out during the ill-fated take off. Moore recounted his own experiments on a DC-8 flight simulator, which showed that a crash was inevitable under the assumed conditions. The discharged engine fire extinguisher suggested that the crew had been fighting a fire at the same time.

The "Journal" documentary passed unremarked in the CASB boardroom but the atmosphere became even more charged, with Bill MacEachern muttering darkly about "scurrilous saboteurs" trying to undermine the work of the professional investigators and Thorneycroft determined to "get the report on the street" by the third anniversary of the crash.

Boag's hints to the media suggested that the new draft would stick with the ice theory but admit problems with the number-four engine. We would be forced to write a dissenting report. But we couldn't just keep railing against the ice theory; we would have to propose an alternative.

Ross Stevenson had always believed that an explosion had brought down the aircraft, and we had discussed this possibility many times. But Ross' belief was based on intuition, not deduction from evidence, and there was no way I was going to supplant one wrong theory with another. The hole Irving Pinkel had found in the fuselage suggested a small explosion in the cabin, but witness testimony and the system failures would place any blast in the cargo hold. And no one had come up with an explosion hypothesis that even began to explain the origin, the links with the failed number-four engine, the hydrogen cyanide in the victims and so on. The problem of course, was that evidence not needed to support the ice theory had been overlooked, discarded or discounted. The investigators hadn't even bothered to keep decent copies of Pinkel's photos.

Once, I had asked why the investigation didn't pay more attention to the possibility of sabotage. Oozing sarcasm, MacEachern had said, "I suppose you think they should have interviewed Yasser Arafat.'' The others pretended to find this funny and refused to discuss the subject. The potential for sabotage was so horrendously self-evident that their indifference — real or feigned—begged explanation.

Stevenson was convinced that there was hard evidence to prove an explosion, if only we could find it. From the outset, he'd been frustrated by the lack of transcripts of interviews with important witnesses like Judith Parsons. In any event, tape-recorded interviews demonstrated that the investigators hadn't asked the right questions. Of the thousands of photos taken at the crash site by RCMP photographers, only hundreds had been kept in the files—and these were stored haphazardly, without identification or index. Before deciding what to put in a dissenting report, Ross wanted to see *all* the photos. He wanted to talk to witnesses. He became obsessed with revisiting Gander.

Thorneycroft and the supporters of the ice theory scoffed at his demands. To them, such manic desires demonstrated his unfitness to be a member of the board; clearly he really fancied himself as an investigator. It was great sport to deflect his questions into an analysis of his unfulfilled delusions.

But for some reason — possibly the increased publicity about restricting the review of board members— Thorneycroft had a sudden change of heart. He would approve Stevenson's trip if it was supervised by one of Hinton's henchmen. I told Stevenson it would be like visiting a gulag with a guide from the Politburo, but he didn't care. In fact, he was exuberant. He spent three days in Gander.

Stevenson observed aircraft taking off from Runway 22 from the positions of the truckers on the Trans-

Canada Highway at the crack of dawn, to see for himself whether the reflection of low-intensity approach lights could be seen, let alone mistaken for a fire. He talked to witnesses, including Judy Parsons and the helicopter pilots who had been flying on the day of the crash. He had a helicopter fly him along the presumed flight path of the Arrow Air DC-8 and he sighted back to Parsons' position in the Tilden parking lot. He was convinced she had seen an in-flight explosion.

The CASB investigators had not seen fit to interview the fire-fighters who were first on the crash site. For two years Stevenson had pleaded in vain for the next best thing; a copy of the report of Fire Chief Hennigar, submitted a few days after the crash. He finally got one.

He also visited the repository for the RCMP photographs in Grand Falls. The RCMP officers were cooperative but not communicative. Somehow, though, his attention was directed to a specific selection of photos showing pieces of wreckage containing holes with outwardly curled edges, like the ones that had caught Irving Pinkel's attention. The exhibits were carefully photographed from both sides and from different angles. The group included other remarkable photos like those of a section of fuselage containing an empty window surrounded by a narrow ring of soot.

No one explained why these particular photos were grouped together, or why someone had deemed it important to take them. There was no index. But copies were willingly provided.

When Stevenson returned to Ottawa, he again tried to convince Thorneycroft that the investigation might have overlooked evidence of a pre-impact fire or explosion. He followed up with a memo listing items that he believed needed to be considered, for the report to be credible: "Soot marks around window frames seen in photographs obtained from the RCMP suggest the possibility of an internal explosion. . . . Recovered doors, frames and wreckage destruction suggest the possibility of rupture from internal over-pressures characteristic of in-flight explosion. This point should be considered by a complete description and analysis of all doors and removable windows." And a dozen other items.

Thorneycroft countered with a memo dismissing Stevenson's points out of hand. With respect to the doors and windows he said, for example, "All wreckage, including the window frames, door frame and doors was subject to detailed and extensive examination by investigators, as indicated in the Conditional Draft, no evidence of in-flight fire or explosion was found."

He might as well have said, "Trust the professional investigators"—or, for that matter, "Stuff it where the sun don't shine."

* * *

While Ross Stevenson was chasing evidence of an explosion, I wanted to rethink the evidence related to the thrust reversers. The investigators seemed so confident of their diagnosis that all four thrust reversers had been stowed that, up to now, I had accepted their conclusion at face value. But Lee Levenson, Dick Moore and Harold Marthinsen were absolutely adamant that the number-four thrust reverser had been a factor. I needed to reflect on it for myself.

The investigators had not documented their negative conclusion about fire or explosion, but the *New*

York Times had reported that a thrust reverser might have been involved; so they had produced a report to back their contention that all four were stowed. I obtained large glossy photos of the wrecked reverser units and started to mull over their conclusions.

The thrust reversers had attracted attention because the number-four unit seemed to have been extended at impact, but the investigators had deduced that it had opened while being dragged backward during the crash. This may seem plausible if one imagines the crippled aircraft sweeping close to the ground with the right wing tilted down. But the right wing was *not* tilted down. The wings of the DC-8 are set in a dihedral or "V" with the tips up about 7 degrees from the roots. This dihedral was cancelled almost exactly by the seven degrees of right bank. Thus, at the moment of impact, the right wing was dead level—or as close as anyone could hope to determine. So whatever happened to the number-four engine should also have happened to number three, hanging next to it on the right wing. The two engines should have sucked in branches and dragged their thrust reversers in the same way. So why was everything about number-four so different? Why had the number-four thrust reverser been dragged all the way back when the other three units were dragged back only a little bit?

Something was very strange here.

A moment's reflection on impact forces sufficed to show me that none of the thrust reversers could have been dragged backward. The DC-8 had plunged violently into the ground; all four engines had clearly struck nose first. Components would have hurtled forward, like loose objects in an automobile that runs into a brick wall.

The DC-8 thrust reversers open something like a fold-up umbrella. "Translation rings" slide back on tracks, opening when they have moved back far enough. The other reversers were apparently closed at impact but, curiously, not one of the translation rings was in the full forward position. Had they been forward at the time of the crash, they should have stayed forward. If, however, the translation rings had been open at the rear of the tracks, the impact forces would have driven them forward, possibly far enough to close the doors.

The only reason I could think of to explain why any of the reverser units was back even a little was that it had been back farther to start with, and then the violent deceleration had driven it forward. Of course, there might be many reasons I couldn't think of—but they weren't found in the investigators' report.

I packed up the glossy pictures and a copy of the thrust reverser report and sent them to an experienced investigator working as a private consultant. When he called me after studying the material, he said, "I think all four were out."

Later, we studied the deductions about thrust reversers in detail. Based on the evidence at hand, a plausible case could be made to support the hypothesis that all four reversers were forward at the time of impact.

So here, after almost three years, a complication came right out of left field. It fit neither the ice theory nor the malfunctioning number-four thrust reverser theory, nor any other theory—except that of a careless investigation, an investigation aimed at producing evidence to support a preconceived conclusion, an investigation where even evidence close at hand was not examined with a critical eye.

BOOK 6
FINAL REPORT

Part 1
End Game

The tragedy has been so uncommon, so complete, and of such personal importance to so many people that we are suffering from a plethora of surmise, conjecture, and hypothesis. The difficulty is to detach the framework of fact—of absolute undeniable fact—from the embellishments of theorists and reporters. Then, having established ourselves upon this sound basis, it is our duty to see what inferences may be drawn and what are the special points upon which the whole mystery turns.

Sherlock Holmes[1]

We received the new draft of the Gander report during the last week of September 1988, almost exactly a year after we had received the draft purporting to answer the "deficiency list". The second time around was just as much of a downer: the new text was bloated with irrelevant clarifications, but the conclusions remained.

I lost a bet I had made with Bobbitt; the new draft didn't concede that the number-four engine had lost power before impact, but only that it had been turning more slowly than the others. And it maintained that the investigators couldn't determine whether the slower rotation was due to loss of power in flight or to "tree fragment ingestion" during the crash. It actually said it couldn't be "conclusively determined" because there was no "definitive evidence" so nothing could be "completely ruled out". The report was peppered with weasel words like "conclusively", "definitive" and "completely" — part of a trend to even more equivocation and side-stepping. It was no longer claimed, for example, that "engines one, two, and three were determined to be operating at high-power settings at ground impact" this clear (if unproven) claim had been replaced by three hopelessly obscurantist paragraphs.

The text about the number-two engine said, "Although the impact reading of the number two engine indicator was well below take-off EPR, it is possible that the reading, if reliable, indicates that power was removed from the indicator later in the impact sequence, after the engine rpm and EPR had decreased as a result of impact and breakup. This assessment is supported by the examination of the engine which indicated that the engine was operating at high rpm at ground impact." (I can't understand it either.)

The new text on engines also reported that "Independent examination of the number four engine confirmed the assessment of the CASB investigators." The independent examiner—Dr. Gary Fowler, a metallurgical engineer from Gardenia, California — concluded that "observed engine damage caused by tree ingestion and resulting deceleration was consistent with a high power output."

Fowler had been retained late in the game, and the board hadn't got to meet him or even see his terms of reference. Nor could we ask how his metallurgical expertise helped him assess tree ingestion. But why would we want to? Who but interfering government appointees would not be completely satisfied? The appraisal of the "independent investigator" hired by Tom Hinton was, after all, consistent with the story Hinton had been pushing so vigorously. And Fowler, we could be certain, was not inexperienced in theorizing about the causes of an air crash.

In 1984, for example, Fowler had advanced a theory of "heat delamination" in the defence of director John Landis, accused of negligence in the grisly deaths of actor Vic Morrow and two children. An explosion, set deliberately during the filming of *Twilight Zone — the Movie*, disabled a helicopter flying just twenty-four feet above the actors. The right skid crushed six-year-old Myca Dinh Le and the main rotor decapitated Morrow and seven-year-old Renee Chen, while six cameras captured the gory details. Fowler's theory, convenient for the defence, that the rotor blade "delaminated" independently of the tremendous blast under it, was rejected by other expert witnesses and, ultimately, by the court[2].

The new draft did rectify certain omissions. Under the prodding of Arrow Air, it finally acknowledged that hundreds of weapons had been scattered over the crash site. Humphrey Dawson's demands had elicited short lists of weapons found on the site and weapons "believed to have been aboard the aircraft", although the report did not comment on the discrepancy between the lists. It did tell us that "There was no evidence found of military ammunition or explosive device". Elsewhere it noted that the equipment "believed to have been aboard the aircraft" did not include "military ordnance, ammunition or other explosive material" — except, that is, for "one clip each of .45 calibre ammunition reported to have been carried by a Criminal Investigation Division Inspector and the Battalion Commander." Given that the investigation had found no evidence of these clips, we were left to wonder what else might have been overlooked.

The new draft had taken to heart Ross Stevenson's complaints about omitted witnesses. The reference to "three witnesses" who saw a "yellow-orange glow" had been changed to read "several witnesses". Somehow, Stevenson was still not happy. Neither was he cheered that the investigators accepted his demand to account for Fire Chief Hennigar's report of thirty or forty explosions, some of which were large enough to cause "mounds of rubble to lift several feet into the air". The new draft noted that "several small post-impact explosions occurred in the burning wreckage" and attributed these to "normal bursting of pressure vessels". The vessels — presumably without safety valves — were not listed.

But this was all a side show. It was the new draft's account of the hydrogen cyanide and carbon monoxide in the victims' remains that best illustrated how the seeds of neglected evidence had grown into a briarpatch of contradictions.

Many Victims Survived the Crash

The conditional draft report sent to the interested parties in December 1987 had given a simple account of the cause of death: "All 248 passengers sustained fatal injuries as a result of impact and post impact fire."

This simple statement didn't rule out an interpretation that occurred to no one at the time: that some victims lived through the crash and died in the subsequent fire. The possibility of a survivable impact, if taken seriously, would, of course, raise disturbing questions: could lives have been saved by speedier arrival of the firefighters, for instance? But no one asked such questions. Pictures and eye-witness testimony of the horrible destruction belied all hope of survival.

The impression that no one could have survived was borne out by the certificates of death by Dr. McMeekin, which cited instantaneous death from accidental causes. McMeekin, moreover, had told the CASB's public inquiry that the remains showed "no reaction to burning in any of the tissues."

Without the toxicological findings, these conclusions added up to a simple, consistent story: everyone died instantly in the colossal crash. For almost three years, no one had suggested anything else. But then public disclosure of the toxicology results called for some creative thinking.

Four pages of opaque "medical information" supplanted what had been covered in a few short paragraphs. The previous draft stated that "no postmortem evidence was found that was consistent with the detonation of an explosive device." The new, improved version said, "The effects of an explosive blast wave were considered indistinguishable from the effects of trauma from decelerative forces, flying debris and structural collapse of the aircraft." Those interested in word games might note that the new phrasing was logically equivalent to claiming that the findings were consistent with the detonation of an explosive device.

It turned out that Robert Lee's revelation in the *Ottawa Citizen* had been understated. Lee had suggested that about half the corpses had contained elevated levels of hydrogen cyanide; the data showed that 158 of 187 samples — about 85 per cent — tested positive. In addition, 69 of 187 samples showed elevated levels of carbon monoxide.

But only non-professionals would assume that this suggested a pre-impact fire. The new draft said that "a complete review of pathological examination results was undertaken for the CASB by forensic pathologists. . . . The primary purpose of this review was to estimate the time interval from injury to death for each victim." What was this? Didn't Dr. McMeekin's death certificates say death had been instantaneous?

The report went on to tell us that the review had established that 158 out of 250 cases where it was possible to estimate the time of death—coincidentally, the same number as had elevated levels of hydrogen cyanide—had survived for more than thirty seconds, possibly up to five minutes. It followed that if the firefighters had arrived in five minutes instead of seven, they might have found many of the victims still alive. Yet no firefighter reported any sign that these victims had made an effort to save themselves — by opening emergency exits, for example.

The report didn't explain that the "complete review of pathological examination results" had been commissioned in the wake of Lee's article, for the express purpose of countering "erroneous news reports" that "cast doubts on the cause of the accident and generally create adverse public perception"[3]. Nor did it say what Dr. McMeekin thought of the disavowal of his death certificates.

But never mind — here, for the sufficiently credulous, was the secret of how so many victims could have ingested deadly fumes. They had lived and breathed after the crash and *then* succumbed. The less credulous might have wondered how victims who lived for five minutes in that blazing inferno could have escaped injury from the raging fire and numerous explosions.

* * *

On October 12, 1988, the CASB board held its first formal discussion of the Gander investigation since the presentation on behalf of Mrs. Griffin in July. Oblique skirmishing had, of course, continued. At one point I had introduced a motion to ask the government to institute a judicial inquiry. The motion had been defeated, but the list of "whereases" had put some of the deficiencies of the investigation on the record. Another motion seeking a judicial interpretation of the CASB Act had also been defeated, but we hoped it would remind Thorneycroft of the legal niceties.

There had been a steady stream of correspondence with Arrow Air lawyer Humphrey Dawson. His most recent letter had arrived just a few days earlier.[4] Arrow was still waiting for information promised five months earlier, Dawson complained. "At that time you assured us that no report would be issued without a draft thereof being supplied to my clients," he reminded Thorneycroft. "I would ask for an assurance that the Board will not issue a report until at least the pathology and other information for which we asked in May has been supplied and our clients have had a reasonable opportunity of considering same." Fat chance of that. Thorneycroft had already started final review of the new draft, which he insisted must be issued before December 12, the third anniversary.

Thorneycroft was in good form. He began with a ten-page opening statement, the likes of which we haven't heard since Bernard Deschênes jumped ship — a textbook example of preemptive historical engineering, a record of the fairness of the process to come, for the benefit of future historians.

"Two years and ten months have passed since the morning of December 12, 1985," intoned Thorneycroft, presumably rehearsing for a future press conference. "The investigators have examined the site and wreckage in minute detail," he continued — asserting another hotly disputed claim as if it were fact. There followed a revisionist history of the investigation. He noted, for example, that "My predecessor, Bernard Deschênes, assisted by members Lacroix, Pultz and Stevenson, conducted a comprehensive public inquiry into this accident." This is the first time anyone had ever styled the public inquiry "comprehensive" the usual alibi for omissions being that the inquiry was "just one step in the investigation process". And, curiously, Thorneycroft had omitted Frank Thurston from the list of board members presiding at the public hearing.

"The investigation has been thorough and complete," Thorneycroft went on, "and our review exhaustive." All conclusions had been verified by "independent experts". Of course, the board members who disagreed were not independent experts. Irving Pinkel was not an independent expert. Harold Marthinsen was not an independent expert. Gary Fowler was an independent expert.

"Nothing has been withheld," Thorneycroft read. He must still have been smarting from the pasting on "The Journal". "The public discussion of the analysis of the Gander accident has, in my view, caused serious harm to the reputation of the CASB in general. It has harmed the good name and reputation of the majority of board members who have respected the confidentiality provision of the Act. Similarly, it has harmed the staff whose work was publicly maligned and who were least of all able to respond. I have seen no justification for such action and unless justification is produced, I will have to take the steps that any responsible chairman would take.

"While I do not intend to limit or restrict discussion, if a consensus is not emerging, I intend to call for a recorded vote and move on. . . .

"The interested parties had generous opportunity to make representations" but changes to the draft were "primarily due to public interest created by the leak of the initial draft." So much for the promise to circulate another draft.

The chairman had also reflected on the possibility of a dissenting opinion. The final report must be produced by his deadline of December 12 and "I do not intend to unduly delay production of the report waiting for dissenting opinions," Thorneycroft said.

"I understand the board has decided not to impose any restriction on the submission of dissenting opinions," he added, alluding to a recorded decision taken under Bernard Deschênes. "Nevertheless, it seems to me that there must be reasonable limits on what can or cannot be included." When the time came, he aimed to "provide reasonable guidelines" to potential dissenters. But "in any case, dissenting opinion not received in a reasonable time frame . . . will not be published."

He then called on the board members to comment, starting with those who had objected to the ice theory. The four of us did so, at great length. The new draft was as specious as the original. Refusing to let interested parties comment on it was is a breach of trust. The time limits he was trying to impose were unjustifiable and unfair—contrast them with the month the investigators had needed to review our "Critique of the Ice Contamination Hypothesis." Moreover, the new draft included extensive additional medical information, yet we had not received any documentation.

Thorneycroft's cheerful assurance that nothing had been withheld reminded me of a curious visit from the director of investigations. Tom Hinton was assuredly not in the habit of making chummy visits to my office, but a few weeks earlier, he had dropped by unexpectedly. He had suddenly got the urge, he said, to soothe my concerns about the baggage of the two soldiers who had missed the flight in Cairo.

Since Hinton was so uncharacteristically informative, I had taken the opportunity to ask about Thorneycroft's answer to my repeated requests for the passenger manifest. According to Thorneycroft, the manifest given to the investigators "could not be identified" because it disappeared during a "reorganization of the files" that Hinton had ordered.

Hinton was chagrined at the implied mistrust. He allowed that "duplicates" had been removed from the files when they were "reorganized to help the board's review," and on further probing he admitted that the "duplicates" might not have been "exactly the same" as what remained. But he reassured me that "the best information had been left on the file."

"If that is not withholding information, then withholding information has no meaning," I angrily told everyone.

The comments from the proponents of the ice theory were shorter. Thurston congratulated the chairman on his "expedient and statesmanlike" opening statement. MacEachern thought it "well reasoned and responsible" and agreed with it 100 per cent. He excoriated the "dissenters" somewhat more moderately than had been his custom of late; "Scenarios have been presented and cases have been made and allegations have been put on the table," he said. "They have been answered. They have been dealt with. Yet when we think they have been put to rest, lo and behold, they pop up again like mushrooms."

Hinton got his chance to explain why material was removed from the files. "The purpose was to try to bridge the gap between the staff and the board with respect to the availability of information," he said. "To facilitate the members' use of the files we put them in order, and I found out at that point that some information, what we believed to be extraneous information, duplicate information, was taken off the files, and I mentioned this quite frankly in my conversation with Member Filotas. All of it was kept in our own records. I think the very fact that I mentioned it to him would indicate that I'm not trying to hide anything."

The only way Hinton could top this pitch was by offering us all a once-in-a-lifetime deal on choice property in the Everglades—maybe with a free trip to Disneyland for those who signed right away. He didn't. We went on with the review of the latest, and final, draft of the report.

* * *

Thorneycroft would not permit a systematic review of the comments from the interested parties, as had been the practice. We could only ask the investigators to explain changes to the report, and then the board would vote on it, section by section. We started with the engines.

Gary Fowler's "independent" examination had given number four a clean bill of health, and there was no need to do anything on the other engines. The revisions in the new draft were accepted by the usual divided vote.

We moved on to the thrust reversers, and I started asking questions. Apropos of something I can't remember, one of the staff happened to mention the recent tests on the number-four engine nozzles in Montreal. What nozzles? What tests?

It turned out that Carl Hoffman, the lawyer representing the estate of flight engineer Mike Fowler, had hired a consultant to examine the engines. The consultant, retired Pratt & Whitney engineer Don Hammel, noticed at once that the number-four engine combustion chambers and fuel nozzles were covered with heavy deposits of soot, in marked contrast to the other three engines. The "professional investigators" had overlooked this conspicuous clue for three years, and Gary Fowler had also missed it.

Hammel was dumbfounded that the investigation had not sought to narrow the uncertainty about the state of the engines by testing components where possible. A simple bench test on the fuel nozzles could show if they had been working properly. At Hoffman's insistence, the nozzles from number-four engine had been tested at Air Canada's engine test facility in Montreal—on September 27, 1988, just two weeks earlier.

Hoffman subsequently wrote the chairman that the nozzle tests "raise doubts about the condition and power output of the number four engine." He requested further tests. Thorneycroft received Hoffman's letter days before telling us that nothing had been withheld and that all conclusions had been verified by independent experts". Tom Hinton had arranged for the nozzle tests days before saying that "we are working as hard as we can to provide all the information to the Board as fast as we can."

Yet the board had discussed and approved the conclusions about the engines in the draft report without a hint about the tests or Hammel's conclusions. We learned his opinion by chance—as we had learned of the witnesses who thought they saw fire, as we had found out about the hydrogen cyanide.

At this stage there was nothing to do but keep trying. I moved to suspend the discussion until the further tests suggested by Hoffman were carried out.

Thorneycroft was uncontrite and unconcerned. He had a majority, and he could win any vote. But Arthur Portelance was ready to make a gesture, and suggested we defer the vote until we got a report on the nozzle tests. The next morning we were told that unidentified Air Canada technicians believed that, while the nozzles "would not be suitable for installation on a newly over-hauled engine", they would be "acceptable as in-service components". There was no written report.

Why, someone asked, had the tests been done at Air Canada instead of at an independent facility? Well, Air Canada had the right kind of rig and had done us a big favor by letting us use it at short notice. "We rely on the good will of Air Canada, who provided a lot of assistance in the Arrow Air investigation," said Peter Boag.

I wondered, but only to myself, about the potential for conflict of interest in scrounging favors from a past and future interested party. Shouldn't the board know about such an exchange of favors? Here was a real-life illustration of the "exclusive authority" of the director of investigations; it ensured that favors granted or received from interested parties, actual or potential, were beyond outside scrutiny. But this was neither the place nor the time for philosophical reflections.

The all-purpose explanation for anomalies in the number-four engine was broad enough to accommodate a few sooty nozzles. The slower rotation had already been attributed to "tree ingestion" — obviously the same source had deposited soot on the fuel nozzles. Enough nonsense, get on with the vote. My motion to postpone discussion was defeated. The staff would, however, put some information on the nozzle tests into the report.*

Having cleared the deck of fuel nozzles, we were ready to continue our review. But wait a minute — a laggard chicken had just come home to roost. Thorneycroft received a fax from Humphrey Dawson, and this time he distributed copies.

"We find it very difficult to understand how, after this length of time after the presentation to the Board and the leaks to the media, you are unable to produce to us detailed pathology/toxicological reports and the information for which Harry Weisberg and I asked you in May," Dawson wrote. Arrow Air had retained Dr. Richard Thorley Shepherd, of the division of Forensic Sciences, Guy's Hospital, University of London to analyze the Gander toxicology results. "Having regard to all the circumstances I do not think that it can possibly be proper for your Board to issue any further report without full consultation with us."

Thorneycroft was willing to send the toxicology reports to Dawson and the other interested parties, but there was no way we were delaying the report. "The interested parties do not dictate our policy," he

---

*The final report would say, "The sooting in the area of the nozzles was considered consistent with the disruption of the airflow and the resulting fuel/air mixture that would have occurred due to tree ingestion."

said. "We'll look at anything they send us when we get it. If it's pertinent we can revise the report, even after it has been released."

We would, in fact, receive additional insight from Dr. Shepherd before the report was released. "It is obviously erroneous, and somewhat circular, to conclude that there was no pre-crash fire," he reported when he saw the toxicology data two weeks later.[5] One of the more grisly examples in the group assessed to have lived for up to five minutes provoked him to comment, "I find it very difficult to accept that survival in a case such as this would have been for more than a few seconds *at most*."

But Shepherd was neither "independent expert" nor "professional investigator", so his opinion would not influence the perception of the investigation—not until it was broadcast on American network television exactly one year later.

In the meantime, the CASB pressed on with its review. Those against the ice theory continued to comment on as many points as possible, with a view to making the report as correct as they could. Some suggestions on wording were accepted but substantial revisions were rejected out of hand. If the discussion dragged, Thorneycroft just said, "Let her lay where Jesus flung her," and called a vote. A majority of five to four invariably decided to do so.

One issue that got this treatment was Captain Griffin's personal notebook. The draft report sent to Mrs. Griffin had contained long verbatim extracts about a flight from the Sinai two years before the Gander crash. The main hydraulic system had failed on approach and the company had directed the crew to take the aircraft to Amsterdam for repairs. The extracts from Griffin's notes detailed the crew's use of manual aileron control and the auxiliary power system to raise the landing gear during a refuelling stop at Cairo.

Ross Stevenson had pointed out that the passages, quoted out of context, seemed to suggest that there had been something improper about the flight. The investigators had replied, fatuously, that the quotation showed that a failure of the main hydraulic system could not have caused the crash. This point could, of course, have been demonstrated simply by reference to the flight manuals. Putting a section of the captain's notebook after the section on de-icing suggested, by innuendo, that Griffin was prepared to fly with a known deficiency—a conclusion contrary to all reports of his character.

Nevertheless, the quotes from Griffin's notebook were included in the draft "unanimously" adopted by five members of the board. The only concession to Mrs. Griffin's concerns was to shorten the passage and paraphrase it to omit direct quotes. And that's the way it stayed, by a vote of five to four.

* * *

On October 24, we at last finished going through the report. The text, conclusions and recommendations had been accepted substantially as received.

I had tried mightily to convince the board to recommend that service bulletins related to the thrust reversers be made mandatory, as persuasively argued by Lee Levenson. No way. Unaccountably, they accepted a suggestion for a finding about missing side panels in a cargo pit — the missing panels compromised the fireproof integrity of the cargo hold — but the board decided to drop the word "fireproof".

I seized the initiative to move that the modified report be adopted as Conditional Draft Number 2, to be sent to interested parties for comment. Thorney-croft just chuckled. MacEachern made a speech about how the board must always act with generosity and fairness to the interested parties — adding that, as some interested parties had abused our generosity, the report should go out as final. My motion was defeated, five to four. But many of the changes had been left to the discretion of the investigators, and I argued that the board should not issue a report it had not even seen. So we got a few days' reprieve.

A draft with the changes was ready on October 28. And then—just before the vote was called—Thurston had second thoughts about the cause.

The several versions of the cause adopted by the majority over the previous year and a half had all cited some variation of an attempt to takeoff with ice on the wings, leading to a stall. The draft sent to the inter-ested parties had said, ''The probable cause of this accident was a take-off attempted with ice contami-nation of the leading edge and upper surface of the wing. This action resulted in a stall at low altitude from which recovery was not possible. Contributing to the occurrence were flight-crew fatigue and the use of inappropriate take-off reference speeds.'' Now, at the last moment, Frank decided that cause and effect should be reversed. He suggested that the *cause* had been a stall, probably due to ice contamination.

Lord only knows if this was just a whim or some-thing deeper. The logic of the report, such as it was, started with ice. Actually, the report said very little about a stall, other than that the FDR showed that it had happened. The ice argument would not suffer without the word ''stall — they could have just said that, with ice on the wings, the aircraft had too much drag and not enough lift to fly.

In any event, Frank Thurston — the consummate wordsmith — proposed the final wording. As in the rest of the report, the meaning was obscured by what purported to be a clarification. He started with an introductory statement: ''The Canadian Aviation Safety Board was unable to determine the exact sequence of events that led to this accident.'' This was true, as far as it went; CASB hadn't come within a country mile of determining even the *approximate* sequence of events which led to the crash, which might or might not have been an accident.

''The board believes, however,'' Thurston continued, ''that the weight of evidence supports the conclusion that shortly after lift-off, the aircraft experienced an increase in drag and reduction in lift which resulted in a stall at low altitude from which recovery was not possible. The most probable cause of the stall was determined to be ice contamination on the leading edge and upper surface of the wing. Other possible factors such as a loss of thrust from the number-four engine and inappropriate take-off reference speeds may have compounded the effects of the contamination.''

The wordsmithing had puffed the cause statement from 53 to 110 words. The stall had become the sole certainty. Flight-crew fatigue was no longer a contributing factor, and inappropriate takeoff reference speeds were now just *possible* factors. The possibility of low thrust had made a last-minute cameo appearance.

There was nothing to say. They were, after all, professionals. Bobbitt, Mussallem, Stevenson and I voted against; the others voted for.

* * *

There remained the matter of a dissenting opinion. Some members argued that there should be no dissent, but our lawyer, Rowland Harrison, had recently written Chairman Thorneycroft outlining our rights as board members. It wasn't a propitious time to rescind recorded decisions; there were other ways of dealing with dissenters. Thorneycroft said that if a dissenting opinion was on his desk by the morning of November 15, it would be included in the final report.

We haggled over the conditions. The board's staff would provide typing and editorial support; the dissenting opinion would be produced to the same editorial standards as the majority opinion and would be contained in a section, equal to "Factual Information" and "Conclusions", and be included equally in the table of contents. There the chairman drew the line.

No, there would be no mention of a dissent in the summary. No, we couldn't have two more weeks. "If you were going to do another investigation, you would need more time," he said. "But you are not investigators. You have two weeks."

That wasn't a lot. Dave Mussallem had participated in this final meeting by conference call, and Norm and Ross and I went straight to my office and called him. We agreed right away that we were going to have a single dissent, and that we were going to meet Thorneycroft's deadline. We also agreed that neither crew error nor ice had had any bearing on the crash. But a dissenting opinion couldn't just deal with what hadn't happened. We knew that the power had been down on number-four engine before the crash, and that the number-four thrust reverser had been deployed, but the other three engines had not been examined adequately to tell what shape they were in. None of us believed that the origin of the crash had been an engine problem.

So what had been the initiating event?

Stevenson was certain that it had been an explosion of some sort, maybe more than one.

"What about you, Norm?"

"Explosion in the cargo hold."

"Dave?"

"Yeah, it must have been an explosion in the cargo hold."

That was what I thought, too. It was agreed. I'd write the minority opinion, checking with the others by phone as necessary. We'd use the pictures Ross had got from the RCMP. Our report would be on Thorneycroft's desk in two weeks.

Double Cross

November 1, 1988. The International Society of Air Safety Investigators had just opened its nineteenth international seminar at Le Meridien Hotel in Vancouver. Ken Thorneycroft was giving the keynote address on the theme "Accident Investigation and Prevention — Working Together".

It was a fine theme, he said. He believed we must "work together in agreed-upon ways, otherwise our different and varying interests will tend to cause us to be secretive and competitive, rather than open and cooperative." His speechwriters had resurrected one of Bernard Deschênes' favorite chestnuts: "The basic guiding principles of an accident investigator, or an accident investigation agency, must be integrity, competence, openness and fairness."

Dave Mussallem, Ross Stevenson and I were in the audience, but we hadn't come to listen to recycled platitudes. Over the weekend I had pulled together a crude draft of our dissent on the Gander report. We would be seeking advice from experienced investigators before finalizing our minority report.

I was particularly interested in the reaction to the photos Stevenson had obtained from the RCMP archives, especially the one showing the section of fuselage with the smoke ring around the open window. I showed the photos to experienced investigators without telling them where they had come from or what I thought they might signify. Several guessed at once that they were from the Arrow Air DC-8. No one dismissed them as prosaic. On two separate occasions, the picture with the smoke ring elicited the same first reaction: "Holy shit!" The invariable second reaction was, "I need to look at the actual piece" — and then incredulity that the item had not been examined in the lab.

One investigator mentioned that Jerry Lederer had been talking about a bomb that was supposed to be on board the Arrow Air DC-8. This was indeed news.

Jerome F. Lederer was the legendary dean of aircraft accident investigators. He had started his career in aviation safety in the pre-Lindbergh era as an aeronautical engineer with the U.S. Air Mail Service, and gone on to become the first director of the Safety Bureau of the U.S. Civil Aeronautics Board. As commercial aviation developed, Jerry Lederer shared in just about every institutional effort to improve aircraft accident investigation. In the wake of the Apollo I fire in 1967, Lederer was induced out of retirement to set up the Office of Manned Space Flight Safety at NASA. Here was someone who might impress even the CASB's "professional investigators".

Although I was familiar with Lederer's papers on accident investigation, I had never met him. But he wasn't hard to spot. He was the smiling little white-haired leprechaun bouncing all over with questions and comments.

The second morning of the seminar, there was a long line waiting for breakfast at the hotel cafeteria. Stevenson and I walked to a little restaurant down the street, and we spotted Lederer at the counter. Ross insisted we go right over and introduce ourselves.

Lederer looked over the pictures of the wreckage. "Have you spoken to Irving Pinkel about this?" he asked. Well, we'd read Pinkel's report, but we hadn't actually spoken to him. "You really ought to talk to Irv Pinkel," Lederer said. We tried to keep the conversation going, but he just repeated, "You talk to Pinkel." There was no doubt he thought we were on the right track.

But there seemed to be a problem back at the office. Norm Bobbitt had stayed behind to get copies of photos and to keep his eye on the production of the majority report. He reported that the photos had not been delivered as promised. The proof copy of the majority report was back from the printer but the table of contents had not allowed for a section on the dissenting opinion. Indeed, the report contained no hint that there might be a dissenting opinion.

I found Thorneycroft during the coffee break and reported Norm's call. "What do you want me to do?" he asked.

"Call Johnson and tell him to get Bobbitt the cooperation he needs so that we can hand over our dissent by your deadline." Thorneycroft agreed to make the call. I told him I was also concerned that the preliminary copy of the report made no accommodation for the dissent.

We were in the middle of a room crowded with people trying to balance coffee cups, eat doughnuts and converse at the same time. We had to stand close to hear each other over the din. "I am committed to include the dissent as a section of the final report," Thorneycroft said. "None will go out without it. You have my word."

I thanked the chairman — and paused just long enough to write his exact words in my notebook before calling Norm.

* * *

Back in Ottawa. I called Pinkel in San Diego. He was very gracious and agreed to discuss the Gander case if the principals who had engaged him agreed. He gave me the number of a Wall Street law firm.

Half an hour later, I was speaking to Pinkel again; he had received clearance from New York to discuss his findings without restriction. He was incredulous that there were no clear copies of his photos in the CASB files. He would rush copies directly, while I sent him copies of photos we were thinking of including in the minority report.

Over the next few days, as I talked to Pinkel and other investigators, everything I heard reinforced the minority conclusion. I worked at home, keeping in contact with my co-authors by phone. Our story finally boiled down to fourteen pages of text under three main headings — No Ice Contamination, Pre-Impact System Failures, In-Flight Fire/Explosion — with a dozen pictures to illustrate our points.

* * *

On Monday, November 14 — just in time to meet the deadline — I plunked a copy of our completed dissenting opinion on Thorneycroft's desk. The chairman thanked me and I walked out.

The stage was set, I foolishly thought, for issuing the CASB's report on the Gander investigation. The announcement of the date of release was pending confirmation of the delivery of the completed reports from the printer.

Negotiating the production of our dissenting report had been a bummer, to say the least. Having met Thorneycroft's deadline, we expect some reaction. I even harbored a glimmer of hope that our dissent would provoke further discussion and another round of comments from interested parties. But there was nothing. I told Thorneycroft that I needed to proofread and correct some typos. Sure, he said, call the editor.

A young woman I hadn't met came to my office with the proofs. She seemed nervous and awkward, surprised to see no horns or fangs, as she handed me a typeset copy of the dissenting report.

The page numbers started at one. The pagination would be adjusted just before production, she explained. What about the table of contents? She didn't know what I was talking about — no one had said anything about a heading for the dissenting report. Okay, I'd talk to the chairman.

I looked through the report. The copyediting was fine, but the pictures were half-page black and white. Our strategy had been based on the supposition that the photos would be reproduced in full-page color. This was just temporary, right? The final report would have full-page color photos?

Nope. This was how it was going to be.

I stormed off to Thorneycroft's office. He had repeatedly promised that the dissenting report would be treated on a par with the majority report, and that meant headings and sub-headings in the table of contents, and full-page color pictures.

Thorneycroft was amused to see me. Well, he said, we were going to go with full-page color, but we had to pull back—production costs and so forth.

But the minority report was predicated on the use of color, I insisted. Our pages could be printed separately and bound in. No way, he said. The largest photos in the majority part were half-page black and white, and all he had promised was equal treatment. Well, dammit all, what about the table of contents?

"I'm not about to let your headings in the table of contents." The chairman smiled at my frustration. He wasn't vindictive, he just held the cards. I was mad as a hornet, but there wasn't much I could do. I started to talk about the arrangements for the press conference.

"I presume you're going to be the spokesman for the dissenters," Thorneycroft said. Norm Bobbitt was about to leave on a long-planned trip to Australia, and Dave Mussallem didn't want publicity. "Stevenson and I are both going to be spokesmen," I said. Thorneycroft chuckled. Only one of you can speak, he told me, take your pick. I protested—he had Hinton, Boag, the whole damn public relations office to back him. And we represented almost half the board.

Too bad again. "My job is to sell the majority report," Thorneycroft informed me. "I'm not about to give you guys a leg up with the dissent. I could be bloody-minded about this and orchestrate you right out." Bloody-minded? I didn't know just how bloody-minded he and his cohorts were planning to be.

It was Friday afternoon, and just before the end of the day a short notice was delivered to my office. The chairman had called a special meeting of the board for the following Tuesday afternoon "at the request of Member Thurston". A call to Thorneycroft elicited no further information; he said he didn't know what Thurston had in mind.

* * *

Tuesday morning was taken up by a representation from interested parties on a helicopter accident. Mussallem knew about that meeting but didn't attend. At noon I called him in Vancouver; he'd heard nothing about a special meeting. MacEachern refused to let him participate by conference call and Bobbitt was off in Australia, so Stevenson and I faced the five champions of the ice theory.

Frank Thurston was wearing the lop-sided smile signalling he was about to engineer something he deemed especially clever. He cleared his throat and declared that various colleagues had asked him to speak. He might have to be critical of the chairman, he said sadly, but he hoped to be constructive.

There had been a "serious divagation" of late, he explained, "from what should be good practice." There was simply "no precedent or protocol on what constitutes the chairman's privilege." Thurston objected to the "chairman's decision to dump a twenty-page discursive treatment of what amounts to a separate investigation" into the board's official report, saying this "parallel report has no precedent and has not been subject to the principles governing the official report."

Thorneycroft tried to look as if he didn't know what was up. The others grinned openly.

Thurston went on for a good fifteen minutes before getting to the point: he had two resolutions to offer for the board's consideration.

First, the dissenting report should not be part of the official report. But there can't be any appearance of a cover-up,'' he added hastily, the dissenting opinion should be ''made available to the public as a stand-alone document.''

Secondly, ''the press conference to release the Gander report is to be treated solely as the chairman's presentation of the official board report.''

Thorneycroft cut in, assuring Thurston that he took no offence at this affront to his leadership. Perhaps he had been presumptuous, he said, ''in making it incumbent upon myself to orchestrate the release of the report.'' Gosh, he sure wasn't comfortable about the arrangements, but he felt he should lean over backward to give all sides an opportunity to express their views. He had given his word. But of course, he was only one member of the board; under the circumstances he had no choice but to abstain from the vote.

I didn't believe the unctuousness could get any thicker. Wrong again; Bill MacEachern congratulated the chairman for his forbearance and statesmanship, saying, ''The mind boggles at the minority report'', which implied the palpable nonsense that ''seven respected entities are either incompetent or have colluded.'' It was unthinkable, he said, that such ''odious and mischievous intervention'' should be attached to an official report of the CASB. Yet he was magnanimous — he wouldn't object to a separate minority report. ''It's so outlandish it can only discredit the authors.''

They smiled and waited for my reaction, a row of Sylvesters who had finally caught Tweety. I began by pointing out that, in fairness, the absent members must be allowed to vote. I'd spoken to both Bobbitt and Mussallem that day and it would be no problem to get their votes over the phone.

The others smirked. But Thorneycroft was well primed. "When we decide on policy, all members must have a say. But this is procedure."

I made one last appeal to due process. Our regulations demanded "reasonable notice" for all meetings, and the absent members had not received it. Don't be silly, they told me. Mussallem had known of the morning meeting, and had he come to that he'd be here now. As for Bobbitt, a notice had been placed in his mailbox the Friday before. The CASB's legal counsel judged this reasonable. Get on with the vote. Stevenson and I refused to participate, so Thurston's two resolutions were "unanimously" approved.

No one had spelled out how the minority report would be "made available to the public as a stand-alone document", but when the majority report came back from the printer I thought I could figure it out. The sole indication of any contrary opinion was at the end of a supplementary section, "Safety Recommendations", sandwiched between the conclusion and the appendices. The last text page mentioned in the table of contents, page 100, was followed by the first entry of the appendix, on page A1. Anyone who stumbled upon page 101, in between, would find only the following:

> This report and the safety action therein has been
> adopted by the Chairman K. J. Thorneycroft, and Board
> Members:
>
> W. MacEachern
> A. Portelance
> B. Pultz
> F. Thurston

> Members N. Bobbitt, L. Filotas, D. Mussallem and R.
> Stevenson dissented. A report of their dissent is available
> on request from the Canadian Aviation Safety Board.

The back of page 101 was blank. If this single sheet
should become disassociated from the report — for
whatever reason—there would be no way of knowing
that it, or a minority report, had ever existed.

The announcement issued the next day on "Release
of Occurrence Report on Arrow Air Inc. DC-8-63 etc."
implemented Thurston's second resolution:

> The Chairman of the Canadian Aviation Safety Board
> will present the public report, its findings and
> recommendations. He will be available to answer
> questions and for interviews at the end of the news
> conference."

The dissenters and the dissenting report had been
deep-sixed.

Press Conference

When Stevenson and I told Roger Lacroix about the
double-cross on the minority report, he got as mad as
during his worst battles with Bernard Deschênes.
"They're not going to get away with this," he told
us. We next saw him on the evening news, denouncing the unprecedented decision to muzzle dissent in
the CASB.

A surprised CASB public relations officer confirmed
Lacroix' story, saying that the decision to withdraw
the minority report "was taken because of the different kind of opinion being expressed by the four dissenters." But when the news item propagated and
Members of Parliament and the Canadian Air Line
Pilots Association denounced the decision, a CASB
spokesman issued a clarification: "No one is being

muzzled, anyone is free to discuss their opinion." Thorneycroft even sent us a memo "in order to dispel any misconceptions." If we wished to attend his press conference we might do so. Also, "Members may respond to queries from the media following the formal Press Conference."

Everyone knew that the majority attributed the crash to ice on the wing, and the reporters knew which members were against the theory, but they didn't know the basis of our dissent. Rowland Harrison had sternly warned us against disclosing our position until the official release. It was good advice and we followed it religiously. Nevertheless, leaks and speculations abounded. By December 7, the day before the scheduled release, both the majority and the minority report had been leaked.

December 7, 1988 — D-day minus one. Ross Stevenson was waiting to review some pictures from the files, but they still hadn't arrived. He called the chairman to complain but Thorneycroft wasn't available. Neither was Hinton. Nor Boag.

It turned out that Thorneycroft, the majority board members, Hinton, Boag and their camp followers were holding a dress rehearsal for the press conference. Ever since Thorneycroft's disastrous debut on "The Journal", he'd been taking lessons on how to strut his stuff. A firm of media wizards had been engaged to mastermind the campaign to "sell the majority report". The gurus and spin doctors were taking Thorneycroft, Boag and the others for a final check ride before the real trip.

* * *

The next day, Thorneycroft's handlers had arranged a two-hour "lock-up" before the press conference, so that reporters could read the report and talk to the investigators in preparation for the main event at eleven a.m. Ross Stevenson went to observe. I'd join him in time to hear the chairman.

But when Stevenson tried to join reporters and staff in the lock-up, the media counselors wouldn't let him in. Ross marched to Johnson's office and demanded to know on what basis he dared to block him from a board function. Hey, there must be a misunderstanding. Johnson escorted him back and he was allowed in.

Thorneycroft had told the press that the minority report would be readily available, but just to make sure, I was bringing extra copies. For good measure, I also took a bunch of copies of our "Critique of the Ice Contamination Theory" in a canvas satchel. As I turned a corner in the corridor, walking lopsided with my satchel, I just about smacked into Boag and Hinton. "Look," I said cheerfully, indicating the bulging bag, "I'm already packed—in case your plan works."

I found a chair near the back of the room. Stevenson was somewhere near the front. I could hardly see the podium through the forest of tripods for the TV cameras; it bore the letters C.A.S.B in black over a Canadian flag. Bits of wreckage, photos of the crash site and a model of a DC-8 served as visuals. Hinton, Boag and Thorneycroft's senior communications adviser were seated at a table to the left of the podium, under a four-foot photograph of a DC-8 in flight. Hinton's jacket was bunched up and his bald dome glinted under the bright lights, but Boag was a picture of elegance and composure.

The senior communications adviser welcomed everyone and spelled out the ground rules: the chairman would speak, and then only one question and one supplementary each, please. Thorneycroft entered stage right, accompanied by an aide.

He wore his usual grey suit, but with a light blue shirt instead of white. He cut a good figure, looking confident; the only thing missing was the band playing "Hail to the Chief". He read a statement describing the circumstances of the crash and the review of the evidence, as seen by the majority. No recycled platitudes about "integrity and openness" today — this was a well-crafted, well-rehearsed statement.

Then he took the bull by the horns—sort of. "It has been reported in the media that not all board members are in favor of the final report," he said. "Five members are in favor of the report and four do not accept at least parts of it." Sure—and Margaret Thatcher and Ronald Reagan didn't accept "at least parts" of the Communist Manifesto. Who was he trying to kid?

But I had to admit that he did a good job—not flashy, but serviceable. He praised the staff, "trained experts who have the experience and knowledge". The possibility of sabotage was investigated "by the experts, the RCMP", who gave the matter "a completely clean bill of health". He stressed that the majority report was "passing a very important safety message to the aviation community" about the danger of trying to take off with a small amount of ice on the wings. "The NTSB," he remarked, "has commented most favorably on the investigation."

A number of questioners goaded the chairman to take potshots at the dissenters. "It has been suggested to me by some of the investigators," one reporter remarked, that three of the dissenting board members were "aggrieved" and "trying to wreak some kind of revenge." Thorneycroft wouldn't take the bait.

When the questions to the chairman were over, I waited for some reporters I had promised to see. But a crowd surrounded me, and another surrounded Stevenson. After some minutes of confusion, the reporters decided among themselves that it would be most efficient to question us together at the microphone Thorneycroft had just vacated.

I wasn't really prepared for this, but Ross was pumped up. He started answering questions at once, blasting away with masses of detail about the evidence — the bug ring, the witnesses, the meteorology . . . all extemporaneously. It sounded sincere and knowledgeable. For the most part I just stood there like a hood ornament. But I was well placed to observe both Stevenson and the questioners, and it seemed to me that he was doing as well as Thorneycroft.

The *Globe*'s Paul Koring still had this thing about ice on the windshield. Stevenson tried to explain the DC-8 heating system, but Koring wouldn't settle for "it depends". Stevenson finally said ice would form on the windshield before it formed on the wing, and Koring's story the next day quoted the reaction of a CASB investigator — Stevenson, with 30,000 hours as pilot in command, was "lucky to be as old as he is". It was about as thoughtful a reaction as any critic of the ice theory ever got.

Of course, a press conference is not the place to sort out a complex technical controversy. To get to the bottom of the Gander crash, we needed a new inquiry — one where evidence could be taken under oath and cross-examined, where professional credentials could be established and challenged. I had made one blunder in preparing our dissent: we had agreed to include a recommendation for a judicial inquiry, but in the rush I had inadvertently left it out. Nevertheless, I was reluctant to join Stevenson's clamoring for an inquiry.

This was partly because opposition Members of Parliament had embraced the concept of a judicial inquiry as a means of hounding the government over what they termed "a major cover-up in the Gander crash affair".[5a] If there was indeed a cover-up, the government certainly wouldn't initiate a judicial inquiry. If there wasn't, the government was still unlikely to heed a call associated with taunts and jibes from the opposition.

In any event, the government would have time to see which way the wind blew. The CASB Act allowed it ninety days to respond to the findings and recommendation. Benoît Bouchard's office announced that a decision would be taken after studying both the majority and the minority report.

* * *

The split decision of the CASB made a good story, but fate conspired to overshadow it. A devastating earthquake took some 50,000 lives in Armenia, and "The Journal" ran a special on that country instead of rerunning its Gander documentary as planned. Reporters on aviation and military matters had to cover the story of a U.S. fighter that ploughed into an apartment building in West Germany killing five and injuring forty—it was the twenty-second crash of a NATO aircraft in West Germany that year.

Most accounts of the Gander report followed the "one side says this, the other side says that" format. Our minority conclusions were generally labelled "speculation". One interesting exception was in the speciality publication *Counter Terrorism Security Intelligence*, which cited unnamed intelligence sources as supporting the hypothesis of sabotage.[5b] Despite the obstacles to the minority and the resources promoting the majority, we managed at least a draw. But no one was content to leave it at that.

Thorneycroft's press conference was only the opening shot in his campaign. Later the same day, Boag and Hinton held a private briefing for key client organizations — Air Canada, the Canadian Air Line Pilots Association, the Air Transport Association. The purpose was to head off a public inquiry, as I found out from a participant who called to tell me of Peter Boag's "anti-dissent tirade".

The briefing was not a complete success. In fact, it boomeranged with the Canadian Air Line Pilots Association, which wrote the prime minister to recommend "an independent audit of the Board's structure, procedures and staffing." CALPA wanted the prime minister to know that "our experts reject the 'official' icing supposition as unproven and cannot accept the speculations of the 'dissenters'." The pilots suggested that "it may be the very quality of the investigations that is leading to the split on the Board".[6]

The American Air Line Pilots Association was less diplomatic. "What the Board did was to conjure up an icing theory immediately after the accident and perpetuate it to the exclusion of other theories," Harold Marthinsen said, adding, "I was appalled by the bias in the investigation."[7] Marthinsen didn't comment on the minority position, but said he suspected the number-four thrust reverser.

For the next phase of their public relations offensive, the majority and their backers returned to the scene of the crash. We learned from the newspapers that Peter Boag had spoken at a public forum in the Hotel Gander "to allay certain myths".[8] These myths, as

Boag saw it, included any theory contrary to the dogma of ice. But he would not stoop to dignifying heresy with specific rebuttal; "These other points of view are not supported by the facts," he assured a crowded room of Gander residents. "As investigators, we deal with facts, we cannot speculate."

Members of the audience loudly demanded to hear the other side. Again, Boag's performance had bombed.

* * *

On December 22, 1988, the big headline in the *Globe and Mail* — "258 killed as jet crashes into Scottish town"—brought me an eerie flashback to the headline of December 13, 1985, "258 from U.S. die in Gander crash".

The sense of *déjà vu* continued as I read the story of the in-flight disintegration of Pan Am Flight 103 over the town of Lockerbie: "several witnesses said the jet was on fire before it hit the ground. . . . large number of U.S. servicemen heading home from West Germany for the Christmas holidays. . . . the presence of the servicemen fuelled speculations that sabotage may have been involved."

In contrast with Gander, the possibility of sabotage was not dismissed out of hand. Michael Charles, "the top British investigator on the scene", commented on the "fragmented and scattered nature of the wreckage". By the second day after the crash, the papers were reporting "investigators increasingly focusing on the theory that a terrorist bomb had exploded in the front cargo hold just below the flight deck, instantly crippling the plane".[9]

During the next days and weeks, news reports revealed that tens of thousands of items of baggage were to undergo laboratory analysis and thousands of witnesses were to be interviewed. The wreckage was painstakingly reconstructed on a giant scaffold. The difficulty of detecting traces of plastic explosives and the poor security at West German airports were widely noted. We learned of a detonator so small it fit into the lock of a suitcase, of a bomb that was "extremely thin and looks like computer printout paper so that it is easily concealed".[10]

I was struck by the fact that both the Arrow Air and the Pan Am flights disintegrated shortly after takeoff from the first stop after an original departure from West Germany; the flights could have been destroyed by a bomb triggered by a timer set ticking by a full compression-decompression cycle.

While speculation about specific causes was premature, the investigators' approach stood in remarkable contrast with the response at Gander. Who could fail to note the contrast with the Gander investigation, where circumstantial evidence was dismissed out of hand, where only three or four pieces of wreckage were taken for forensic examination, where the investigators couldn't say what had been loaded into the cargo hold and made no effort to analyze security measures or reconstruct the airframe?

* * *

With the Gander report out at last, Ross Stevenson could catch up on some delayed holidays, so for the next few days I took all calls to authors of the minority report. One was from a Mrs. Phillips in St. Petersburg, Florida, the step-mother of one of the Gander victims. Sergeant James Douglas Phillips, Jr., had been twenty-three years old when he died. For three years, the U.S. Army had been telling Zona Phillips and her husband that the Canadian investigation would provide a full account of the circumstances of their son's death.

Since the CASB didn't rank passengers' next-of-kin as "interested parties", the Phillipses had been left in the dark about the progress of the investigation. Now they had heard a brief news report about the split verdict. What did it all mean?

Zona Phillips was an engaging, sympathetic person, and it was clear that the crash weighed heavily on her and her husband. She talked of efforts to form a support group with other families who had lost loved ones at Gander; for some reason she couldn't fathom, the army had discouraged these efforts. Nevertheless, some two dozen families had kept in touch. They wanted to know everything about the investigation.

I outlined the majority and minority positions and offered to send copies of the reports and newspaper clippings. I wasn't sure what else I could properly send, but I told Mrs. Phillips that, as an interested party, Theresa Griffin had access to a wealth of material. I could also send along some information on the public-interest groups who had helped secure information for relatives of the victims of the 1983 Korean Airlines Flight 007 disaster.

Mrs. Phillips sent a poignant letter thanking me for the material. "The families that I am in contact with are very upset," she wrote. "We have waited very patiently over the past three years for some information that we felt would be accurate and precise. We had a high degree of faith that the Canadian government would leave no stone unturned and that nothing would be covered up. . . . We all feel that it simply can not end this way. . . . an incomplete finding for such a senseless and tragic event. . . . our grand-daughter will never know her daddy and experience his unconditional love. . . . ''

Many families wanted to know how their loved ones had died, she added. What could they do to complete the investigation? Could I send copies of the reports to the others and let them know what they could do? Zona Phillips enclosed a list of about twenty-five names and addresses.

What could I say? The CASB claimed it would reopen any investigation to consider new evidence, but it was unlikely anything further would be done about Gander. Many Canadians were urging their government to order a judicial inquiry, but so far the government had shown little interest, possibly because the crash involved an American airplane and American citizens.

By this time the United States General Accounting Office had informed the CASB that it was investigating ''certain issues'' relating to the Arrow Air DC-8 crash at the request of a United States senator;[11] maybe the families could get help from their own government. I mentioned this hope in a short letter I enclosed with the copies of the majority and minority reports I sent to the other families.

By now, too, the possibility of sabotage had been made much more plausible by comparisons with the disintegration of the flight over Lockerbie, so I also wrote that ''As an author of the minority report, I believe that the official investigation did not adequately pursue the possibility of deliberate sabotage.''

But one of the letters could not be delivered. It was returned to the CASB, where someone in the mailroom opened it before giving it back to me. Unbeknownst to me, a copy was sent to Chairman Thorneycroft. Weeks later he would condemn me for writing to the families, calling my letter a "ghoulish step" that was "morally reprehensible if not illegal".

Part 2

Rebound

Battle for the Hearts and Minds of the Media

> What you do in this world is a matter of no
> consequence. The question is, what can you
> make people believe that you have done?
> *Sherlock Holmes*[12]

As the year turned, the proponents of the majority
report redoubled their drive to contain the damage
from the dissenting report. The secret briefings to special interest groups were supplemented by a broadly
based public relations campaign. With full use of the
CASB budget and complete control over the public
relations staff, they had little reason to fear failure.

Their strategy called for diverting attention from the
technical evidence by leaking stories about the alleg-
edly malicious motives of the "Gang of Four". They
found a good listener in the *Globe*'s Paul Koring, who
had cultivated sources among the CASB investigators
while covering the Air India 747 crash for Canadian
Press.

Koring's first story inspired by the media offensive,
a page-one article on January 13, dredged up some
detritus from the earlier attempt to use the media. The
story revealed that there had been a "secret addendum" to the report from Hickling Management Consultants in October 1987, savaging four members of

435

the CASB who were "seriously eroding the morality and credibility of the board."

As far as I could see, Koring's revelation came as a genuine surprise to the whole board. Thorneycroft obtained copies of the four-page addendum. The vitriol was astonishing. Without giving names or citing cases, the consultants claimed that "A minority group of four has attacked the authority of the Chairman, has been disloyal to him, created potential conflict of interest situations for both the Minister and the Board, abused the CASB staff and generally perceive the CASB and its operations very negatively."

Roger Lacroix had not resigned until February 1988, so at the time of this addendum five of us had perceived the CASB's operations "very negatively". Koring's story went on to say that "the four disaffected members issued a separate report and held news conferences to underscore the differences with the agency's professional investigators." Lacroix hadn't signed the minority report, but he was the only one who had held a news conference.

"Two of the four applied for positions as CASB investigators and were judged by the PSC [Public Service Commission of Canada] as unsuitable," the addendum noted. "One served for a protracted period at CASB and departed. One person, who may be a member of the group, was removed as a pilot from two different commercial operations." If so, the government's method of choosing board members was, to be charitable, deficient. But not one of the derogatory descriptions fit any of us.

Why would a respectable consulting firm transmit such unfounded, possibly libellous innuendo to the Transport minister? The only one who might know was Ken Johnson, who had been on the advisory committee directing the consultants' study. But Johnson wasn't saying anything.

In any event, the secret addendum was a red herring. John Crosbie had recognized the inherent conflict of interest; the middle managers who had hired the consultants had done so in the past and could be expected to do so again. Indeed, the study had recommended a blatant explosion of power for these same middle managers. Crosbie saw that this compromised all the conclusions, and asked John Sopinka to take an independent look. The *Globe*'s own stories showed that Sopinka's engagement had mooted the Hickling report and, along with it, its "secret addendum." There could be only one reason for dredging up its remains now — to discredit the authors of the minority report.

The assault continued the next day, taking up the theme that the investigators suffered abuse at the hands of the dissenting members. This abuse, Tom Hinton said, was expressed in the form of "very close and aggressive questioning", and caused "serious morale problems and the resignation of [unnamed] senior investigators."

The *Globe*'s stories were picked up by other Canadian newspapers. When reporters contacted me, I reminded them of Sopinka's conclusion that "the CASB Act as it has been interpreted in practice and by the Hickling Report has fragmented this agency to such an extent that it has crippled its effectiveness." Since then, Sopinka had been appointed to the Supreme Court of Canada. Who, I thought, would dare question his credibility? I soon found out.

Specialists in character assassination by innuendo can find something odious under the most innocuous of stones. Prior to his appointment to the Supreme

Court, Sopinka had been counsel in the Toronto office of Stikeman Elliott. By coincidence our counsel, Rowland Harrison, worked with the Ottawa branch of the same firm. The association garnered a front-page story in the *Globe*.[13]

We had come to know Harrison because he had analyzed the CASB legislation for the Canadian Bar Association while still a professor at the University of Ottawa; Sopinka had been counsel to the Dubin Commission, which had recommended the CASB in the first place. It had never occurred to any of us to think of either lawyer in terms of the firm he worked with. In fact, I hadn't even realized they worked in different branches of the same firm. But according to the *Globe* chairman Ken Thorneycroft knew more than I did.

"Gosh, that seems odd," he recalled saying to himself. Frank Thurston recalled his own amazement: "I was shocked at the time that Sopinka should have been appointed to do the job," he said. "The whole operation was, shall we say, unusual, to be charitable". But no one would come right out and accuse the Supreme Court Justice of influence peddling and the story went nowhere.

Meanwhile, the rest of the Canadian media showed little inclination to explore the issues raised by the country's worst-ever air disaster, and one effort I made to interest a well-known journalist boomeranged badly. John Burns, the *New York Times'* stringer in Toronto, told Ross Stevenson that he would like to do a feature story on the attempted suppression of our minority report. Burns was about to leave on assignment to Moscow, but he had high regard for the *Toronto Star*'s national reporter, Carol Goar, and recommended we interest her in the story.

As it turned out, Carol Goar had close contacts with Bill MacEachern, and her story centered on my guileless admission that I had been appointed to the board because of my friendship with Conservative Cabinet Minister Tom Siddon.[14] She didn't add that Siddon and I had shared a close interest in aviation accident investigation since the Dubin Commission a decade earlier, or that Siddon had been the only active politician to make a submission to the commission. But it didn't really matter. Any contact I had had with politicians had long since ended. And the focal point in the search for the truth about the Gander crash had now left Canada.

* * *

In the United States, Zona Phillips lost no time in urging the victims' families to publicize the split decision of the CASB. Stories about the Gander controversy began to appear in smaller papers across the country, and American reporters started calling.

The CASB's public affairs officers adopted a strategy of dismissing our report as an expression of meaningless malevolence. But the American media were not interested in the office politics of an obscure Canadian agency. They wanted to talk about the evidence. The CASB wouldn't, but we were only too glad to go into details. I tried to focus attention on unanswered technical questions. Who, for example, had examined the burned-out window over the right forward baggage compartment? How could it be that none of the thrust reversers was latched?

I also emphasized the curious similarities with the Pan Am crash at Lockerbie. Stories about Pan Am flight 103 put the minority report in perspective: large airliners are vulnerable to sabotage; Semtex is difficult to detect; the search for evidence of sabotage calls for meticulous laboratory examination of hundreds of thousands of tiny shreds of wreckage. Television coverage of the Pan Am crash had graphically captured the investigators' obsessive care in preserving bits of clothing and upholstery. The painstaking reassembly of the 747 from tiny fragments served as a standard for assessing the casual disposal of the Arrow Air wreckage.

Moreover, the Pan Am investigation had unveiled West Germany as a haven for Middle Eastern revolutionaries, and deficiencies in airport security had been publicized. The stopover of the Arrow Air flight in Cologne suddenly seemed more sinister — as did the fact that both flights were bringing American troops home for Christmas. To complete the picture, it was now beyond dispute that the Pan Am 747 had been disabled by an explosion in the forward cargo hold—the very means of destruction postulated in our minority report.

The echo of our point of view bounced back to Canada in consequence of a letter I wrote to the editor of *Aviation Week and Space Technology*, pointing out a misconception in their brief story on the Gander report.[15] A reference to the similarities between the flights tweaked the editor's interest, and instead of printing my letter *Aviation Week* carried a story, "Safety Experts Cite Similarities Between DC-8, 747 Crashes",[16] quoting Irv Pinkel and Harold Marthinsen's support for the minority position. Being termed "safety experts" by America's foremost aviation trade journal gave our egos a boost, and took a lot of steam out of the campaign to paint us as ill-motivated malcontents.

The next issue of *Aviation Week* carried an editorial urging the Canadian government to convene a judicial inquiry to clear up ''troubling discrepancies'' about the investigation. That really made my day.[17]

The respectability given to the minority report by the *Aviation Week* articles also revived the interest of the Liberal Party. Stymied by their minority in the House of Commons, the Liberals threatened to use their majority in the Senate to hold hearings to ''shed light on the doubts and suspicions raised by unanswered questions about the crash''.[18]

Thorneycroft repeatedly promised that all information would be dispensed freely once the final report was issued. Roger Lacroix and Ross Stevenson took this to heart and provided extensive briefings to researcher Marc Laframboise and other Liberal representatives. (I remained skeptical of political involvement of any stripe, and said so publicly, much to Roger's consternation.) The CASB staff retaliated by inviting members of the Liberal caucus for briefings from Boag and Hinton ''to explain the background to the investigation supporting the Arrow Air report.'' So it was somewhat amusing to be accused of the crime of speaking with Liberals by fellow board members Bill MacEachern and Arthur Portelance. Ironically, the only times I had knowingly communicated with anyone associated with the Liberal Party were when I had spoken to former Liberal politicians MacEachern and Portelance—and then only to try to set them straight in our boardroom.

I took it more seriously when Thorneycroft sent a long appraisal of my performance to the Clerk of the Privy Council Office. His appraisal was right in line with the strategy of focusing on personality and secondary issues.

''I am utterly appalled at the insensitivity he has displayed by communicating with next of kin, and suggesting that they solicit assistance of their Congressmen and Senators in applying pressure on the Canadian Government," he wrote. "This approach, in my view is particularly morally reprehensible and distasteful to the extreme." In fact, the "insensitivity" had been toward his unspoken wishes, not toward the families—who had contacted me in the first place and had replied with uniformly kind calls and letters.

This time Thorneycroft made his aims abundantly clear: "Mr. Filotas has clearly demonstrated that he is incapable of performing his assigned role as a Member of the CASB. He is a serious liability to the Board and an embarrassment to the government. . . ." Thorneycroft would only write such drivel if he was assured of a receptive audience. It was clear to me that he was merely recording the conclusion of whoever really called the shots at our "independent" agency — and that the end of my short and inglorious career as a board member was at hand.

Ironically, Thorneycroft's vendetta placed me in an enviable position. My only hope of redress was through finding the truth about the Gander crash. I was left no choice but to do the right thing.

The Great Impostor

The board meeting of February 17, 1989, finished the scheduled agenda around noon. The chairman said he had another item of new business for the afternoon. "Could you indicate the nature of the item?" I asked.

"Yes," he answered, "I want to talk about the contract for Frank Thurston."

Well, it had been a long time.

About a year before, after Thurston had let slip that he had had a contract to act as a board member when his first appointment had terminated in June 1985, he had immediately regretted his remark and clammed up. But a check of the *Canada Gazette* had shown that he had been appointed to the CASB for a one-year term on June 1984 and reappointed for a three-year term in June 1986. Yet in the intervening period he had continued to act exactly as before — attending meetings, voting on issues, chairing meetings in the absence of Bernard Deschênes and even serving as a presiding officer at the CASB's public hearing on the Gander crash in April 1986.

Oddly, the CASB's annual report for 1985 didn't list the board members' names or the names of the presiding officers at the public inquiry into the Gander crash, nor were the presiding officers named on press releases or most other documents pertaining to the public inquiry.

At the time I had not yet been named to the CASB. Those who had been on the board attributed the omissions to Bernard Deschênes' aversion to sharing the limelight. But a check of board documents seemed to indicate a systematic exclusion of members' names in the interval from June 1985 to June 1986. It was hard to avoid the conclusion that the omissions were meant to mask Thurston's nebulous participation. What if Frank was only pretending to be a member of the CASB during that year? What would that make of his role in the public inquiry and other decisions about the Gander investigation?

We knocked the implications back and forth among ourselves until, in the midst of a tumultuous meeting in October 1988, Norm Bobbitt brought the question into the open. Thorneycroft's opening statement before the final review of the Gander report purported to summarize the history of the investigation, and Bobbitt immediately spotted an omission. "Member

Thurston's name is conspicuous by its absence from the list of members who assisted Chairman Deschênes with the public inquiry," he told Thorneycroft.

Bobbitt reviewed the dates of Thurston's appointments to the CASB and asked for clarification of Thurston's status at the time of the public inquiry, suggesting that the uncertainty of his status raised questions about the fairness of the public inquiry and whether Thurston might have had a conflict of interest.

Thorneycroft was nonplussed. "Communicate your concerns to me in writing," he said.

"Maybe Member Thurston would like to comment on it, Mr. Chairman," Bobbitt replied.

"Mr. Chairman, I have no intention whatsoever of commenting on it," Thurston curtly assured him.

MacEachern decided to put in his two cents' worth. "I find this intervention odious, to say the least. This is a mischievous intervention, symbolic of other attempts that have been made to derail the board and more particularly the Gander investigation. And I repeat my earlier expression—odious."

"I agree," said the chairman.

Tempers flared and I appealed to the chairman to answer Bobbitt's question and put the matter to rest once and for all. He refused, and said again that Bobbitt would have to ask in writing.

Bobbitt did, and got his answer two weeks later. Thorneycroft castigated his "unacceptable behavior" and "seriously deficient judgement" for "insulting fellow Board Members." Bobbitt repeated his request for the information and it was ignored. He raised the subject again at a board meeting in January but Thorneycroft refused to discuss it.

It was then three months since Bobbitt had asked a simple direct question. Canadian legislation stipulated that contracts let by public agencies had to be available for inspection by the public. On the way out of the meeting I picked up an "Access to Information Request Form", addressed it to the CASB, and requested "All contracts and agreements with Mr. F. Thurston of Ottawa effective during the period June 1985 to June 1986." I attached a cheque for five dollars made out to the Receiver-General of Canada and found the staff member who handled such requests. I even waited by his desk to verify that my submission was logged in. The law required a response in thirty days. Now the time was just about up.

When I arrived at the boardroom, everyone else was already seated and a stack of papers had been distributed. A grim-faced Frank Thurston was talking. As I took my place beside Bobbitt, he gestured toward the papers and handed me a note: "THIS IS DYNAMITE."

"I do have to regard this as a pretty gross invasion of my privacy," Frank was saying in a trembly voice. "I resent it and would oppose it if it were possible. . . . If we are going to examine one another's credentials and performance and all the rest of it, let's do it all. Just call in the RCMP if necessary . They might be induced or even directed to look into the various leaks we've had. . . . This transaction is abominable. I believe it calls for reciprocal action. If we are to raise the level of harassment, I consider that I have the right to reply in kind. And I will." Frank breathed heavily. He took his briefcase and walked out.

"This is a sad bloody situation," Thorneycroft said, and called on Ken Johnson to "walk us through the sequence of events, as best as we can reconstruct them, from the time that Frank became a special adviser to the present time."

Special adviser? What's this special adviser?

Well, for the first and only time during my tenure as a member of the CASB, we received an adequate answer to a sensitive question. The package of papers traced the machinations behind Frank Thurston's reappointment to the board. Evidently it had been prepared for someone else, and that someone else must have told Thorneycroft and Johnson that they had no choice but to hand over the complete package.

It turned out that Frank Thurston had not, in fact, had a contract between June 1985 and June 1986. As the end of his first term approached, Bernard Deschênes had tried unsuccessfully to secure his reappointment by the Cabinet. The memos and letters tracked the maneuvering and manipulation. They also exposed Ken Johnson's major role. In February 1986, seven months after Thurston's appointment to the CASB had expired, Johnson had signed a letter to Transport minister Don Mazankowski on behalf of Chairman Deschênes. "You will recall that Mr. Frank Thurston was appointed as a part-time member of the Board and that this appointment was not extended beyond the one year period ending June 1986", the letter noted. Approval was sought to enter into a contract with Thurston at a per diem rate greater than that which could have been granted by Deschênes.

The contract was "to provide guidance and advice in the review of aircraft investigation reports" — presumably to Ken Johnson. But the minister was slow to respond. The behind-the-scenes manoeuvring became even more frenetic. On one hand, efforts continued to get the Cabinet to reappoint Thurston to the board. At the same time, efforts continued to secure a contact so that he could be paid in the interim at the desired rate. All the while, Thurston attended and chaired board meetings and served as presiding officer at the public inquiry on Gander. The minister's approval was eventually obtained for the interim contract, and Frank was reappointed to the board. The contract was used retroactively to pay for his services between appointments.

So now we knew — Thurston had been, retroactively, a special adviser to Ken Johnson during the time of the public inquiry. It seemed to me that one might legitimately wonder if gratitude for Johnson's exertions to get him reappointed could have influenced his judgement on the ice theory.

I consulted a top-notch lawyer. "The very idea of a retroactive appointment boggles my mind," he wrote. "Probably, everything done when Mr. Thurston acted as a Member of the Board and when he voted as a Member of the Board when he was actually not a member (during approximately a one year period) is invalid. However, in so far as the Gander crash is concerned, this may not be of any great impact since it would appear that he was properly reappointed back to the Board by the time the decision on the Gander crash occurred."

In the end, there was nothing we could do. But the bureaucratic machine kept churning. A few days after the meeting about Thurston's contract, my cheque to the Receiver-General was returned with a memo: "We are cancelling the above mentioned request since the requested documents were distributed to all Board Members." Traces of my request under the Access to Information Act had been expunged. It wasn't quite the memory hole at the Ministry of Truth where Winston Smith dispatched inconvenient documents. But it wasn't not a bad approximation.

447

Part 3

Transport Follies

Coordination

"The appropriate Minister shall, within a
period of ninety days after he has been notified
of the findings of the Board . . . reply to the
Board, in writing, advising the Board of the
action, if any, taken or proposed to be taken
in response to those findings and
recommendations."

CASB Act, Sect 23(5)

As the majority and minority factions waged their
public relations battle during the spring of 1989, the
Canadian Aviation Safety Board's schizophrenic
report on the Gander investigation was building into
a king-size headache for Benoît Bouchard. Controversy over the investigation had nagged the minister
ever since he had taken over the Transport portfolio
in April 1988, but he had stuck to the pressline suggested by his officials at the outset and stolidly refused
to comment. Interference from the minister, Bouchard
kept saying, would compromise the board's independence. He would have his say in due course.

Due course was fast approaching. A few days before
March 8, tickler messages would be popping up on
computer screens all over the country, reminding edi-
tors and reporters that the law required Bouchard to

respond by that date to the board's findings and recommendations on the Gander crash.

"Benny" Bouchard, a 49-year-old former high school teacher from the French-speaking town of Roberval in the interior of Quebec, had been named Minister of Transport just in time to announce the replacement of Bernard Deschênes by Ken Thorneycroft. Commentators generally agreed that Bouchard's elevation to the important Transport portfolio was a positioning maneuver for the coming federal election — to show Quebec clout in the Cabinet. But his preoccupation with the political problems of his home province cut down on the effort he could devote to mastering the technical aspects of his new department. To make things worse, the Deputy Minister of Transport, Ramsey Withers, left for a post with the private consulting firm run by Gary Ouellet, and was replaced by a career bureaucrat from the Privy Council Office. Thus both the minister and the deputy minister were new on the job.

By the spring of 1989, Bouchard was beset by more than his share of problems. The government's high-priority push to secure ratification of the Meech Lake Accord demanded more and more attention. The Transport department was deluged with complaints about delays and overcrowding at major airports. Commentators castigated the department's short-sighted policies for training air traffic controllers, and publicity about the near collision of two airliners raised the question of safety versus convenience and economics. The minister was sidelined for a time with quadruple bypass heart surgery. But the work of the huge Transport bureaucracy ground on, and the well-oiled administrative mechanism churned out the minister's responses to the hundreds of accident reports produced each year by the CASB.

The board's reports were formally addressed to the minister, but were routed directly to the Aviation Group, for analysis by technical staff. Aviation Group staffers prepared a response, in the form of a letter from the minister to the CASB's chairman, which bubbled up through the usual bureaucratic chain of command — first the director, then the office of the Assistant Deputy Minister for Aviation and then the Deputy Minister. As a final step, the minister's personal staff reviewed letters for political sensitivities. Those that passed muster were recommended for the minister's signature. In practice, most of these letters were thoroughly routine — thanking the CASB for its good work, noting that the minister had accepted the board's recommendations, perhaps adding that the department had already taken action. Normally such letters could be signed by machine and transmitted to the CASB within the mandated ninety days without ever coming to the minister's personal attention.

Before attempting to draw up recommendations for consideration by the CASB board, staff members went over the possibilities with counterparts in the Transport Department. The two staffs, which often interchanged members through the normal course of advancing careers, generally worked hand in glove.* Recommen-

---

*As John Sopinka observed in his report, "Since the investigators have not been firmly placed under the Board it is more difficult to view them as having been separated from Transport. Investigators depend on Transport Canada for career progression inasmuch as the CASB is not large enough to permit them to rise in the ranks. They may therefore be seen as keeping one eye on their own advancement and therefore tending to soften their criticism of Transport Canada."

dations or wording deemed unpalatable need never see the light of day. And if any of the departmental officials suspected political sensitivity, the minister's staff could be consulted as well. Interactions could take place over the phone or by personal meetings—without leaving a potentially embarrassing paper trail.

The CASB's recommendations often urged Transport to "review documentation", "liaise with the FAA", "strengthen procedures" or "clarify regulations". Such pablum was convenient to both sides: the department could comply with a token gesture, and the CASB could point to another implemented recommendation to show how well it was doing its job. Even when a recommendation looked tough at first sight, there might be less to it than met the eye; the Transport department might have been on the verge of adopting the measure in any event.

All this cooperation and coordination might seem a sensible way to keep safety recommendations flowing efficiently. But soon after I joined the CASB I became aware of a hidden process behind the flow of recommendations. Seemingly trivial wording changes to recommendations proposed by the staff drew fierce resistance. At one point, someone gave the game away by artlessly suggesting that the wording couldn't be changed because "coordination" had already been done with the Transport department.

My efforts to explore the distinction between coordination and collusion would be one of the sins to brand me a dissenter. But the insight I gained from thinking about the interface between the two staffs would come in handy when it became necessary to read between the lines of the minister's response to the Gander report.

* * *

The CASB Act did not distinguish between majority reports and minority reports, which were rare but not unprecedented. Ross Stevenson and I had written a minority report, for example, after serving as presiding officers at a public inquiry into the ditching of a twin-engine turboprop aircraft into the ice-cold waters near Toronto Island airport in January 1987. To us, the official cause that an experienced, competent pilot "undertook the flight with insufficient fuel" was not particularly enlightening. We thought it significant that "the pilot believed that his job would be jeopardized if he did not run the tanks nearly dry when the aircraft was to be left for long-term storage after a short flight." We also noted that "Transport Canada's surveillance and audit procedures allowed safety margins to be eroded in the operation of this company"[19] (The Minister of Transport dealt with our findings by ignoring them.)

Bouchard's office had distanced him from the uproar over the alleged attempts to suppress the minority report on Gander by assuring reporters that the minister would comment on both reports after his officials had studied them. Whoever the nameless officials might be, their labors would be conducted under the authority of the head of the department's Aviation Group, Assistant Deputy Minister Claude LaFrance.

LaFrance and Thorneycroft were old comrades — first as generals in the Canadian air force, later as senior executives in the Transport department. LaFrance had spent thirty-four years in the air force before retiring as a major-general in 1981. Between 1981 and his appointment as Assistant Deputy Minister of Transport for Aviation in October 1985 he handled a variety of sensitive assignments in the civil service. A framed letter of appreciation from the prime minister proudly displayed in his study attested to LaFrance's pride and success in this part of his career. So when the controversial reports on the Gander accident landed on his desk in December 1988, he was well aware that he was "expected to lean over backward to accommodate the technical analysis to the concerns of the political bosses"[20]

It might be supposed that he would discuss such a sensitive balancing act with his boss, the deputy minister. But LaFrance's old boss, former chief of defence staff Ramsey Withers, had been replaced by a former ambassador and deputy secretary to the Cabinet, Glen Shortliffe. LaFrance said he neither expected nor received specific instructions from Shortliffe. Instead, he called together his best analysts and told them that the Aviation Group "needed to look into [the majority and minority] reports in some depth, because of the complexity and the dissenting element." The time bomb set ticking by this apparently conscientious attention to duty would cost LaFrance his job.

So, after the release of the Gander report, behind the scenes, out of the public eye — while the CASB staff was holding meetings and briefing journalists, while opposition politicians were denouncing the government's cover-up, while I was corresponding with the families of the victims — LaFrance's group in Transport Canada set out to analyze the widely divergent views presented in the majority and minority reports.

* * *

On March 7 — the day before he was due to respond to the CASB's report — Bouchard attended a meeting of the "inner Cabinet" at the prime minister's Meech Lake retreat in the Gatineau hills north of Ottawa. As usual, a throng of reporters skulked outside, in hopes of a scoop or at least a good quote before the departing ministers were stuffed into the waiting limos.

Usually the pickings were slim, but that day there was a rare nugget. An impromptu question to Bouchard about the Gander investigation elicited the response that the minister had "totally accepted" the conclusions of the majority report and "totally rejected" the minority report. Bouchard had scooped his own press conference, scheduled for ten a.m. the next morning.[21] The premature revelation set the stage for a wild week of confused, overlapping events, and ensured that the Gander investigation would stay in the headlines for months to come.

Chairman Ken Thorneycroft had thoughtfully scheduled a press conference of his own following the minister's, to give journalists time to hustle across the bridge to CASB headquarters. But Bouchard's premature revelation forced a change in the battle plans. Not only that, but—like a flight being put off by successive half-hour delays — Bouchard's press conference kept getting put further and further back. Officially I was told nothing about either Bouchard's or Thorneycroft's plans, but throughout the day reporters kept calling in the latest postponement. Those with deadlines were getting frantic and asking for a reaction ahead of time.

Finally, at about four in the afternoon, the minister was ready. His strategy was based on the narrowest possible interpretation of the CASB Act. He said nothing about the findings, but confined his remarks to the recommendations. The majority report proposed three: that the Transport department sponsor a safety campaign about the dangers of ice; that it establish more detailed de-icing procedures; that it establish better procedures for checking flight data recorders. The minority report had findings but no recommendations, since I had forgotten to include our recommendation for a judicial inquiry.

Bouchard simply announced that he was endorsing the CASB's recommendations. Yes, pilots should be aware of the dangers of ice on the wings, and his department would continue to ensure that the dangers were understood.

In answer to reporters' questions he still professed support for the ice theory, but somewhat more cautiously than in the remarks he had tossed off the day before. There might not be "total evidence", he allowed, to support the icing theory, but he was prepared to support it anyway. He judged the minority report by more demanding standards, saying that it had failed to substantiate the hypothesis of an onboard explosion with the "absolute evidence" needed to order a judicial review.[22]

Copies of his response were delivered to the members of the CASB at 5:15 in the afternoon. Chairman Thorneycroft met the press shortly thereafter, declaring himself well satisfied. With the threat of a judicial review lifted, he could loosen up. He expected the four dissenters to continue their "guerrilla warfare" to reopen the investigation, he said. "The dissenters will never be satisfied. They are using Gander as a vehicle to voice their objections to the way we function."[23]

What was he going to do with the dissenters, now that the minister had rejected their foolishness? "It would be nice to cut the cancer out," he said, "but I don't think that's possible, quite frankly." When the new legislation came to replace the CASB, the dissenters "can try to get on." Thorneycroft smiled. "But

I don't think some of them will make the team.''[24] The Gander controversy had, it seemed, been laid to rest.

But not for long. While no one in the media seemed to notice, the stalling over the press conference and the cautious wording of the minister's prepared statement signalled heavy-duty doubt and gaping schisms in the Transport department.

Tom Hinton and the professional investigators had pulled out all the stops to "coordinate" a ringing declaration of the insidious dangers of tiny amounts of "ice contamination". Only a high-profile campaign by the Department of Transport to alert pilots of this peril would persuasively endorse their conclusion. But the minister's support was, at best, lukewarm. "There is no record of an accident to a Canadian-registered DC-8 that has been attributed to ice contaminated wings in the last 20 years," said his written text. "Transport Canada officials have reviewed the detailed procedures that are required. . . . These procedures are effective in ensuring that flight crew can detect the presence of ice on the wings. . . ." In short, the Transport Canada officials were thumbing their noses at the ice theory — and from within the bowels of the department, someone was sending a cryptic message.

* * *

Two days later — Friday, March 10 — a Fokker F-28 operated by Air Canada's regional subsidiary Air Ontario crashed into a wooded area off the end of the runway while trying to take off from the northern Ontario town of Dryden. The aircraft broke into three sections and caught fire. Twenty-two of the sixty-nine on board, including three of the four crewmembers, died. It was the worst accident to a scheduled flight in Canada since the Cranbrook disaster eleven years earlier. The CASB launched an investigation.

That evening, I received a call from an acquaintance who had once worked for the Transport department. My friend had some interesting information related to the Gander investigation, something I had to see for myself. Could we meet tomorrow, at an office in suburban Ottawa?

My confidant was waiting at the door and ushered me into a deserted room. A floppy disk was loaded into the drive of a computer and a few keystrokes brought up the title page of an internal Transport Canada review of the CASB's report on the Gander investigation.

"You can take all the time you want to look, but don't ask how I got it. And I can't allow you to make a copy, either." I spent the next two hours reading a detailed critique of both the minority and majority reports. It was thorough and obviously genuine. Its scathing criticism of the ice theory and the work of the CASB investigators led to an inescapable conclusion — the CASB investigators had embraced the ice-on-the-wing theory at the outset, and then stretched supporting arguments, often past the breaking point, and ignored all clues and evidence pointing in other directions. I was surprised to see it put so baldly. The "coordination" between the CASB and the Transport department had evidently broken down.

To be sure, the Transport Canada report was also critical of our minority report, but these criticisms seemed more like legalistic nitpicking; they objected, for example, to our conclusion that the crew had carried out its duties without apparent fault, pointing out that the crew had under-reported the takeoff weight by a few thousand pounds. I realized that this was technically ·correct and that our conclusion should have repeated the wording from the text, namely that no act or failure to act by the crew had contributed to the accident.

How could the minister have been induced to endorse the ice theory in face of such a devastating report by his own officials? Could he have been misinformed, or had he acted deliberately? Who had advised him? My friend professed not to know. I did learn that the report had existed in various drafts for several weeks and that its contents had been commonly known and accepted by technical staff throughout the department — and that a few days before Bouchard's press conference all printed copies of the report had been collected. Apparently, those responsible for the recall had forgotten about copies on disk.

The report was political dynamite. I wondered how it could be publicized. But there was no way I could get a copy, and I had been sworn to such secrecy that I didn't even tell the others on the minority side.

Two days later I was still mulling over what to do when I received an excited call from Mark Kennedy of the *Ottawa Citizen*. Kennedy asked if I was aware that Transport Canada had produced a long internal report on the Gander investigation. "Uhmm," I said. He then read several passages that I recognized as being from the text I had read off the computer screen. Mark had obtained some extracts from the report but was wary of a setup. I didn't tell him or anyone else that I had read the complete report, but I did tell him that I was morally certain the passages were genuine and not taken out of context.

The next day, March 15, Mark Kennedy's headline story revealed that an internal Transport Canada review "blasts board investigators for mismanaging" the Gander investigation and "focusing on a preconceived icing theory at the expense of other possible causes." The story revived the controversy and propelled it to record heights.

Flap over Gander Crash Cooks Deputy's Goose (Headline from *Edmonton Journal* March 22, 1989)

Benoît Bouchard's flat-out rejection of a judicial inquiry into the Gander affair averted Chairman Thorneycroft's worst nightmare, but the hidden subtext of the minister's announcement infuriated Tom Hinton, Peter Boag and the other architects of the ice theory. They had called for an all-out war on ignorance about the peril of ice on the wing. Rejection of this call must be the result of enemy plotting, and the ringleader of the enemies had to be former colleague and fellow "professional investigator" James Stewart. As Director of Safety Programs, Jim Stewart perched one rung below Claude LaFrance on Transport Canada's bureaucratic ladder. Everyone at the CASB knew Stewart was in charge of drafting replies to the Safety Board's recommendations and launching them *en route* to the minister's office.

In many ways, Jim Stewart's career had paralleled that of Peter Boag. He had joined Transport Canada's Aviation Safety Bureau after leaving the air force in 1982, at the same time as Boag. He was a few years older and his flying experience was considerably more extensive, but as new civil servants the two men were rivals for positions in the accident investigation agency about to be created in the aftermath of the Dubin Commission.

Both men were bright, articulate and ambitious, but Jim Stewart was more independent-minded and less inclined to deference to authority. This did not sit well with Ken Johnson and Tom Hinton, who were drawing up the organization charts. So when the CASB was formed in 1984, Stewart remained with the Transport department while Boag tied his fortunes to the new agency. They both prospered. But Stewart rose higher, to a directorship, which, in bureaucratic terms, was equivalent in rank to Boag's boss, director of investigations Tom Hinton.

The CASB investigators knew, more than anyone else, that the devastating Transport Canada critique of their work had been done under the direction of Jim Stewart. The paranoia that drove them to assail the motives of any critic now turned on Stewart—and the crash of the Fokker F-28 at Dryden provided them an opening to strike at his perfidy.

* * *

Investigators from the nearest CASB regional office at Winnipeg rushed to the crash scene at Dryden, but only to hold the fort. This was a big accident, "high profile" in the parlance of the CASB, and it called for professionals from the board's headquarters.

The headquarters contingent arrived the next day, Saturday March 11. While the confusion and organizational effort paled in comparison to that at Gander, the situation still called for major mobilization. The investigators, including Peter Boag, worked at a feverish pace over the next several days.

Surviving passengers and other witnesses had noted a heavy coating of wet, fluffy snow on the F-28's wings before takeoff. Snow on the wing? There you were — had Transport Canada not shrugged off the warnings about the dangers of wing icing, the Dryden accident might not have happened.

No one had observed a thick layer of heavy, wet snow on the wings of the Arrow Air DC-8. Nor had the Arrow Air DC-8 lumbered to the very end of the runway, but had lifted off normally with runway to spare. The wreck of the Air Ontario F-28 at Dryden looked like the aftermath of a crash on takeoff, not like "what you'd expect if you threw a grenade and the pieces flew everywhere." But this was no time for nitpicking, this was an opportunity to take the offensive. On March 15, the day Mark Kennedy broke the story of the Transport department's internal report in the *Ottawa Citizen*, Paul Koring reported in the *Globe and Mail* that CASB investigators "had identified wing icing as the prime suspect" in the Dryden crash and were linking the accident with the Arrow Air crash at Gander.

The unnamed investigators denounced Transport Canada for failing to launch the anti-ice campaign they had demanded. They could even name the prime culprit — "James Stewart, who as Transport Canada's director of flight safety programs is responsible for drafting the mandatory replies."

The next day, most news reports about the CASB focused on the leaked transport department report. Parliament had recessed for Easter and Bouchard was vacationing in Florida. "Bouchard's embarrassed officials acknowledged that the leaked report had failed to reach the Minister's desk," said a *Toronto Star* editorial, and the *Ottawa Citizen* quoted Bouchard's spokesmen as saying that the minister would "take another look" at the possibility of a judicial inquiry.

The *Globe and Mail*, however, continued to highlight Transport Canada's dereliction of duty. When accepting the recommendations of the majority report, Bouchard had said that all airlines had been warned about

461

the dangers of ice. But the *Globe* had learned that the warning had not reached Air Ontario until after the crash at Dryden.

"CASB officials who spoke on the condition that they not be identified, suggested that the Air Ontario accident, should it be found to be caused by wing icing, would have been preventable if Transport Canada had not fumbled CASB's call for warnings." The proviso "should it be found to be caused by wing icing" was evidently superfluous; the suspicion of ice had grown into the certainty that the "Fokker F-28 jet crashed on take-off with ice encrusted wings."

The pilot would have been no less aware of the risk of taking off with ice on the wings than of taking off with alcohol in his bloodstream; there was little reason to suppose that additional warnings would have changed anything in either case. The *Globe*, however, seemed more taken with the political than the practical. The minister had said that a cautionary letter had been sent to all airlines on the first of March. In fact, the letter had only gone to national carriers then. Regional carriers such as Air Ontario had not been notified until March 16. Paul Koring, apparently overdosing on the theatrics of parliamentary question period, was determined to get to the bottom of this outrageous misrepresentation. And for all the investigators' animosity toward Jim Stewart, he was not a big enough fish. Assistant Deputy Minister Claude LaFrance must answer—and LaFrance's superiors left him twisting slowly in the breeze.

In what Koring described as a "testy exchange with reporters", LaFrance, ever the good soldier, fell on his own sword. "We provided the wrong information to the minister," he confessed. The "testy exchange with reporters" was, in fact, a testy exchange with reporter

Paul Koring—the product of that reporter's persistent efforts to trap LaFrance into embarrassing his superiors.

To Claude LaFrance, the delay in sending a redundant letter to Air Ontario was a "bureaucratic error". The mail had been slow. But failure to analyze the Gander report would have been an inexcusable dereliction of duty. His people always analyzed CASB reports; that was the way the ball started rolling on the minister's legally mandated responses. It was routine, and the Gander crash was like the rest only more so — the worst and most controversial air crash in Canadian history. As LaFrance had told his staff, the Gander report had to be examined "in some depth, because of the complexity and the dissenting element."

But Koring's cross-examination was relentless. "Despite repeated questioning," he reported, "[LaFrance] could cite no ministerial or legislative mandate for second-guessing the independent agency that had been created for the purpose of taking accident investigation out of the hands of Transport Canada." LaFrance should have had his lawyer with him.

Paul Koring's front-page story of March 17 named his CASB sources for the first time. Peter Boag believed that "Transport Canada failed to understand the grim lesson" of Gander. Boag also believed that Transport Canada had deliberately set out to "undermine the credibility" of the CASB's professional investigators.

Who could have masterminded such malevolence? Well, Assistant Deputy Minister Claude LaFrance had ordered the "secret critique" of the CASB investigators, but the critique had, in fact been "written under the direction of James Stewart, who has sought CASB positions on at least two occasions without success."

The implication seemed to be that a spiteful Stewart, with the possible complicity of his boss, Claude LaFrance, had conspired to mislead the minister in order to wreak revenge for rejection by the CASB. The idea was wacky enough to fit right in with the addendum to the Hickling report, and the insinuation against the dissenting members at the December press conference. Perhaps no one had told Paul Koring that both Stewart and Boag liked to forward their names whenever a senior job became open in either the CASB or Transport Canada, a practice not unusual for ambitious bureaucrats.

But Jim Stewart had committed an even greater sin. He had fraternized with dissenters, a transgression so abominable that it had to be confirmed by the highest authority. So Ken Johnson too momentarily shed his cloak of anonymity to confirm that "one of the CASB's dissenting board members, Les Filotas . . . had sent Mr. Stewart his critique of the investigation last year."

Johnson was reliving his indignation over the copy of our "Critique of the Ice Contamination Hypothesis" I had passed to Stewart a year earlier. Although Thorneycroft had stopped pursuing the matter after I submitted a legal opinion that rejected his claim that it had been an "illegal act", Johnson had never forgotten. Choosing his words carefully, he told Koring that "the sending of that document was certainly not within the range of expected behavior" for a member of the board. So here we had Peter Boag and Ken Johnson shooting the breeze with a reporter about the Dryden accident barely a week after it happened, and complaining about a technical analysis being passed to a designated interested party three years after an accident.

Ironically, Stewart hadn't received our critique with great enthusiasm, and had only said he found it interesting. The irony didn't end there. Pressed about his contacts with CASB board members, he admitted to Paul Koring that he had "some personal and social contacts" with unidentified board members. Much later, he told me that he had been referring to going out for lunch with one of the majority.

The attempt to discredit Transport Canada by linking the Gander crash with other cases continued on the front page of the *Globe* on April 1. "CASB sources" had drawn Koring's attention to a hard landing of an Air Canada DC-8 in Edmonton on March 28, and — lip service to confidentiality now out the window — the "professional investigators" had told him that "ice build up on the aircraft's right wing apparently caused the plane to stall and dip suddenly." The pilot had been lucky, they explained, "in that the stall occurred just above the runway", otherwise it would have been Gander all over again.

April Fool. The eventual report on the DC-8's hard landing made no mention of a stall and said the incident occurred because "the pilot experienced disorientation during the transition to visual flight".[25] It didn't really matter; by that time no one was listening.

* * *

On Monday, March 21, Transport minister Benoît Bouchard had come back from his Florida vacation to find the opposition demanding his head. Opponents of the ice theory were mad because Bouchard supported the majority report in the face of criticism from his own officials. Advocates of the ice theory were mad at the delay in sending the ice advisory to Air Ontario. Those who didn't care about ice were mad because a shortage of aviation inspectors and air traffic controllers was threatening air safety.[26] Everybody was mad at the overcrowding and congestion at Canadian airports.

Pressure was also mounting in the United States. Zona Phillips had finally got some political support: Representative Bill Young, Republican Member of Congress from Florida, had asked the secretary of state to request a new Canadian investigation of the Gander crash. Young had released a January 1986 cable from the U.S. embassy on Mauritius about the claim that the Arrow Air flight had been brought down by ''a cold-blooded, premeditated act.'' He told reporters that the previously top-secret diplomatic communication ''raises the possibility that terrorists may have played a role in the Arrow Air crash.''[27]

Within twenty-four hours of Bouchard's return it was announced that the minister had accepted the resignation of Assistant Deputy Minister Claude LaFrance. ''Circumstances surrounding the minister's statement of March 8'' were cited. ''Sources in the government indicated that the idea Mr. LaFrance should resign came from Mr. Bouchard.'' LaFrance was the first senior Ottawa bureaucrat to lose his job in a political uproar in recent memory. ''The wrong guy quit,'' said the opposition.,[28] Bouchard promised further measures.

BOOK 7

DAMAGE CONTROL

Part 1

The Estey Report

> The right to search for the truth implies also a
> duty: one must not conceal any part of what
> one has recognized to be true.
>
> *Albert Einstein*[1]

On March 29, 1989, Benoît Bouchard faced the television cameras once again, ready to announce the further measures he had promised "to restore the credibility of accident investigation."

"Firstly, I have asked an eminent authority, Mr. Justice Willard Estey, to review the entire record of the Gander investigation," Bouchard read slowly from his prepared text. "He will be asked to provide a report to me within ninety days as to whether any further investigation or inquiry is warranted." Bouchard explained that when he had accepted the CASB recommendations on Gander, "I had not been advised of the internal technical review conducted by my department." He had now seen the review and was making it public; it would be part of the "entire record" studied by Estey.

Secondly, the CASB was stripped of responsibility for the Dryden investigation. Virgil Moshansky, a sitting judge from Alberta, had been appointed to conduct a full judicial inquiry.

Finally, the government would introduce legislation to replace the CASB with a new agency as a matter of urgent priority.

Well, fast-tracking the legislation to dump the CASB was no surprise, and a judicial inquiry into the Dryden crash should curb the unseemly campaign to use the investigation to promote the ice theory. But I had mixed feelings about the review of the Gander investigation. The minister had carefully avoided defining its scope and depth. Also, despite the honorific "Mr. Justice", Estey was no longer a judge.

The 69 year-old former Supreme Court Justice had taken the surprising step of returning to private law practice. He was now deputy chairman of the Central Capital Corporation, chairman and director of Central Guaranty Trust Company, and a lecturer at both the University of Toronto and Osgoode Hall Law School, and he held numerous directorships. He was also counsel for the giant Toronto law firm of McCarthy & McCarthy, and it was in that capacity that Bouchard had secured his services. Bouchard had not launched a judicial review; he had retained counsel to help him out of a tight spot, as had his predecessor, John Crosbie, when he retained Sopinka.

There was no doubt of Estey's distinguished record, but Ken Thorneycroft had a distinguished record too —I was reminded of that by the chummy reference to "concerns raised by . . . Ken Thorneycroft" in Bouchard's statement. We were not about to get a sympathetic ear. But then, weren't after sympathy. Estey had been a justice of the Supreme Court. The majority wouldn't be able to bamboozle him.

Had I paused to reflect on Estey's appointment, I might have recalled the appointment of Chief Justice Earl Warren of the United States Supreme Court to "ascertain, evaluate and report on" the investigation of the assassination of President Kennedy. Justice Warren had also been an extremely distinguished and incredibly busy man. He too had accepted the appointment on top of other onerous commitments. In attempting to head off controversy, rather than to expose the truth, the Warren Commission had added to the mystery and ensured that the controversy would continue, if not in perpetuity, certainly for more than twenty-five years.

At the time, though, it seemed that we had gained about as much as we could hope for. I praised Bouchard's measures. For once, Thorneycroft and I agreed—but the chairman had different expectations. He predicted that Estey's review "will show very clearly that the accident investigators are truly professional and that they do excellent work"[2].

Thorneycroft's theme was echoed by the CASB's public affairs office and Bouchard himself. Representatives of the CASB praised the "professional investigators" and openly mocked the dissidents, who were not protesting against "the Gander report as such" but were using the Gander incident as a means of protesting "their role and responsibilities on the board"[3]. Bouchard told an interviewer that "the problem we now have is that the definition of tasks is confused. The board members, certain board members, want to be investigators." Why didn't the minister just fire the rascals? "The governor in council can only fire them for cause," he confided. "You can imagine that if I tried to fire the current board members for cause, I'd be tied up in court for ten years".[4]

Well, at least the minister didn't think he had a legally acceptable cause to fire me. It wasn't too reassuring, though, to learn that he would do it except for the inconvenience.

* * *

During the next two days a flood of calls came in from acquaintances and strangers. Some offered congratulations for a great triumph in securing an independent review, others warned that Estey's review was just a delaying tactic. Either way, there couldn't be any harm in taking the initiative. Over the weekend I summarized where we stood on the Gander investigation in a long letter to Estey:

> With over 20 years' experience as an aeronautical engineer I soon realized that facts known about the Arrow Air DC-8's trajectory were incompatible with the icing hypothesis. . . .
>
> On a more intuitive level, I was disturbed by the resistance to discussion of certain aspects of the evidence. I sought some indication or hint that our Board's review might be circumscribed by national security or some other overriding consideration. . . .
>
> Since the release of the Board's report I learned that Arrow Air may have participated in the arms-for-hostages exchange conducted secretly on behalf of highly placed individuals in the American administration. If so, there would have been strong motivation for attempts to confound aspects of the investigation. . . .
>
> Congressional hearings into the so-called "Iran-Contra affair" revealed, *inter alia*, that some of the clandestine arms deliveries were made by "privately chartered DC-8". Efforts to contain leaks of information with regard to the ransom of American hostages, arms sales to Iran, diversion of funds to Nicaragua and plans to overthrow Colonel Khadafy of Libya came to a head during the first weeks of December 1985. . . .

Efforts to prevent the leak of information involved deliberate deception of both foreign governments and American legislative oversight committees. Thus, a reasonable person could infer that attempts would have been made to suppress information related to the transport of materials and personnel in support of such activities—whether or not this transport was illegal in itself or related in any way to the Gander accident. . . .

Suggestions of an incomplete investigation were also made to me by fire fighters who were first on the scene but who were not interviewed by CASB investigators. . . .

You have undoubtedly heard of the mysterious illness among workers who were on the accident site. I have been very concerned that these illnesses could be somehow related to some unidentified material aboard the aircraft. . . . the contents of large, coffin-size wooden crates (loaded in Cairo at the cost of leaving soldiers' personal gear) was not determined.

Our minority report postulated an explosion in the forward cargo hold and disintegration of the Arrow Air DC-8 similar to what later happened to the Pan American B-747 at Lockerbie, Scotland. . . .

. . . the fuselage section with the ring of soot around an empty window illustrated in our minority report was not shown to Mr. Pinkel. On the basis of photographs I sent him he agreed that this piece of wreckage strongly suggests a pre-breakup fire. . . .

. . . I have heard, in particular, of three former Arrow Air pilots who told reporters of personal involvement in arms deliveries. I have also heard of an internal report from the Miami office of the Federal Aviation Administration apparently at odds with the official acceptance of the ice contamination theory. This suggests similar internal division between technical experts and administrators in the FAA as in Transport Canada. . . .

When I arrived at the office on Monday morning, I had my letter sealed and addressed to Estey at his law office in Toronto and left it with the receptionist for delivery by courier. The fall-out came sooner, and from a different direction, than I expected.

The next morning the board met to formally relinquish control over the Dryden investigation. We agreed to terminate the CASB investigation and turn over all files to the judicial inquiry. There was almost no discussion and the vote was unanimous. But before we adjourned, Bill MacEachern had a question for Thorneycroft. Would the chairman think it a good idea, he asked in his most syrupy manner, if board members sent Justice Estey material they thought he should see? Thorneycroft looked horrified. After a few moments of grave reflection, he opined that it would be most improper for any board member to attempt to communicate with the Estey.

For a while MacEachern and Thorneycroft continued their little charade, but I wouldn't bite. At last MacEachern got to the point and demanded a copy of whatever I had sent to Estey. He knew the exact time my letter had been picked up and how much the delivery had cost. Sure, he could have a copy, I told him. If Estey wanted to let him have one, that was fine with me.

MacEachern and Thorneycroft were not amused, but it amused me to see them so upset that I should presume to write a letter. There wasn't much they could do about it — or so I thought.

Two days later Thorneycroft sent us copies of a "self-explanatory" letter, dated that day, from Willard Estey to David Bell, who the chairman informed us was the "transport weenie" assigned to work with Estey. Actually, David Bell was the same senior official who had forwarded the pressline advising Bouchard on how to stonewall questions about the Gander investigation. Estey's letter referred to a telephone call from Bell "concerning the intention of members of the CASB to make submissions to me concerning their views as to the matters which I must consider."

> After reflection, it strikes me that I must conduct my review within the four corners of the evidence and exhibits before the Board and, of course, the majority and minority decisions. If I were to entertain submissions from Board Members, I believe that would be both improper in law and would oblige me to hear submissions from all interested parties on an equal footing.
>
> For that reason, it is my intention to return to you all material received from any party or person wishing to make a submission to me. I would be grateful if you would then forward along to that person the material so returned, together with a copy of this letter.
>
> I will not read or review any such material, but will return it in the state in which I receive it.

I could hardly believe this. On the basis of one phone call, Estey had decided to ignore those who made his review necessary. We couldn't even ask what he meant by "improper in law". At the very least, Estey's letter implied that the majority had instant access and influence over his procedures. At most . . . I shuddered to think what it suggested at most. For the first time since joining the CASB I was thoroughly and utterly discouraged.

But the game was far from over. More wild cards would be dealt — off both the top and the bottom of the deck — before the final scores were tallied.

* * *

Bouchard's damage-control strategy got a fair reaction the first day, then rapidly unravelled. Journalists and opposition politicians soon pounced on discrepancies between the texts of the Transport department review distributed at Bouchard's press conference and the leaked version that had provoked the furor two weeks earlier.

According to the leaked version, ''The Transport Canada review does not substantiate the CASB conclusion that the aircraft stalled probably due to ice-contaminated wings.'' In the minister's version, the conclusion had undergone a transformation: ''The Aviation Group review does not dispute that the wings may have been contaminated with ice; however, the substantiated findings could also be consistent with at least one other scenario''.[5]

Opposition members saw the alterations as ''a clumsy attempt at misleading the Canadian media and the Canadian public'', and called for the minister's resignation—again.

''We wonder just what Estey is going to be asked to review . . . the version of the document prepared by officials in the department . . . or is he going to be asked to review the final document whose final text has been doctored?'' asked the deputy leader of the Liberal Party.

''To respond to that kind of slander would dignify it and Mr. Bouchard has no intention of doing so,'' retorted a representative of the minister. But the minister's staff had a handy scapegoat; all difficulties and subterfuges were attributed to departed Assistant Deputy Minister Claude LaFrance. LaFrance refused to comment. Bouchard weathered the attacks, but his credibility suffered and editorial support for his measures faltered.

The Estey review lost even more credibility when the media revealed that the former justice had refused to hear representations from Theresa Griffin and Zona Phillips. Estey's letter to the Phillips provided a clue to the limitations on his review. "I have been instructed by the Minister to study the record," he wrote. "My report is limited to determination as to whether the record placed before the Board supports the conclusions reached. . . . Many persons have volunteered information . . . but by reason of the nature and the limits of my retainer, such requests and offers have been respectfully declined."[6]

Estey also turned down Roger Lacroix's offer to give evidence. "My terms of reference do not authorize me to take any evidence," he wrote. "I am simply instructed to review the record as compiled . . . ."[7]

"Estey set his own terms of reference," Bouchard maintained. "He has a mandate to do what he wants."[8]

Incensed, the victims' next of kin cranked up the clamor in the United States. Mrs. Phillips and Mrs. Griffin came to Ottawa to plead for a full judicial review, but Bouchard refused to see them. "It wouldn't be relevant to meet them at this time," he told the press. "To meet them would prejudice the inquiry."[9]

Supporters of the majority believed the next of kin were butting in because of the letter I had written at Mrs. Phillips' request, the one Thorneycroft called a "ghoulish step" that was "morally reprehensible if not illegal." A copy found its way to the *Globe and Mail*. The *Globe* hadn't seen fit to carry even the wire service reports on the scuffles in Parliament over the letter from Bouchard demanding "full and unqualified support" for Thorneycroft, but it did carry an item on "the solicitation, by a board member, of the views of the next of kin after an official report is issued'.[10]

The story didn't make a great splash, but it did have some effect; copies of the article were distributed in the aircraft accident investigation community, and I got a lot of critical feedback.

In the meantime, Bouchard launched the final solution to dissension on the CASB — "legislation to enhance transportation safety by establishing a new independent Board to investigate air, marine and railway accidents". The new legislation stipulated a maximum of five members. The new agency would take over investigations in progress, but would not be concerned with "closed" investigations such as Gander.

The strategy wasn't difficult to discern — force the legislation through, before Parliament adjourned for the summer and before Estey submitted his report. Presumably Estey's report would discredit the dissenters and set the stage for granting Thorneycroft's wish to "cut the cancer out". The new streamlined board could be introduced during the summer break, without fanfare or embarrassing questions in the House of Commons.

At first, the opposition fought back. "The government's urgent desire to get the bill passed by July stems from its desire to silence four members," Liberal Transport critic Maurice Dionne told the *Ottawa Citizen*. "The motive of the government is all too apparent. It is an attempt to sweep under the rug the botched investigation into the crash at Gander."[11] But the government was adamant and had a majority. Bouchard saw no need to wait for Estey's report. "I don't see any link between the bill and the investigation," he said. "We know what the problem is at the CASB and we're fixing it with this bill."[12]

The opposition rumbled about using its majority in the Senate to force changes, but apparently the lure of the summer sun was too strong. Resistance crumbled and on June 20, 1989, the House of Commons passed the bill to supersede the CASB with the new streamlined, five-member, multi-modal agency. Estey's report was not due for another week. By that time, Parliament would have recessed for the summer.

The new law did not go into effect immediately. The government could "proclaim" it at its leisure. When the original CASB had been passed five years earlier, the government had proclaimed it in two stages. Members of the former AARB and new CASB had worked together during the six-month transition period and had continued to join in discussions on drawn-out cases for much longer. This time the government did not reveal its plans for implementing the new legislation, but no one expected a six-month transition.

Thorneycroft sent a reassuring memo to the CASB staff: "I would like to emphasize that this new legislation should not cause concern in any way for your future employment. You will simply be doing the same kind of work in a new and larger organization. . . . In the meantime enjoy the summer period."[13]

Fabricated Evidence

> History shows that deceit in the annals of science is more common than is often assumed. Those who improved upon their data to make them more persuasive to others doubtless persuaded themselves that they were lying only in order to make the truth prevail.
>
> *William Broad, Nicholas Wade*[14].

Back when Theresa Griffin had been preparing for her representation, the CASB had promised to let Harold Marthinsen, director of accident investigation for

ALPA, analyze the FDR tapes at ALPA headquarters. When the CASB then changed its mind and said Marthinsen would have to do the work in Ottawa, the plans had fallen through; Marthinsen just hadn't the time. He got to examine the tape that September, but the CASB's final report soon slammed the door on further debate.

Nevertheless, the FDR data had continued to nettle Marthinsen. Eventually he compared the traces recorded at Gander with those from previous takeoffs — a task I hadn't been able to persuade the CASB to do in two years of trying.

The altitude traces on some twenty earlier takeoffs showed fluctuations much like the ones the majority report interpreted as the signature of a stall. In the previous traces, the fluctuations disappeared as the aircraft gained altitude. Marthinsen's work plainly confirmed the contention in our minority report that "fluctuations in the altitude trace near lift-off are characteristic of the installation and not of the accident flight."

During a telephone conversation on June 8, 1989, Marthinsen told me he had corroborated this conclusion through comparisons with metal FDR tapes from previous crashes. He had found similar fluctuations, for example, on the altitude trace from the Capitol Airways DC-8 that crashed during takeoff at Anchorage in 1970.

Marthinsen had also reviewed the analysis that purported to work backward from the FDR traces to establish that only the ice theory fit the observations. I had dismissed this fantasy almost two years earlier, on the grounds that the myriad assumptions in the computation amounted to assuming the results. Harold agreed, but he was not ready to concede that the misconception was entirely innocent.

Wishful thinking or self-delusion might explain a distorted analysis, especially one by zealous henchmen of an airline or manufacturer. But using the FDR to smear the reputations of dead pilots went beyond the pale — especially when the deed was perpetrated by official investigators. Marthinsen wanted to expose this scandal, and told me he was thinking of publishing a technical paper.

"Well, that's nice, Harold," I replied "but by then the curtain will have come down on the CASB and the Gander chronicles will be gathering the same dust as the *Hindenburg* chronicles. Why don't you write up what you've got in the form of a letter to Chairman Thorneycroft, so he can have the benefit of your insight?" Thorneycroft had, after all, maintained that he'd like to be the first to know of new evidence.

Marthinsen rhymed off a list of things he was busy with and said he'd think about my suggestion. He didn't think too long; on June 20, the day Parliament axed the CASB, he called to tell me he had set out his thoughts in a letter to Thorneycroft. He was also sending copies to the other board members, interested parties and others he thought might want to know.

* * *

I had underestimated the depth of ALPA's indignation. Mild-mannered, grandfatherly Harold Marthinsen had not so much sent a letter as launched a 21-page nuclear warhead.

Marthinsen began by reviewing some of the technical aspects of the FDR readout. He then systematically annihilated the arguments advanced to show that the aircraft had stalled, because of ice or anything else. "In summary," he wrote, "none of the parameters contained on the FDR either taken alone or 'when viewed together' indicate that the aircraft was in a stall." So much for the majority report.

But Marthinsen reserved his main salvo for a study "that purports to analyze the flight data recorder" using a "readout of the FDR which they claim is the basis for their analysis." He was referring to the study at the University of Dayton Research Institute (UDRI) using data supplied by the CASB investigators. Some of the numbers used by UDRI had come from a preliminary report that was later withdrawn. "Perhaps the UDRI did not know," Marthinsen allowed, "that this readout contains uncorrected data and bears no resemblance to the final readout." But this benefit of the doubt didn't extend to a second group of numbers that he believed had no basis at all. I would have said that the authors were forced to make questionable assumptions to compensate for missing data. But Marthinsen didn't see it that way. "It should be clear," he fumed, that these other numbers had been "conjured up" using "faulty or perhaps more to the point, fabricated data." He went on to inform the chairman that:

> Someone within the CASB should have seen the study for what it was worth. But instead the results became a part of the final CASB report.
>
> We have over 20 year's experience of reading out hundreds of metal foil flight recorder tapes. In all these years, we have not seen such abuse of FDR data as in this investigation. The flight recorder was put in airplanes as a tool to assist accident investigators to arrive at the true causes of accidents. They were not intended to allow "conjecture" which suddenly changes to "factual conclusions".

Marthinsen urged Thorneycroft to do everything in his power "to correct the erroneous statements and conclusions in the CASB report."

The first thing in Thorneycroft's power would have been to inform Estey of ALPA's conclusions, so during a board meeting on June 22 I noted that the material we had all received from Mr. Marthinsen the previous day was relevant to Mr. Estey's study. I moved that the board forward ALPA's analysis to Estey at once. Bill MacEachern dismissed Marthinsen's missive as "mischievous, spurious and abusive", and my motion was defeated.

But the considered opinion of ALPA's director of Accident Investigation couldn't be kissed off quite so easily. Not only had he denounced the majority report, he had done so in language that made terrific copy. "U.S. pilots condemn Gander probe," said the headline in the *Ottawa Citizen* of June 24. The story was carried by the wire services and many Canadian and American newspapers.

Benoît Bouchard's office announced that Estey's report, due on June 27, would be delayed for at least a week. "It is not known," the *Ottawa Citizen* reported on June 27, "if Estey's decision to delay is connected to the unexpected broadside from the U.S. group representing 41,000 pilots." It was known, however, that Parliament would adjourn on June 29. Whatever Estey concluded, the government wouldn't be answering questions about it in the House of Commons.

A Fundamentally Legal Process

The release of the Estey report slid past the original deadline into the peak of the summer doldrums. A spokesman for the Transport minister finally announced that the report would be released at a press conference on the afternoon of Friday, July 21. The timing was typical for announcements the government wanted to bury—too late for weekend editions, stale by Monday. But — fittingly, in view of its antecedents—first public disclosure came by way of a leak.

On July 18, a front-page story in the *Globe and Mail* by Paul Koring announced that "sources in the CASB and the Department of Transport" had revealed that Estey's report "generally exonerates the CASB's professional staff investigators who found that a sandpaper-thin layer of ice sufficiently reduced the lifting capacity of the wings of the ageing DC-8, which crashed on take off." Koring's sources had also informed him that the judicial review by Estey "found no need for any further public inquiry", and that "Judge Estey rejects the dissenters' allegations as being in the realm of gossip and unfounded innuendo."

To demonstrate how low the dissenters would stoop to promote their calumny, Koring also disclosed that "one of the dissenters. . . tried to get in touch with Judge Estey during the review", and that, fortunately for truth and due process, "Mr. Filotas' letter was returned unread by Judge Estey, whose mandate was to review the massive amounts of existing evidence collected by police and investigators."

It seemed that Estey's report was a grand victory for the majority and a humiliating rout for the minority. But something didn't ring true. Why leak such a great story in advance of the triumphant announcement by Mr. Estey himself? Why dilute the impact and preempt the story line? And why schedule such a long-awaited exoneration for release on a Friday afternoon?

The absence of direct quotes showed that Koring had not actually seen Estey's report. Furthermore, his sources had misled him on minor matters. While Estey had said he would return submissions unread, for example, he had not, in fact, returned my letter, and I had no idea whether he had read it or not. More significantly, Thorneycroft and the others were not gloating.

The *Globe*'s report of Estey's vindication of the majority was carried on the wires and widely reported on Thursday, and many reporters called for interviews. I said I wouldn't comment until I saw the report, but sure, I'd answer questions after the press conference.

* * *

Bouchard's short press release and Estey's 37-page report were distributed to reporters ahead of the press conference the next day. I picked up copies as I went in, a few minutes before Estey arrived.

The press release was to the point. "Mr. Bouchard stated that he accepts Justice Estey's recommendation that no further investigation of this accident is warranted."

Estey's report began by outlining how he had confined his study "to transcripts of evidence taken by the Board at its public hearings, together with many exhibits, reports and reviews received or conducted by the Board in the course of its deliberations." In short, he had only considered the evidence gathered in support of the ice theory.

Estey then gave a brief summary of the accident circumstances. The very first page contained a factual error; the troops on the Arrow Air DC-8 had not been "on duty in the Sinai Desert as part of a United Nations Force", they had been part of a "multinational" force organized by the United States after the

United Nations declined to sponsor the force. It wasn't a big mistake, just a bad omen.

Estey's description of the short flight before the crash also popped a few circuit breakers. Rotation and lift-off, he wrote, "occurred at forty-seven and fifty-one seconds respectively from the beginning of the take-off roll." This was what the majority report said, all right, but the times had been deduced—fabricated, Harold Marthinsen might have said — from the disputed FDR traces. While our minority report challenged the spurious precision, Estey took the numbers for gospel. Things were not looking good.

I continued to read as Estey came in. Evidently, press conferences didn't figure among the former judge's favorite pastimes. No, he didn't have an opening statement. "The meeting was the idea of the press and I'm prepared to answer any questions," he said. "There's no point in me going through the report. You've read it. I haven't changed my mind. It's self-explanatory."

Paul Koring's first question grabbed all my attention. "You find that the evidence does not support the conclusion of ice contamination as the cause or even the probable cause of the accident," he began.

What was this? Estey had rejected the ice theory? Then Koring's sources had suckered him by withholding the most important conclusion. Estey's report had been selectively leaked for damage control; had it been leaked that Estey repudiated the ice theory, *that* would have been the headline. As it was, the recommendation against further inquiry had had a free run for two days.

Koring asked Estey's reaction to the notion that "ice is almost impossible to find as a cause particularly in an accident where there's an ensuing fire."

"Well, you're quite right," Estey replied. "Ice doesn't survive a fire. You even know that in law school." He waited for the laugh. "But the presence of ice can be discerned by indirection, the same as we can speak with assurance of what's on the other side of the moon. . . . There isn't a shred of evidence which indicates that ice was the controlling factor in this crash." He might be enough of a lawyer to narrow his terms of reference to exclude facts inimical to the interest of his client, but he was enough of a judge not to stand for bullshit. When Koring persisted in trying to get him to admit that "there is never absolute evidence of icing," he was not amused.

"You can have a survivor who said, I saw it flake off. . . . You can have passengers look out and see ice on the wing."

Estey still dismissed the minority report with scathing generalizations, and followed the established strategy of ignoring specifics. He didn't comment, for example, on the photos of the fuselage panel. "There is no evidence in the transcript [of the public hearing] or in the studies undertaken during the investigation," he said, to substantiate the minority conclusions. That, of course, was the essence of the minority position — but the supplementary evidence we presented was not within "the four corners" Estey had considered. He dismissed our "Critique of the Ice Contamination Hypothesis" with the comment that it "was responded to by an equally substantial document generated by the staff of the board."

Estey complained that "the minority refers to some pathologist without a name from somewhere else, who did not testify. . . . That kind of thing is not helpful in any kind of process, particularly one which is fundamentally a legal process". It might be fundamentally a legal process to him, but it was fundamentally an investigative process to me. If he had only asked, we would gladly have furnished the name of Dr. Richard Shepherd — or he could have obtained Shepherd's name from "the entire record of the investigation", had he taken that record to include correspondence from interested parties. We could also have given him the name of Dr. P. E. Bessant-Matthews of the Institute of Forensic Sciences in Dallas, which appeared only in the record of our work on the minority report.

But it seemed that Estey was not really interested in evidence. His main point had nothing to do with evidence. "The primary purpose of all this is to promote air transportation safety," he said, echoing the old refrain. "You can whip milk for a while and get butter, but you can't get butter by doing it fifteen times over." In his opinion, "no reasonable hope exists that we will find a cause for this enigmatic tragedy." So stop beating a dead horse, and leave my client alone.

It was a magnificent example of a circular argument and a self-fulfilling prophecy wrapped into a single package. We shouldn't go on with the investigation because we wouldn't be able to find the cause, and dropping the investigation would ensure that we never did—thus proving Estey right in the first place.

Estey left the microphones and departed the room, and I took his place. I hadn't had a chance to read his report so I'd have to wing it. I thought of a line that had got Jerry Lederer a laugh at the Vancouver Air Safety Symposium. "My first reaction to what we heard here today is a bit like how you might react to a bikini bathing suit. What Judge Estey has uncovered is very important, but what remains covered up is vital," I said. "And what he has uncovered is that ice was not the cause of this accident. . . . What still remains covered up is all those circumstances that lie outside the four corners considered by Judge Estey."

Despite the putdown of the minority report, I was buoyed by the categorical rejection of the ice theory. Had we rejected the ice theory without suggesting an alternative, we wouldn't have got this far. So what if Estey said there was no basis to support our speculations"? We'd stopped a false cause being assigned to one of the world's worst aviation disasters — that had to be worth something.

"This labyrinth has many more than four corners," I said. "There's all kinds of angles to this accident that haven't been examined up to this time. It's very unfortunate that the investigation is going to be closed."

"Are you saying then, Dr. Filotas, that you don't agree with Justice Estey when he says we'll never know the answer?"

"Well, we certainly will never know if we don't look any further. There's no question about that."

There was also no question that, in Canada at least, all official efforts to look further had ground to a halt. But then, the deepest motivation for finding the cause of the crash did not lie in Canada. As Estey himself had said, "Human beings for some reason have in their genes an instinct that says, we want to know how our friends and relatives die." One of the reasons he cited against further inquiry was compassion for these human beings: "To expose the next of kin to an expensive and unproductive extension of this inquiry would, of course, be wrong."

But those whose friends and relatives had died in the Gander crash had had very little to say in the proceeding. The next move would be up to them.

* * *

Improbable Cause

When I returned to my office from Estey's press conference I was greeted by an upset secretary, who told me that an RCMP officer had been looking for me. A member of the CASB's administrative staff had let him into my office where they had gone through my papers. The secretary had been forced to produce a copy of a letter I had recently written to Harold Marthinsen.

On Monday the RCMP officer returned and I met him in person. It seemed that the letter I had written to Estey had not exhausted Bill MacEachern's interest in my correspondence; observing that I had written to Marthinsen, presumably another "scurrilous saboteur" who dared to criticize the ice theory, he had accused me of betraying secrets.

The officer politely informed me that a complaint had been laid and that he had had to investigate. He had examined my innocuous letter and determined that the complaint was groundless. He apologized for the inconvenience and left.

The RCMP officer was Staff Sergeant Richard Jordan, the officer who had refused to lay charges against reporter Doug Small a year earlier "to please elected officials".*

---

*Jordan wouldn't reveal who had laid the complaint. Neither would Thorneycroft, who had approved calling in the law. I found out the details later through an Access to Information request to the RCMP.

490

Part 2

Exeunt CASB

Willard Estey's putative concern for the next of kin infuriated Zona Phillips. She had been working herself to exhaustion for over seven months to extend the inquiry. What did Estey know about the sacrifices the families would willingly endure to establish how their loved ones had perished? How dare he speak for the families whose testimony he had spurned?

Zona managed the office of her husband, Dr. Doug Phillips, Sr., a pathologist and lab director at the Edward H. White Hospital in St. Petersburg. The Phillips' resolve to learn more about the death of Doug Jr. dated back to Christmas 1985 — spent, not with their son as expected, but with the horrible knowledge that his charred remains lay in the mortuary at Dover.

Although the families' support group had folded in the face of army disapproval, Zona, an indefatigable telephone talker and letter writer, had kept in touch with the families. Many had a desperate need to know what had happened, a desperate need for reassurance that the lives of the young soldiers had not been lost through negligence or inattention.*

_____

*A similar need surfaced at the height of the Vietnam War, when American soldiers were being killed by misdirected artillery from their own side — "friendly fire". The U.S. government should have learned that, if grieving parents are misled about the circumstances of their loss, anguish can turn to outrage — outrage and a fierce determination to learn the truth.

The CASB's split decision spurred Zona Phillips to renew her effort to locate the families. By mid-spring 1989, some fifty families across the United States had joined a group dubbed "Families for Truth about Gander". Their stated purpose: "to inform, educate and support the members toward perpetuating a meaningful investigation into the true causes and circumstances surrounding the Gander air disaster." Since American politicians and government officials claimed they had no power to hold an investigation, the families pinned their hopes on a judicial inquiry in Canada. They turned to the newspapers and airwaves to help promote their cause.

Zona Phillips was a 49-year-old dynamo, angry, articulate and emotional. Her engaging and sincere manner came across well on radio and television, and her story of a four-year-old granddaughter who wanted to know how her daddy had died was filled with human interest. So were dozens of similar stories around the country. The major media showed little enthusiasm but throughout the first half of 1989, stories about the grass-roots campaign to find out the truth appeared in local papers and on local radio in places such as McKeesport, Conyers, Daleville and North Huntingdon.

The Phillipses and the Families for Truth about Gander caught the ear of reporter David Ballingrud of the *St. Petersburg Times*. Ballingrud's beat covered aviation and military affairs, so he was well prepared to probe the snakepit of technical and political issues under the human-interest story.

He reported on the families' disappointment with the CASB's split decision, their determination to work for a judicial inquiry and the fiasco unfolding in Ottawa. At the same time, he pored over the evidence

and interviewed protagonists, bit players and kibitz-
ers. Much more than Willard Estey or any other outsider, Dave Ballingrud sought "to review the entire
record of the Gander investigation."

But Ballingrud didn't just write about the evidence;
he added to it. An interview with former Arrow Air
pilot Steve Saunders established that at least some FBI
agents had worried about sabotage. Chuck Foster,
director of the FAA's seven-state north-western region
at the time of the Gander crash, confirmed that at least
one senior official of the FAA had credited the hypothesis of an onboard explosion[15].

Ballingrud also grasped the potential significance of
the Arrow Air crash in relation to the Iran-Contra
chronology. Having covered American military matters, he could conceive that troop movements between
the United States and the Middle East might be useful
for covert action—and, accordingly, that it was important to clarify the revision of the victim count from 258
to 256. His search through the Iran-Contra files uncovered the Arrow Air arms deliveries to the Contras in
January and March of 1985. He also found the link to
the Iran arms deliveries through Arrow pilot Jake Bolivar, and was the first to point out the significance of
Oliver North's warnings about the Islamic Jihad.

While none of this added to the direct evidence, it
piled motive upon motive for obscuring the circumstances — and it established a connection with those
who had proved ready to lie and hoodwink government agencies. Regardless of what might have caused
the Gander crash, elements of the American government had the interest, the means and the will to keep
Canadian investigators from digging too deeply.

As Henry David Thoreau put it, "Some circumstantial evidence is very strong, as when you find a trout

in the milk.''[16] While neither Dave Ballingrud nor anyone else had found a smoking gun, a school of fish was certainly swimming around in our milk — and some of them looked much more like trout than red herring.

As a pathologist, Doug Phillips had a professional interest in the medical evidence surrounding his son's death. After speaking with Dr. Richard Shepherd and an army pathologist, he became convinced that the majority report was wrong about the times of death and, consequently, about the absence of a pre-crash fire. Zona and Doug Phillips had been so frustrated by Estey's refusal to hear new evidence that they had travelled to Canada in May of 1989 to lobby for a full judicial inquiry, but neither Bouchard nor any other representative of the government would give them the time of day. They did meet members of the Liberal opposition who pledged support, and the trip generated a good deal of media coverage, which increased the membership in Families for Truth. But the primary mission proved an expensive and exasperating failure.

By the time Estey issued his report, Families for Truth about Gander had grown to about ninety members. Estey's recommendation against an extension of the inquiry was a solid blow but not a knockout. If the Canadians were not willing to track down the cause. or at least expose all the circumstances, then it would be up to the United States. Families for Truth about Gander incorporated, and sold bumper stickers and T-shirts to raise money for stamps and telephone calls to lobby for hearings by the U.S. Congress.

When the controversy erupted over the Transport Canada report charging that the Gander investigation had been mismanaged, the Phillipes' representative in Congress, Republican Bill Young, asked Secretary of State James Baker to press Canada for a new investigation specifically noting "the possibility that terrorists may have played a role in the Arrow Air crash".[17] Young's initiative cooled when the Republican administration signalled its reluctance to get involved.

But Democratic Congressman Robin Tallon of South Carolina felt no compunction about upsetting the administration when the family of Captain Kyle Edmonds of Hartsville, South Carolina, sought help to learn more about the crash. Tallon said the family was "distraught that the United States government was not responding to their questions and concerns about the crash." At first he expected a relatively easy job of cutting thorough some departmental red tape. Instead, he got "a bureaucratic runaround unlike any I have experienced in my nine years as a Congressman."

Tallon simply couldn't understand how the army, the State Department and agencies such as the FAA, NTSB and FBI could all dismiss the deaths of some 250 American soldiers as "a problem for the Canadians" — especially when the Canadians had proved themselves inept and indifferent. Tallon got his dander up and became the tireless champion of the Families for Truth about Gander. "It's incumbent on our government to issue a report and look at some of the information that wasn't considered in the much disputed Canadian report," he said. He held a press conference, put a list of unanswered questions on the congressional record and asked the House Armed Services Committee to hold hearings.

The story about the Gander crash finally started to catch the interest of the major media. To many Americans, Estey's statement that "no further investigation of this accident is warranted'' meant that the Canadians couldn't be bothered to find out how hundreds of Americans had been killed on their territory. Zona Phillips bitterly contrasted the week-long public inquiry into the Gander crash with the full judicial inquiry into the Dryden accident, which had taken twenty-four Canadian lives; the Dryden hearings were expected to last a year, and the families of the victims were full participants.

For those who believe that an event becomes news only when it's broadcast nationally on major networks, the controversy over Gander became news on October 13, 1989. On that date Barbara Walters told viewers of ABC's news show *20/20* that a five-month investigation had uncovered a story "that could rock the nation". The two-part documentary summarized the evidence in graphic images—ground handler Ted West saying he had put his hand on the airplane and felt no ice; Judy Parsons tracing the track of the "bright orange object" that became a fireball, from where she had stood in the Tilden parking lot; Irving Pinkel pointing to the hole in the fuselage that was "hard evidence of an explosion on board"; Dr. Shepherd showing a computer printout and saying, "It means the fumes of a fire were inhaled before the plane crashed."

Congressman Tallon also appeared on the program and suggested a motive for a cover-up. "I think it would have been a catastrophic problem for Ronald Reagan into his second term if he had been faced with the American public finding out that we had just botched an arms-for-hostages deal."

But perhaps the most surprising revelation was CASB chairman Ken Thorneycroft's explanation of how witnesses were selected for the hearing. "Let's have a round-table discussion to decide what evidence we want to come out at a public hearing and then select a group of witnesses who will provide that evidence."

After the *20/20* documentary finally brought the Gander controversy to a large audience in the United States, an "angry and frustrated" Tallon was able to persuade 103 other Members of Congress to cosign a letter petitioning President George Bush to "initiate a formal United States investigation to explore all possible crash theories."[18]

Back in Canada, Benoît Bouchard wasn't impressed. It would take more than "just a TV program" to get the case reopened, also no, the government would not ask U.S. authorities about the legitimacy of the allegations. Willard Estey wasn't impressed either. "It's the same old recycled American garbage," he said. "The U.S. journalists know we're suckers up here. We'll print anything they say or churn out."[19]

And indeed the stories churned out. Syndicated columnist Jack Anderson added a new and bizarre twist with the suggestion that covert operatives in the U.S. Army had perpetrated a cover-up within a cover-up "to scare the Army brass into steering any investigation away from the embarrassing prospect of an explosion caused by negligence".[20] Anderson claimed that the Chairman of the Joint Chiefs of Staff had ordered an internal investigation, and that in consequence "the army quietly retired" General John Crosby.

Even more bizarre allegations surfaced with respect to the contents of the "coffin-size" crates loaded on the Arrow Air DC-8 in Cairo. A former Israeli intelligence agent claimed that the crates contained the bodies of soldiers who had died in a failed rescue mission in Lebanon. Another investigator claimed he had proof that the boxes contained expendable anti-tank weapons that had been written off the inventory in training exercises in the Sinai.[21]

Government agencies could have scotched such stories with a credible account of the contents. Instead, they added to the speculations by giving out contradictory and incomplete information in response to a torrent of letters and requests under the Freedom of Information Act. President Bush didn't even reply to the letter from almost one quarter of the members of the House of Representatives. As in Canada, the authorities in the United States hoped it would all just go away.

* * *

By the time Zona Phillips called to tell me that a report of the Sub-committee on Crime of the House Judiciary Committee had condemned the involvement of American agencies in the Gander investigation, I was no longer a member of the Canadian Aviation Safety Board. The CASB had, in fact, ceased to exist some eight months earlier.

The government waited for almost a year after Parliament passed legislation to replace the CASB before actually pulling the plug. Frank Thurston's appointment was extended for a year, and Thorneycroft made sure we got copies of the paperwork showing that all was in order. Those of us who had signed the minority report were, in effect, cut off from everything but formal communications related to board meetings. Yet the lame duck CASB waddled along, going through the motions of meetings and considering the steady stream of reports on light aircraft accidents, hostility between opposing factions somewhat subdued but never far below the surface.

The board had stopped discussing the Gander case, but it remained a constant ghostly presence. Reporters kept up a regular stream of calls from all over the United States and Canada, asking for interviews or an opinion on the latest conspiracy theory, or just passing on rumors or good wishes. One rumor had it that the Americans wanted the CASB kept on until the post-Estey news stories had run their course. Other stories said the controversy had made it difficult to recruit a credible chairman for the new agency. If Thorneycroft or the majority members had any inkling of the government's intentions, they weren't telling.

As stories about the Families for Truth about Gander proliferated in the American media, Thorneycroft acquired a sudden obsession about appeasing the families. He had called Zona Phillips right after Estey issued his report, offering to bring Peter Boag and Dave Elcombe to Florida to explain the majority report. When she hesitated he sent a letter by courier urging her to think it over and call him or Ken Johnson with any questions.

"We do not understand what you and your associates could say to us at this point," Zona replied. "Since Mr. Estey found the majority and minority reports unacceptable, it would seem the issue is closed." She enclosed a four-page list of open-ended questions, saying that a meeting might be productive if Thorneycroft could supply written answers as a basis for discussion and if representatives of the minority came to give their side.

Thorneycroft called her, suggesting it would be better if he answered the questions in person, but she insisted on written answers. In case it might be something personal about Thorneycroft, Bill MacEachern weighed in with an offer to come and speak to the families. No thanks. Appeasing the families evidently had high priority. In due course Thorneycroft sent a detailed twenty-page response to the questions, adding, ''I would still be pleased to come to Florida with the investigator-in-charge of the investigation and our physician to give you further clarification.'' He didn't mention the request for representation from the minority, or explain that his job was ''to sell the majority report and not to give a leg up to the minority.'' But of course he didn't need to.

''We are in receipt of the answers to our questions,'' Zona Phillips shot back. ''In general they fall short of being satisfactory. They are vague, ambiguous, evasive, contradictory, incorrect and ignore details in many instances.'' She gave examples. ''Your interview on ABC *20/20* was not well taken by our families and friends,'' she added. ''To say that we and they were extremely upset would be a gross understatement! It was inflammatory, perverse and insecure, reflecting an insolent attitude about a very grave issue.'' Thorneycroft's offer to meet the families was declined.

Thorneycroft never mentioned any of this abortive missionary excursion, but Zona kept us informed. It was mildly amusing to hear about his frenetic efforts to influence the families, when he had been ''appalled by the insensitivity'' I had displayed ''by communicating with the next of kin.'' Why, I wondered, would Thorneycroft seek to flagellate himself before the families when the government's decision gave him a perfect pretext for putting the whole Gander mess behind him? Was it possible that placating the families was a litmus test to determine his future on the new board? If so, he failed as miserably as he had failed in selling the majority report.

* * *

March 29, 1990, would be the last working day of the fiscal year. Rumors had been circulating that the government would bring the CASB to its overdue end on that date in a neat actuarial stroke. But a week before Thorneycroft announced that he had been requested to submit a budget to extend operations for ninety days into the new fiscal year, and he circulated an agenda for a board meeting in two weeks.

It made sense. Ninety days would extend into the summer recess. The CASB could then be dispatched without a lot of fuss — no question period, no regular press scrums, just like the release of Estey's report. It seemed we had another three months, and I'd continue going through the motions of being a member of the CASB, although my real interest now focused almost exclusively on solving the Gander mystery.

I spent the morning of March 28 going over the reports that were scheduled for the next board meeting. It was a heavy agenda — as was our practice, we were ripping through the large backlog of accidents and deferring policy items or controversial topics. I was twice interrupted by calls related to the Gander case. One was from Zona Phillips, to tell me that Congressman John Conyers of Michigan had asked the General Accounting Office for answers to some specific questions about the role of American agencies. The other was from a friend at the FAA who had traced the tail number of an American-registered DC-8 that landed in Tel Aviv on September 15, 1985, "reportedly out of Iran".

It had been revealed during the Iran-Contra hearings that this DC-8 had just delivered a shipment of 486 TOW missiles to Tabriz as part of the developing arms-for-hostages initiative. The DC-8 had been chartered from a Florida-based company, but the title had recently been transferred to the Brussels-based Air Tours of Nigeria, which really operated out of Miami. The pilot was Herman Duran, partner of Arrow Air pilot Jake Bolivar. Most of the specific information had been censored out of the Iran-Contra transcripts, but not consistently, and I was trying to piece together what was left. But recently I'd had what I thought might be a stroke of luck.

The arrival of an American-registered plane from Iran was sufficiently unusual to occasion speculation in Israel. A curious Israeli noted the U.S. registry — N421AG — before the DC-8 was hustled off to the Israel Aircraft Industries hangars at Ben-Gurion Airport for repainting. I had found this tidbit in the September 18, 1985 issue of the *Jerusalem Post* in one of my early-morning sessions in the microfilm room of the National Archives.

During the time I had left in a position with relatively easy access to aviation information, I was trying to trace the DC-8s involved in the Iran arms deliveries. Unfortunately, the registration noted in the *Jerusalem Post* turned out to be the tail number of a Cessna decommissioned in 1983. Well, I had three more months—or so I though.

In the early afternoon there was another call; a friend had just heard that five members had been named to the new safety board. ''Their names are not household words. None of them is from the existing board. Word is that the new board will be in operation within days.'' So what? I went back to reading reports until the next call.

It was from Thorneycroft. Could I come to his office right away? As soon as I walked in I knew the tip had been right. Norm Bobbitt was already there, sitting at a table holding a letter. Thorneycroft was holding another one for me. He looked neither happy nor sad.

The letter was from a functionary in the Privy Council Office. I didn't have three more months, after all. Referring to me as "a former public office holder", it said that the new act had been proclaimed as of that day and that "the said Act terminates the Canadian Aviation Safety Board and consequently your public office. . . . The government is under no obligation to compensate for loss of office. . . . I want to take this opportunity to thank you for the time you have devoted to the service of the Canadian public and I wish you well in your future endeavors."

Thorneycroft said that he had only known for about half an hour. He had to clear out of his office because the new chairman was coming in the next day. We smiled but didn't shake hands.

When I left my office later that afternoon, it was for the last time. Some days later, employees of the new board packed my personal belongings in cardboard boxes and shipped them to my home.

Congressional Hearings

Two days of Congressional hearings held almost exactly five years after a chartered DC-8 crashed at Gander, Newfoundland, produced virtually no significant answers to a swarm of questions. The impression strongly persists that a coverup of massive proportions obscured many aspects of the Arrow Air crash. . . .

Counter-Terrorism & Security Intelligence[22]

On December 4, 1990, the audience assembled in Washington for the opening session of the House Judiciary Subcommittee on Crime oversight hearing into the fatal plane crash in Gander. The session had been moved to a larger room to accommodate the crowd of spectators and media. The fifth anniversary of the crash was the next week, and this was the first public inquiry of any kind since the controversy had broken out.

At the time of the Arrow Air crash, and while the Canadian investigation was under way, the Subcommittee on Crime had been drafting legislation to correct a perceived shortcoming in the United States' power to stop terrorist attacks on Americans abroad. The resulting "long-arm" law licensed the FBI to hunt down terrorists who murdered Americans outside the United States. But the U.S. State Department and other agencies blocked the FBI's first notable attempt to apply that law, in an episode involving a politically sensitive plane crash in a foreign country — the one that killed General Zia, American ambassador Arnold L. Raphael, American military adviser Brigadier-General Herbert Wassom and twenty-nine others in Pakistan in August 1988.[23]

Despite the State Department's attempts to deny the possibility, the Pakistani crash was eventually attributed to sabotage. There was no doubt in this instance that some American authorities had tried to block the investigation.

Who tried to block the investigation into the possible terrorist murder of American troops at Gander? If no one did, then why didn't the FBI investigate? If it didn't investigate because of misplaced confidence in the Canadians, why was the decision not reversed after the Canadian investigation was discredited? Where was the documentation to show that anyone gave a damn?

Congressman Robin Tallon prodded the Subcommittee on Crime to conduct hearings, and shortly after ABC aired the *20/20* documentary in October 1989, the subcommittee announced it probably would. But by mid-summer 1990 nothing had happened, and a spokesman for the chairman, Democrat Bill Hughes of New Jersey, said hearings were unlikely.[24] It seemed like another fizzle.

But behind the scenes, the subcommittee's staff continued to beaver away, and in late October the subcommittee suddenly announced that it would hold hearings in December.[25] The unexpected decision was based on a sensational report compiled by the research staff. In keeping with the Canadian precedent, the staff report was promptly leaked.[26] It verified that "professional pilots and experienced investigators within the FAA" had rejected the ice theory, exactly as former FAA executive Dick Skully had told the CASB two years earlier. But it also contained real news.

George Seidlein, the NTSB's senior on-site investigator, had also rejected the ice theory. This had created "a public relations problem" for the NTSB, so Seidlein had been "pushed aside". The report portrayed the NTSB as "grossly negligent", adding that "a layman, after reviewing the Canadian reports, could determine that there were obvious flaws in the investigation and conclusions . . . it appears that the NTSB was predisposed not to find fault."

The subcommittee staff found no evidence of terrorism. But, the reports said, "we do not believe that concerned Americans can find any comfort in this fact. We say this because we found no evidence that there was an investigation of possible terrorism." The FBI's account of its role in the investigation was described as "unacceptable if not also unbelievable."

This sounded very interesting. and I accepted the invitation to testify; Norm Bobbitt and Ross Stevenson went with me to observe and give moral support. We learned that someone had been invited to speak for the majority but had declined on the grounds that "the hearings will amount to a political sideshow".[27]

We got to the hearing room a few minutes before ten a.m. The committee members had not yet arrived. They would be sitting in front of a deep blue velvet backdrop on a gently curving dais on the long side of a rectangular room with oak panels and a forty-foot ceiling—a much classier setting than that of any hearing of the Canadian Aviation Safety Board. A crowd of several hundred spectators and reporters was milling about, and technicians were still adjusting lights, taping wires to the carpet and racing around with canisters of film.

I recognized a number of journalists who had been covering the story for years. Harold Marthinsen was there. Theresa Griffin had come up from Miami, and introduced her brother-in-law, Bill Griffin. Bill Griffin later reminisced about his brother, the man whose carelessness had supposedly caused his own death and the deaths of 255 people entrusted to his care, and his anecdotes reinforced the picture of the painstaking perfectionist

Zona and Doug Phillips were there, of course. Doug was going to testify on behalf of the Families for Truth about Gander. He, Lee Levenson and I were on the first panel, along with retired army colonel Lewis L. Millett. Millett, an imposing old man with a bristly white moustache, wore a chestful of medals on his full uniform. The Second World War veteran had lost his son in the Gander crash. Our panel could be considered the case for the plaintiffs, setting the stage for testimony from the army, the NTSB, the FAA, the FBI and the State Department.

Chairman Bill Hughes opened the meeting. He said that "the National Transportation Safety Board and other agencies of the U.S. government chose to sit back and watch as the Canadian efforts became embroiled in controversy and dissension." The hearing would attempt to determine why the participation of American government agencies "was as limited as it seems to have been."

Doug Phillips said that "our government's apparent lack of concern for 256 Americans who may have been murdered" was "incomprehensible and un-understandable," and that the families demanded a thorough investigation of the causes and circumstances. I outlined the minority position, suggesting that the subcommittee could at least find out what had been loaded on the Arrow Air DC-8 in Cairo and re-examine the medical evidence. Lee Levenson summarized the deficiencies in the investigation and blasted Canadian and American authorities for ignoring problems with the DC-8 thrust reverser system.

Colonel Millett closed the plaintiffs' case with a highly emotional appeal for a new investigation. "All around the country Christmas decorations are being hung," he said, "holiday plans made, gifts wrapped. This will be the fifth Christmas that . . . 248 sons of liberty will be missing from the family hearth." Millett wanted just one gift for Christmas — to know the true cause of his son's death. He wiped tears from his eyes, providing the visual for all the next day's news articles.

Of course, none of us had said anything new. But the next panel, representing the United States Army, might be able to clear up many outstanding questions.

First on the list was General John Crosby, now retired and looking vaguely sinister in a grey business suit with a black eyepatch.

Dr. Robert McMeekin and Ronald Carpenter were with Crosby, both in full uniform, looking much the same as they had when testifying at the CASB public inquiry in 1986. McMeekin, no longer with the Armed Forces Institute of Pathology, seemed a shade greyer and more stooped. Carpenter, now a lieutenant-colonel, still looked like an Eagle Scout.

Crosby said that the Gander crash had "wrenched the souls and torn the hearts of the U.S. Army like no other peacetime event" (I recognized the phrase from the account of an army historian). His own torn heart went out to the family members who had lost their loved ones. But he and the army had played "only a support role in the investigation of the cause of death." He hadn't received intelligence briefings. No members of the 160th had been among the passengers. Bulldozing the crash site "was never raised as an issue of supreme importance" while he was in Gander.

"Didn't it bother you, General Crosby, that they ruled out all possibility of an explosion in two days?" A dutiful soldier wasn't bothered or not bothered by such things. Crosby just gathered information and passed it back. And, when all was said, Crosby had maintained his story and added nothing.

But McMeekin quietly devastated the medical conclusions of the majority report. He had been responsible for the autopsies, "focusing on identification"—not for forensic studies. He was not familiar with the toxicology reports and had not inquired about them. He was not familiar with Dr. Elcombe's report. Yes, he had signed the death certificates. What did the word "instantaneous" mean? It meant what it said, "certainly less than one minute." No, he didn't rule out the possibility of an explosion in the cargo hold.

This was sensational testimony. The only medical expert to testify at the CASB's public inquiry, the one Estey had praised to the skies, had contradicted the medical findings in the majority report. If death had come "less than one minute" after the crash, the toxicology results meant an in-flight fire. McMeekin had maintained his own story, but he had kicked the stuffing out of Dr. David Elcombe and the majority report.

Ronald Carpenter, U.S. Army liaison officer in Cairo in December 1985, added a key piece of new evidence. He had not been able to arrange a ramp pass for Captain Gerald De Porter, the accredited customs supervisor, and so De Porter had refused to sign off the customs documents. In short, the Arrow Air flight had left without clearing customs. De Porter had submitted a report to this effect, but the fact had been obscured in a cloud of testimony and documents about "the rigorous pre-flight inspection." This gave a concrete example of contrived testimony. Revealing that the doomed flight had not cleared customs would have been embarrassing, and might have led to disciplinary action and even sensationalist speculation. Since this had nothing to do with the crash, which we all knew was due to ice on the wings, why make a big deal about it? So the testimony had been contrived to evade the point.

Carpenter had skewered the "professional investigators" at the CASB. If they knew that the Arrow Air flight had not cleared customs, they withheld key evidence contradictory to their conclusion. If they didn't, they were inexcusably sloppy in checking basic documents. Since they didn't keep complete records, we could at least be certain of the sloppiness. As Senator Warren Rudman said with respect to some of the Iran-Contra activities, "I tend to believe it was a case of gross incompetence. I guess it's better to be dumb than crooked.''[28]

The army testimony left all kinds of unasked and unanswered questions, but it shattered any illusion that nothing more could be discovered about the Gander crash. There was only one more day scheduled for the hearings. Tomorrow we would hear from the NTSB, FAA, FBI and State Department.

Ross, Norm and I had dinner with Zona and Doug Phillips. After today, we all agreed, the Subcommittee on Crime couldn't avoid recommending a full-blown congressional investigation. We looked forward to the next day.

* * *

The following morning; the hearings had shifted to a smaller room. The atmosphere was less imposing and fewer reporters were present. The chairman and his retinue arrived at 9:40, ten minutes after the appointed hour, but there seemed to be some conflict of schedules—the other nine members of the committee were still not there. Subcommittee staffers made hurried calls. Representative Larry Smith of Florida arrived at ten o'clock and Chairman Hughes gavelled the meeting to order.

The first panel was ready to testify. Ronald Schleede was Chief of Major Investigations at the NTSB. George Seidlein, now retired, was "the investigator from the National Transportation Safety Board Assigned as the United States representative to Canada to participate with the Canadian Aviation Safety Board as a party to the investigation".[29] The deputy general counsel of the NTSB was with them. Members of the subcommittee drifted in and out as Schleede and Seidlein read prepared statements.

Schleede disclosed that he had collaborated with the CASB investigators on the report that was presented to the board. And, of course, he agreed with the ice theory. But it was Seidlein we wanted to hear. Everyone knew from the leaked staff report that Seidlein, the nominal NTSB investigator, had been removed for refusing to endorse the ice theory.

In contrast to former fighter pilot Schleede, who looked every inch the part, Seidlein was rumpled and decidedly low key. He wore an old blazer, and reading glasses hung from a cord around his neck. He had indeed been at the crash site in Gander, he said. He had kept himself up to date with the CASB's investigation and attended debriefing sessions, and he volunteered that he had seen the *20/20* television program. He would like to comment on the allegation of conspiracy and cover-up.

"Would I or any member of my team have been a willing and known partner in a conspiracy?" he asked rhetorically. "The answer is a categorical no." Apparently he hadn't weighed the possibility of unwilling or unknown partners. "Forensic technicians from the Royal Canadian Mounted Police," he explained, "found no evidence to support the claim of an in-flight fire or explosion, as reported on the U.S. television program." But he was not as expansive about the other conclusions. "I did not see the final CASB report nor was I ever asked to express an opinion."

Well, what about the ice theory? "There was absolutely no evidence that ice was a factor," Seidlein said. It turned out that Seidlein had been sidelined just after he formed this opinion. He had been an accident investigator with the NTSB for thirty years and had been investigator in charge on six or seven major accidents. No, he had never been removed from another case. He had signed off on them all.

If there was "absolutely no evidence that ice was a factor", why did the Canadian investigators attribute the crash to ice? "Sometimes it's difficult for investigators to overcome speculation they have seen in the media," Seidlein suggested.

Why would the NTSB support the ice theory? "The NTSB was predisposed not to be critical of the Canadians. So the draft report was rubber-stamped," he replied impassively.

Chairman Hughes wanted to make sure he had all this straight. "Did ice cause the crash?" he asked.

"No."

"Did you express this to anyone?"

"No."

"Were you asked?"

"No."

Incredulous, Hughes turned to Seidlein's former boss and asked, "Why wasn't he asked to participate?"

"Well, he was saying he was going to retire. . . ." Schleede began. But the chairman cut him off.

"Oh, come on. Don't give me a snow job. Why wasn't he brought in?"

Even though it was December, Schleede had a nice tan. He turned red under it. "He wasn't intentionally excluded," he replied.

"Who are you trying to kid? He's your man. Why did you send him up there? I find that very curious. He's your top guy, and you would have us believe you don't care what he thinks?"

Schleede steadfastly maintained that he had had no idea his investigator in charge didn't agree with the ice theory.

"Give me evidence of ice," the chairman demanded.

"There was a freezing mist," Schleede said truculently. He couldn't recall who else in the NTSB agreed with the ice theory. He himself agreed with it and he took sole responsibility for recommending that the NTSB support the majority finding.

Did Mr. Schleede and the NTSB still support the ice theory in spite of the rejection by the Canadian Transport Department, a Canadian judge, the American Air Line Pilots Association and his own investigator in charge?

Yes, they did, said a defiant Schleede. But he kept his eyes down.

"You're all nuts," interjected Representative Larry Smith, expressing what most of the audience must have been thinking by that time.

Schleede wasn't used to looking like a fool and a liar, but evidently it was better to absorb the flak than to suffer the alternative, whatever that might be. In fact, he was displaying the same intransigent and perverse devotion to the party line that I had witnessed in the CASB boardroom throughout the review of the Gander investigation. The difference was that, here, Schleede was exhibiting his truculent illogic in public, before a hostile chairman — and being thoroughly humiliated in the process.

"How could you be so derelict in your duty?" the chairman asked, as he adjourned until the afternoon.

* * *

Bob Cook, who had been the FAA representative on the Gander investigation, was now deputy U.S. representative at the International Civil Aviation headquarters in Montreal. Cook too looked like a pilot —

not a hot-shot jet jockey like Ron Schleede, but a transport driver trucking DC-3s on eastern routes in "Terry and the Pirates"—tubular, with thick black hair, dark jowls and a moustache.

Cook said he had arrived at Gander with NTSB investigator in charge George Seidlein at about seven p.m. on the day of the crash, and complained that the FAA had had to play second fiddle to the NTSB. The FAA had not been allowed to go to Cairo, for example. He had told his bosses that the FAA was being frozen out, but they supported the NTSB. He readily admitted that the manager of the FAA's Flight Standards Division in Miami hadn't agreed with the ice theory —he, himself hadn't liked a number of things about the investigation, but he had supported the ice theory anyway. Conclusive proof of ice or lack of ice would never be available, he said. Cook gave the impression of the good bureaucrat reluctantly doing what was expected. He got off lightly.

In fact, the steam seemed to have gone right out of Chairman Hughes and the two or three other members who showed up in the afternoon. The examination of the final two witnesses was spiritless and perfunctory.

Nicholas O'Hara, deputy assistant director of the FBI's Criminal Investigative Division, had had responsibility for oversight of the Gander investigation, but it didn't seem he had done much. Three forensic specialists were deployed to Gander on an FBI aircraft, he said. They stopped off in Ottawa for "legat" (legal attaché) John Terrio. The RCMP "declined FBI assistance" so, after cooling their heels for twenty four hours in Gander, the FBI agents went home.

"I'm at a total loss to understand why you would fold your tent after two days," the chairman said.

"There was no indication that the crash was caused by an explosive or destructive device or other criminal act," O'Hara explained.

"Did you or anybody in the FBI review the Canadian forensic or pathological data?"

O'Hara said he had, but he didn't sound convincing. Some sparring with the committee elicited that he had talked to Tom Thurman, one of the agents who had flown to Gander, just the week before, after receiving the request to appear before the subcommittee.

Tom Thurman, "one of the Bureau's best explosives technicians", had been dispatched to the crash scene immediately after Pan Am 103 disintegrated over Lockerbie. There he had taken a leading role in combing the wreckage for clues of an explosion. But Thurman had had no contact with the RCMP with regard to the Gander crash. His assistance had not been needed, O'Hara explained, since there was no physical evidence of an explosion. Thurman had looked at photos but, as far as O''Hara knew, he found no sign of an explosion.

At the close of a hearing such as this, the chairman usually gives witnesses a week or two to submit supplementary written comments. I made a note to myself for a recommendation for the next round of hearings: "Ask Thurman for opinion on Pinkel's report and minority photos."

O'Hara's testimony continued. No, the FBI had not prepared a report, just kept a few working papers. All this was moot anyway; the RCMP had investigated and found no evidence of an explosion, and everyone knew the RCMP was a top police force. No, the FBI had not obtained the RCMP report — that was unnecessary.

The final witness was Robert Pines, deputy assistant secretary for European and Canadian Affairs in the State Department. His evidence was, in a nutshell, that the State Department had had nothing to do with the Gander investigation. Pines and some of the subcommittee members made fleeting insider allusions to friction between the subcommittee and the FBI over the General Zia crash, but that had been in the bad old days. Relations had picked up. I couldn't for the life of me glean anything from the exchange.

The chairman released the witnesses; it was time for him to sum up. I was expecting the usual formula: thanking the witnesses, promising a mulling over, asking the witnesses to submit any further thoughts in writing before such-and-such a date. But Hughes didn't do any of this. Instead, he announced that the committee had unanimously agreed on three recommendations. The whole audience was stunned as he began to read from notes that had to have been typed, at the latest, during the lunch hour. So the testimony and questioning of the last three witnesses, at least, had been a sham. Somehow, between yesterday and this afternoon, the Subcommittee on Crime had had a change of heart about getting to the bottom of the Gander mystery.

"In June 1990 the subcommittee sent a request to the Canadian embassy requesting specific information about the RCMP investigation into possible terrorist or other criminal involvement in the Gander crash," said the preamble to the first recommendation. "We have received no response . . . we recommend that, in the 102nd Congress, this committee continue to pursue this avenue of additional information."

The second recommendation should have pleased Lee Levenson: "Due to preoccupation with icing on the airframe as the cause of the crash, inadequate analysis has taken the place of other possible causes of the crash. This is particularly true of mechanical failure and systems failure as possible causes. We recommend the committee focus particular attention on additional analysis in these areas."

Finally, the committee recommended that "the Attorney General review the history of both Gander and Pakistani air disasters . . . to insure there is high level review and full coordination at the executive level."

The recommendations served to wash the Gander file from the subcommittee's deck. It would be useless to lobby committee members or submit additional testimony.

Hughes gavelled the meeting closed, picked up his papers and left the dais. He had shattered the hopes and expectations of the Families for Truth about Gander, and their supporters. The sudden termination was completely at odds with the introductory remarks he had made just the day before. Yet it was completely in keeping with the official reaction that had characterized the Gander case from the beginning. The media people crowded around Hughes, but he had nothing to add.

The rest of us found it difficult to answer the questions. The Phillipes' disappointment was palpable. Doug Phillips—usually a gentle, scholarly man—was close to rage. Ross Stevenson denounced the proceedings angrily. I really didn't know what to think; I said that I was grateful for the crumbs, which were more than we'd been able to get for the past two years —that was true but it was of little help.

As we gathered up our papers and left for home, American troops were massing in Saudi Arabia, preparing to teach evil Saddam Hussein a lesson. Operation Desert Storm would soon sweep the Persian Gulf and dominate the news. Reminders of the American military machine's inability to protect its own would become unseemly and unpatriotic. The Gander chronicles, it seemed, were destined to rest beside those of the *Hindenburg* and other great disasters where efforts to uncover the cause were drowned in a sea of bureaucratic self-interest, shameless incompetence and dogged, inexorable deceit.

NOTES

Correspondence and other internal government documents cited in the notes have been made public through the media or through the Freedom of Information or Access to Information Act. I have attempted to cite the most accessible source.

The following abbreviations are used:

AW — *Aviation Week & Space Technology*
CP — *Canadian Press*
IHT — *International Herald Tribune*
G&M — *Toronto Globe & Mail*
NYT — *New York Times* (section, page and column cited)
WP — *Washington Post*

Arthur Conan Doyle —
Sir Arthur Conan Doyle, *Sherlock Holmes: The Complete Novels and Stories*, Bantam Classic Edition, 1986.

Bradlee —
Ben Bradlee Jr., *Guts & Glory: The Rise and Fall of Oliver North* (D.I. Fine) 1988.

chrono—
The National Security Archive, *The Chronology: The Documented Day-by-Day Account of the Secret Military Assistance to Iran and the Contras* (Warner Books, New York, 1987).

Dubin—
Report of the Commission of Inquiry on Aviation Safety, Commissioner: The Honourable Mr. Justice Charles L. Dubin, Volume 1, May 1981 (available from Canadian Government Publishing Center Hull, Quebec).

Emerson—
Steven Emerson, *Secret Warriors: Inside the Covert Military Operations of the Reagan Era* (G.P. Putnam & Sons, New York, 1988).

Final report—
CASB Report No. 85-H50902, Arrow Air Inc. Douglas DC-8-63 N950JW, Gander International Airport, Newfoundland, 12 December 1985. (Consits of a majority report and separately bound Dissenting Opinion.)

I-C (Iran-Contra report)—
U.S. Government Printing Office, *Report of the Congressional Committees Investigating the Iran Contra Affair with supplemental, Minority, and Additional Views*, H. Rept. No.100-433, S. Rept. No.100-216, Washington, 1987.

I-C, Deposition—
One of the volumes of Appendix B of the Iran-Contra report cited above, volume and page numbers are cited.

Ledeen—
Michael A. Ledeen, *Perilous Statecraft: An Insider's Account of the Iran-Contra Affair* (Scribner 1988).

Segev—
Samuel Segev, *The Iran Triangle: The Untold Story of Israel's Role in the Iran-Contra Affair* (The Free Press, New York 1988).

Senate hearings—
U.S. Government Printing Office, *Airline Safety*, Hearings before the Permanent Subcommittee of the Committee on Government Affairs United States Senate, March 6-13, 1986, S. HRG 99-1049.

Subcommittee on Crime—
United States House of Representatives Committee on the Judiciary, Subcommittee on Crime, Oversight Hearing on the Fatal Plane Crash in Gander, Newfoundland, on December 12, 1985 (held in Washington Dec. 4-5, 1990)

Transcript—
Canadian Aviation Safety Board (CASB), *Transcript of the Public Inquiry into the Accident which occurred near Gander Airport, Gander, Nfld. December 12, 1985.*

Veil—
Bob Woodward, *Veil: The Secret Wars of the CIA 1981-1987* (Simon & Schuster) 1987.

Prelude

1. Michael M. Mooney, *The Hindenburg* (Dodd, Mead, 1972).
2. A.A. Hoehling, *Who Destroyed the Hindenburg?* (Little,Brown & Company, 1962).

Introduction

1. Arthur Conan Doyle, *The Adventure of Westeria Lodge.*

2. Conversation with Ross Stevenson, January 5, 1986.

3. "'Ball of Fire' lit sky for 5 seconds witness says", *Toronto Star*, Dec. 13, 1985.

4. "258 from U.S. die in Gander air disaster", *G&M*, Dec 13, 1985.

5. Subcommittee on Crime, Exhibit 33.

6. "Claims by 2 groups discounted", *NYT*, Dec. 14, 1985, p. 4; datelined GANDER, Dec. 13 (UPI).

7. Conversation with author, March 16, 1989.

8. "Canada's Worst Aircrash Kills 258", *Ottawa Citizen*, Dec. 13, 1985, p. 1.

9. Subcommittee on Crime, Exhibits 6 and 7. The log for December 14 twice notes that the investigator in charge had determined that "there was no explosion or fire prior to impact".

10. "Aircraft's wings not de-iced", *Edmonton Journal*, Dec. 13, 1985, p. 1.

11. "U.S. Autopsies set for Crash Victims", *NYT*, Dec. 15, 1985, p. 23.

12. Canadian Aviation Safety Board, Annual Report 1986, p. 12.

13. Veil, p. 85.

14. Long Commission Report, Congressional Record, Jan. 30, 1984.

15. Speech to American Bar Association, July 8, 1985.

16. I-C, p. 252; Ibid, Note 139, p. 265.

17. "President Issues Veiled Warning", *NYT*, June 17, 1985, p. 1.

18. Emerson, p. 211.

19. "Reagan in Appeal: After third incident he calls on West to help combat terrorism", *NYT*, June 21, 1985, I 1:3.

20. Bradlee, p. 305.

21. "27 Injured in Copenhagen in Mideast Terrorist Blasts", *NYT* July 23, 1985, I 3:5.
22. *NYT*, June 21, 1985, I 1:3, op. cit.
23. Emerson, p. 146; Peter Mass, *Manhunt: The Incredible Pursuit of a C.I.A. Agent Turned Terrorist* (Random House 1986).
24. "Gander Crash Puts Spotlight On Aviation Group and its Regulatory Crashes", *NYT*, Dec. 24, 1985, I 9:1
25. John Nance, *Blind Trust* (William Morrow, 1986) p.159.
26. Senate hearings, p. 94.
27. "Active Liberal named to head safety board", *Winnipeg Free Press*, Feb. 11, 1984.
28. Lloyd Axworthy, "Control of Policy", Policy Options, April '83, p. 17.
29. See for example text of an address given by B. M. Deschenes at the International Society of Air Safety Investigators (ISASI) Annual International Seminar, Atlanta Georgia, October 7, 1987.
30. Transcript, p. 21.
31. Made public in "The Gadzos Report", *Canadian Aviation*, December 1988, p. 21.
32. C.H. Rolf, "Report to the Attorney General of Alberta on a Public Inquiry held under the Fatality Inquiries Act", November 24-26, 1986.
33. Canadian Aviation Safety Board, "Propeller Blade Examination, King Air A-100, C-FDOY", CASB Engineering Project EP 187/86.
34. CASB Aviation Occurrence Report 85-H50007, Okanagan Helicopters, Bell 214ST, C-GVZO, Big Lake, Alberta, 23 October 1985. Conditional Draft No. 1, Aug. 1, 1986, p. 24 (modified in the final report).

Book 1: Political and Bureaucratic Imperatives

1. *G&M*, May 13, 1978, p. 65.
2. ATS/Aeradio Operational Interface Study Group
3. Stevie Cameron, *Ottawa Inside Out* (Key Porter 1989), p. 55.
4. "Commission of Inquiry for Aviation Safety", Phase II, Opening Statement by John Sopinka, Feb. 25, 1980.
5. A full account of this accident and its repercussions can be found in N. A. Komons, *The Cutting Air Crash: A Case Study in Federal Aviation Policy* (U.S. Department of Transportation 1973).
6. "Federal Aviation Agency Historical Fact Book: A Chronology, 1926-1963", p. 17.
7. Report of J. R. Booth, p. 20 (reproduced in Dubin, Exhibit II.)
8. Report of the Board of Inquiry into the Accident at Toronto International Airport, Malton, Ontario, to Air Canada DC8-CF-TIW aircraft on July 5, 1970
9. The McLearn report is reproduced in Dubin, Exhibit II.
10. Aircraft Accident Report, Lockheed L188 CF-PAB, Rae Point, Melville Is. *NWT*, 30 Oct., 1974.
11. Dubin, p. 113.
12. *G&M*, Aug. 29, 1978, p. 5.
13. Dubin, p. 127.
14. Dubin, p. 133.
15. Dubin, p. 149.
16. Dubin, p. 155.
17. "That objective look", *G&M*, Jan. 10, 1981 p. 6.
18. e.g. "Makeup of Committee under fire, *Halifax Chronicle Herald*", May 30, 1981, p.4.
19. "Opposition criticizes choosing air official to study safety report", *G&M*, May 29, 1981, p. 8.

20. "Mr. Pepin Wings It", *G&M*, June 24, 1981, p. 6.
21. Letter, AARB Chairman Max Friedl to Associate Deputy Minister of Transport Jaffery Wilkins, February 14, 1983.
22. "MoT offers feeble excuse for fake transcripts", *Winnipeg Free Press*, Dec. 16, 1983.
23. Minister of Supply and Services Canada, *Final Report of the Board of Inquiry into Air Canada Boeing 767 C-GAUN Accident — Gimli Manitoba*, July 23, 1983, p. 17-18.
24. Ibid, p. 18.
25. "Trudeau rewards surprised board", *Ottawa Citizen*, July 9, 1984.
26. "The CASB as a Political Plaything", *Canadian Aviation*, Aug. 1984, p. 2.
27. Transport Canada Press Release, No 326/84 for release November 5, 1984.

Book 2: Corpus Delecti

1. Arthur Conan Doyle, *The Sign of Four*.
2. "Tragedy at Gander", *Macleans*, Dec. 23, 1985.
3. "258 Americans die in Gander airport disaster", *Toronto Star*, Dec. 13, 1985.
4. C. Hennigar, (Chief, Firefighting & Rescue, Gander International Airport), "Casualty Report Arrow Air Crash, 12 December 1985".
5. "Crash cause remains mystery", *Ottawa Citizen*, Dec. 14, 1985.
6. "FBI Expert Joins Crash Probe", *WP*, Nov. 29, 1989.
7. Lt. Col. R.A. McLean (Commander, Canadian Forces Base, Gander), "Report on CFB Gander Assistance", 1010-1 (BComd), Jan. 29, 1986.
8. T.C. Badcock, *A Broken Arrow* (Al Clouston Publications, St. John's 1988).

9. *NYT* Dec. 15, 1985, p. 23.
10. Emerson, p. 181.
11. U.S. Army press release, Dec. 14, 1985.
12. "Doubt Surrounds Probe of Gander Crash", *St. Petersburg Times* Feb. 26, 1989.
13. Ibid.
14. Ibid.
15. Ibid.
16. "Report on CFB Gander Assistance", op. cit., Tab L.
17. CASB Comunique #18/85, Dec. 12, 1985.
18. Transcript p. 2003.
19. Authority to enter into contract dated May 8, 1985. Obtained under The Access to Information Act.
20. Interview, ABC Television Documentary 20/20, "What Really Happened?", Oct. 13, 1989.
21. Transcript p. 23.
22. Transcript p. 52.
23. Transcript p. 24.
24. Transcript p. 53.
25. Transcript p. 73.
26. Transcript p. 80.
27. CASB Report No. 85-H50902, Site Survey/Security, Appendix G. (This supporting document has been omitted from the group of Laboratory Reports "available on request from the CASB" listed in Appendix E of the Final Report.)
28. RCMP Forensic Laboratory Report, File No. 85-OL-01597, Report No. ONE, Submitted by B. W. Richardson, April 1, 1986.
29. Walter Cunningham, *The All-American Boys: An insider's candid look at the space program and the myth of the super hero* (Macmillan, New York 1977), p. 100.

30. "The Evidence", *San Diego Reader*, Sept. 13, 1990.
31. RCMP Forensic Laboratory Report, File No. 85-OL-01597, Report No. TWO, Submitted by B. W. Richardson, July 2, 1986.
32. International Civil Aviation Organization (ICAO), *Manual of Aircraft Accident Investigation*, 4th edition 1970, Part III, Chapter 11, Para. 5.1.
33. Memorandum J. Melson to Peter Boag, July 25, 1986.
34. Memorandum Joe Bajada to Peter Boag, July 7, 1986.
35. Report of the Court of Investigation, Accident to Air India Boeing 747 Aircraft VT-EFO, Feb. 26, 1986; p. 103, Para. 3.2.11.33.
10a. Arthur Conan Doyle, *The Red-Headed League*.
36. Transcript p. 173.
37. Transcript p. 181.
38. Statement included as Appendix A-14 in the report of RCMP S/Sgt. W.R. Fraser, Feb. 23, 1986, RCMP File Reference: Div. 85-B-4337.
39. Ibid, Appendix A-15.
40. Transcript p. 1459.
41. Transcript p. 1462.
42. Transcript p. 1465.
43. Transcript p. 1479.
44. Transcript p. 1489.
45. Report of RCMP S/Sgt. W.R. Fraser, op. cit., para. 101.
46. Transcript p. 1470.
47. Korean Overseas Information Service, "Investigation Findings: Explosion of Korean Air Flight 858", January 1988.
48. Transcript p. 1472 (corrected to match tape).
49. Transcript p. 1474.

50. Diplomatic Security and Anti-Terrorism Act; Title 18, U.S.C, Sect. 2331, Extraterritorial Jurisdiction Over Terrorist Acts Abroad Against U.S. Nationals.
51. Destruction of Aircraft or Motor Vehicles, Title 18 U.S. Code 32.
52. Transcript p. 1481.
53. Unsigned summary of interview taken in Miami on March 3, 1986 on FBI letterhead. Subcommittee on Crime Exhibit No. 24.
54. Letter from R.S. Saunders to Director of FBI, February 2, 1987.
55. Memorandum from John A. Mintz to Mr. Clarke Re: Arrow Air Investigation, March 24, 1986 (with addendum). Subcommittee on Crime Exhibit No. 26.
56. Ibid.
57. Steven Emerson and Brian Duffy, *The Fall of Pan Am 103: Inside the Lockerbie Investigation* (Putnam's Sons, New York 1990), p. 53.
58. Washington Press Conference, Dec. 28, 1988. Reported by Emerson and Duffy, op. cit., p. 143.
59. Mintz letter to Clarke, op. cit. Expanded in testimony of Nicholas O'Hara at the Subcommittee on Crime hearings in December 1990.
60. Minutes, CASB board meeting Jan. 26, 1988, p. 3.
61. James L. Pate, "Bones of Contention: Army Stonewalls on Lab Scandal", *Soldier of Fortune*, Dec. 1986, p. 35.
62. Ibid.
63. Emergency Preparedness Canada, "The Gander Air Crash", Report 87.01.01, January 1987.

64. Letter from the Newfoundland Medical Board to Dr. A. E. Shapter, James Memorial Hospital, Dec. 14, 1985.

65. Emergency Preparedness Canada, op. cit., p. 53.

66. "Fate knocks on door of bereaved families", *Ottawa Citizen*, December 14, 1985.

67. "Clues are Sought to Crash That Killed 256 in Gander", *NYT* Dec. 14, 1985, p. 1.

68. Major P. Shearston, Memorandum for the record, Feb. 24, 1986.

69. Report of RCMP S/Sgt. W.R. Fraser, Feb 23, 1986, op. cit., para 92.

70. U.S. Army Document, DAPE-PSJ, 10 February 1986, annotated "Approved by BG Beavers". Copy included under Tab S, Report on CFB Gander Assistance, op. cit.

71. Sharon B. Young, "Remains Misidentified, Ex-Lab Worker Says", *Army Times*, Sept. 22, 1986, p. 3.

72. Transcript p. 777.

73. Richard A. Gabriel, *Military Incompetence: Why the American Military Doesn't Win* (Hill and Wang 1985), p. 182.

74. Final Report p. 53.

75. Ibid, Item 18, p. 94.

76. Arthur Conan Doyle, *The Boscombe Valley Mystery*.

77. Questionnaire to inbound troops on the December 4, 1985 Arrow Air flight administered on behalf of the Task Force Commander (Lt. Col. Charles R. Huggins) in February 1986.

78. Interview with CASB investigators, Jan. 20, 1986.

79. *Aviation Daily*, Jan. 24, 1983, p. 121.

80. Laurie Taylor, *Air Travel: How safe is it?* (BSP Professional Books, Oxford, 1988), p. 213.
81. A decision record of the pre-hearing conference is Exhibit 1 of the CASB's Public Inquiry.
82. Letter from F.E. Curran (Director of Military Services, Transamerica Airlines) to Military Traffic Management Command, March 4, 1986.
83. Decision record of the pre-hearing conference, op. cit.
84. Transcript p. 348. It was also agreed that Williams would answer written questions.
85. "Gander air crash inquiry may shock", *Toronto Star*, Jan. 28, 1990, p. 1.
86. Transcript p. 364.
87. Transcript p. 369.
88. Transcript p. 459.
89. Transcript p. 475.
90. Final Report, Sect. 1.17.4.6, p. 54.
91. Transcript p. 730.
92. "The Gander Air Disaster: An Accident or Sabotage?", *Ottawa Citizen*, March 25, 1989, p. B16.
93. Transcript p. 1234.
94. Interview with FBI, March 3, 1986. Subcommittee on Crime, Exhibit 24.
95. Transcript p. 1230.
96. Transcript p. 441.
97. *Toronto Star* Jan. 28, 1990. op. cit.
98. Transcript p. 486.
99. Letter from Congressman John Conyers (Chairman, House Legislation and National Security Subcommittee) to Charles Bowsher (Comptroller General of the General Accounting Office), Nov. 28, 1989.
100. United States General Accounting Office, "Military Airlift: Information on Gander Crash and

Improved Controls over Military Charters", Sept. 1990.

101. Transcript p. 555.
102. Transcript p. 462.
103. Final Report, p. 53.
104. Inventory of weapons produced by Pentagon Telecommunications Center, Jan. 23, 1986, Subcommittee on Crime, Exhibit 12.
105. Transcript p. 453.
106. Transcript p. 525.
107. Transcript p. 369.
108. Transcript p. 561.
109. Jean Dolan, "Personal Reflections on the Gander tragedy", *The Free Press*, Quakertown, Pa., p. 5, Sept. 2, 1986.
110. Commander of MTMC, Falls Church, Va. to list of recipients.
111. "Aircraft Accident Investigation" by Robert R. McMeekin in Roy L. DeHart, ed., *Fundamentals of Aerospace Medicine* (Lear & Febeiger, Philadelphia 1985), Chapt. 27, pp. 762-814.
112. "What Really Happened?: The Challenger disaster", *Miami Herald* Nov. 13, 1988.
113. Dr. Joseph Kerwin (NASA spokesman), Press conference, July 28, 1986.
114. Transcript p. 1512.
115. Transcript p. 1514.
116. "News Stories Sometimes Planted in Garden Plots", *Ottawa Citizen*, June 27, 1988.
117. Letter from Harry M. Weisberg to Ken Thorneycroft, April 16, 1988.
118. "The Rime of the Ancient Mariner", Samuel Taylor Coleridge.
119. Transcript p. 1801.
120. Transcript p. 1906.

121. Transcript p. 1924.
122. Transcript p. 1846.
123. "Canada Air Probe Faulty", San Diego Union, June 25, 1989.
124. Transcript p. 1925.
125. For a full discussion of the Air Florida accident see *Blind Trust*, by John J. Nance, op. cit.
126. The four cases are scattered through Brumby's testimony, but are conveniently listed in Appendix C of the final report.
127. National Transportation Safety Board, Report No. NTSB-AAR-78-7, Japan Airlines, DC-8-62F, JA-8054, accident at Anchorage, Alaska, on 13 January 1977.
128. J. Hanjian, "The Effect of Frost on Take-off Performance of F-86 and B-47 Aircraft", Wright Development Centre, May 1956.
129. Air Transportation Association of America, Minutes of the Second Meeting of Working Group on Operation with Accumulated Frost, Ice and/or Snow and Request for additional data, December 5, 1969.
130. Letter from O.R. Dunn (Director of Aerodynamics, Douglas Aircraft Company) to E.L. Thomas (Director of Engineering, Air Transport Association of America), Dec. 2, 1969.
131. Transcript p. 1923.
132. This account is drawn from, Stanley Stewart, *Historic Air Disasters* (Ian Allan Ltd., 1986).
133. Transcript p. 269.
134. Transcript p. 242.
135. Statement included as Appendix A-40 in the report of RCMP S/Sgt. W.R. Fraser, op. cit.
136. "Human Factors Group Factual Report", Public Inquiry, Exhibit 8.

137. Transcript p. 244.
138. His statement included as Appendix A-37 in the report of RCMP S/Sgt. W.R. Fraser, op. cit.
139. Transcript p. 225.
140. Transcript pp. 20002-8.
141. *The Tower Commission Report: The Full Text of the President's Special Review Board* (Bantam Books, 1987).
142. chrono pp. 201-3.

Book 3: "That Oliver North Crap"

1. After Edward P. Boland, chairman of the House Select Intelligence Committee. This was, actually, an extension, with change of language, of the original "Boland amendment" of December 1982.
2. Emerson p. 221.
3. Peter Maas, op. cit. Wilson is now serving a 52-year prison sentence for illegally supplying arms to Libya's Muammar Gadaffi, and for conspiring to murder federal prosecutors and witnesses.
4. Secord was originally indicted on twelve separate criminal charges. Eleven of these charges were dropped in return for cooperation with the prosecutors; he pleaded guilty to the remaining charge of lying to congressional investigators. In January 1990 he was sentenced to two years probation.
5. chrono p. 234.
6. I-C, Testimony of Albert Hakim, p. 1034 etc.
7. I-C; *IHT*, January 19, 1987.
8. I-C, Deposition of John J. Rugg, Vol. 24, p. 1-32.
9. I-C, Testimony of Albert Hakim, op. cit., p. 1034 and 1037.
10. Christopher Robbins, *Air America* (Avon, 1985).

11. Victor Marchetti, John D. Marks, *The CIA and the Cult of Intelligence* (Knopf, New York, 1974)., p. 137.
12. I-C, Deposition of James Bastian, Vol. 2, pp. 894-1064.
13. I-C, Testimony of Albert Hakim, p. 1040.
14. "Contra Arm Deals Tied to Lisbon", *IHT* Jan. 19, 1987.
15. I-C, Deposition of Robert H. Mason, Volume 17, Exhibit RHM 21.
16. Pamela Naughton, personal communication.
17. I-C, Deposition of Richard B. Gadd, Vol. 11, p. 206.
18. See also: Leslie Cockburn, *Out of Control: The Story of the Reagan Administration's Secret War in Nicaragua, the Illegal Arms Pipeline, and the Contra Drug Connection* (Atlantic Press 1987); Contra Arms Crew Said to Smuggle Drugs, NYT Jan. 20, 1987 p. 1.
19. Speech to Cuban-American audience in Miami, chrono p. 610.
20. I-C, Chapt. 8-12.
21. I-C, Deposition of Hugo de la Torre, Vol. 8, p. 1047.
22. Segev, p. 173.
23. "Mystery Plane from Iran", *Jerusalem Post*, Sept. 18, 1985, p. 1.
24. Ledeen, p. 134.
25. I-C p. 169. Buckley's successor as CIA Beirut Station Chief also became a terrorist victim. An Associated Press bulletin reported on December 25, 1988 that he was on board Pan Am Flight 103, blown up over Lockerbie, Scotland.
26. *NYT* September 19, 1985, I, 10:6.

27. The Congressional investigation determined that the scheme was, nevertheless, illegal. I-C p. 178.
28. chrono p. 175.
29. chrono p. 185.
30. I-C p. 178.
31. Ledeen, p. 153.
32. chrono p. 187.
33. chrono p. 187.
34. I-C p. 191, Note 147.
35. I-C p. 191. The Turkish authorities who believed the aircraft was owned by the government of St. Lucia sent a bill for $460 for the overflight. *WP* Feb. 24, 1987.
36. Bradlee p. 320.
37. chrono pp. 281, 284 etc.
38. I-C, Deposition of James Bastian, op. cit.
39. "Arrow Air Delivered Arms for Contras," *South Florida Business Journal*, Vol. 10, No. 33, April 16, 1990.
40. Donald T. Regan, *For the Record: From Wall Street to Washington* (Harcourt, Brace, Jovanovitch, San Diego, 1988), p. 23.
41. The account of operation Snow Bird and the early days of the Special Operations Division is based on Emerson, p. 22 etc.
42. There were eight-hundred soldiers, fifty civilian observers and three-hundred support personnel. *NYT*, Feb. 16, 1984 I, 1:2
43. Segev p. 194-5; I-C p. 175.
44. Veil, p. 493.
45. Emerson p. 212.
46. Bradlee p. 447.
47. Veil p. 496.

Book 4: Draft Report

1. CASB minutes Dec. 5, 1986, p. 2.
2. CASB Act, Sect. 23(3).
3. "Are Air Crash Inquiries Being Impeded?", *WSJ*, Sept. 10, 1988.
4. Mahon, op. cit., p. 243.
5. Conversation with author, Aug. 27, 1990.
6. Mahon.
7. Ibid., p. 283.
8. The authors eventually published a version of their report, "Analyses of Arrow Air DC-8-63 Accident of December 12, 1985: Gander, Newfoundland", James K. Luers and Mark A. Dietenberger, *Journal of Aircraft*, Vol. 27, No. 6, June 1990, pp. 543-550.
9. Robert J. Serling, *The Electra Story: Aviation's Greatest Mystery* (Doubleday, 1963). The jury eventually found Lockheed and Allison equally responsible.
10. Interpretation of the Canadian Aviation Safety Board Act, Robert Decary, Q.C., April 16, 1987.
11. Memo from Deschênes to Thurston, June 16, 1987.
12. Letter from Paul M. Tellier to Bernard Deshênes, June 10, 1987.
13. Letter from John Crosbie to Bernard Deschênes, July 10, 1987.
14. Letter from Bernard Deschênes to John Crosbie, July 14, 1987.
15. Ira H. Abbott and Albert E. Von Doenhoff, *Theory of Airfoil Sections, Including a Summary of Airfoil Data* (Dover, 1949).
16. Regulation 20, Canada Gazette Part II, Vol. 118, No. 25.
17. Minutes, meeting of October 14-16, 1987, p. 5.

18. Letter from Ken Johnson to Frank Thurston, Oct. 21, 1987. Reproduced in minutes.
19. *G&M Report on Business*, Feb. 1988, p. 49.
20. The impression that Hinton's tight control of information extended to the investigators themselves was confirmed later when I saw a copy of an impassioned memo the CASB's director of engineering had written Hinton about "management related problems" on March 10, 1986: "I do not react well to Engineering Branch personnel telling me that they have been instructed not to keep me informed (presumably for fear that the IIC [investigator-in-charge Peter Boag] will not be the sole source of accident related information). . . . I also find my lack of control over what is happening in these situations embarrassing and certainly in contravention of good management practice."
21. Pratt & Whitney Aircraft Group, Dec. 15, 1986.
22. "Board veils pilots' role in crashes, study finds", *G&M* Nov. 25, 1987.
23. Letter from Roland Harrison to Frank Iacobucci, Nov. 23, 1987.
24. CASB minutes, December 8-10, 1987, p. 4. (Finding 29, p. 100)

Book 5: Fear and Loathing at the CASB

1. "Aviation Safety Board plagued by internal rows, chairman says", *G&M*, Jan. 13, 1988, p. A5.
2. Letter from K. Scott Gudgeon to Franz Reinhardt, Jan. 29, 1988. I found a copy of the letter in the files much later.
3. U.S. Senate hearings, Aircraft Chartered by the DoD.

4. Letter from Ray Davis to Ken Johnson, Feb. 10, 1988, attached to minutes of Feb. 23.

5. "Heaven help air safety chief", *Ottawa Citzien*, April 6, 1988, p. A12.

6. Transport Canada press release No. 93/88, for release April 5, 1988.

7. Memorandum to Minister, April 29, 1988.

8. CIWW News, April 15, 1988.

9. CBC, The National, April 15, 1988.

10. Letter from Arrow Air executive vice-president Harry Weisberg to Ken Thorneycroft, April 16, 1988.

11. "Blast caused Gander crash: expert", *Ottawa Citizen*, April 30, 1988. p. 1.

12. CASB Report No. 85-H50902 Site Survey/Security, op. cit., Photo 22.

13. Hans Diehlmann, interview with CASB investigator Jan. 19, 1986.

14. Human Factors Group Factual Report Public Enquiry, Exhibit 8, op. cit.

15. Global TV, April 5, 1988.

16. Commons Debates, June 8, 1988, p. 16, 240.

17. Ibid.

18. "Comments on the Critique of the Ice Contamination Hypothesis", D. J. Langdon, P. Eng., Chief, Systems Engineering, June 13, 1988.

19. "Inquiry demanded in Gander crash", *Ottawa Citizen*, July 14, 1988, p. A1.

20. "Gander Report," CP, Aug. 22, 1988.

21. "Crash Illness," CP, Aug. 8, 1988.

22. *NYT* Oct. 18, 1988, p. A5.

Book 6: Final Report

1. Arthur Conan Doyle, "Silver Blaze".

2. Stephen Farber and Marc Green, *Outrageous Con-*

duct, (Ivy Books, 1988), p. 195 passim.

3. The phrases come from a letter from Ken Johnson to Lt. Gen. Q. H. Becker, April 20, 1988.

4. Letter faxed from Dawson to Thorneycroft, Oct. 7, 1988.

5. Letter from Shepherd to Dawson, Nov. 18, 1988. Copy to CASB Nov. 23, 1988.

5a. Liberals Accuse the Mulroney Government of Major Cover-Up in "Gander Crash Affair", Dec. 8, 1988.

5b. "Arrow Air: Ice or Sabotage?", *Counter Terrorism Security Intelligence*, Dec. 12, 1988, p. 2.

6. Letter from Norman Bindon to Brian Mulroney, Jan. 6, 1989.

7. "Pilots' group disputes report blaming ice for Arrow disaster", *Louisville Courier-Journal*, Dec. 13, 1988.

8. "Performance of Arrow Aircraft consistent with icing, says board", CP, Dec. 20, 1988.

9. "Crash resembles Air India Blast", *Ottawa Citizen*, Dec. 24, 1988, p. A6.

10. "Flight 103", CP Jan. 15, 1989.

11. Letter from David C. Williams to Ken Johnson, Dec. 13, 1988.

12. Arthur Canon Doyle, *A Study in Scarlet*.

13. "Crosbie chose Sopinka to study air safety board after meeting dissidents", *G&M*, Jan. 27, 1989, p. 1.

14. "Air safety board's credibility slips away", *Toronto Star*, Feb. 14, 1989.

15. "Minority Faction Faults Arrow Air Crash Analysis", *AW*, Dec. 19, 1988, p. 107.

16. "Safety Experts Cite Similarities Between DC-8, 747 Crashes," *AW*.

17. "Review the Gander Crash," *AW*.

18. "Leaders of Canada's Liberal Party May Seek Hearings on Gander Crash," *AW*, Feb. 27, 1989, p. 24.
19. CASB Aviation Occurrence Report 87-070004, Trillium Air, Britten-Norman Islander C-GIRH, Toronto Island Airport, January 12, 1987.
20. Interview with the author, July 10, 1990.
21. CP 07-03-89; "Bouchard rejects crash probe", *Ottawa Citizen*, March 8, 1989; p. A4.
22. "Not enough evidence in Arrow bomb claim, says Bouchard", *Ottawa Citizen*, Mar. 9, 1989, p. A3.
23. "Panic-stricken feds lurched between crises in aviation's PR fiasco", *Canadian Aviation*, May 1989, p. 24.
24. "Ottawa rules out holding inquiry into Arrow crash", *St. Johns Telegraph-Journal*, March 9, 1989.
25. CASB Aviation Occurrence Report 89W0066, Air Canada McDonnell Douglas DC-8-73-R C-FTIR, Edmonton International Airport, March 28, 1989.
26. "Air safety report demanded by July", *Windsor Star*, March 23, 1989.
27. "Young urges Arrow air crash be reopened", *St. Petersburg Times*, March 18, 1989.
28. *G&M*, March 22, 1989, p. A1.

Book 7: Damage Control

1. Inscription on statue, National Academy of Sciences, Washington D.C.
2. "Judges to look at two crashes", *London Free Press*, March 30, 1989.
3. *L'Heure Juste*, CHRC, March 30, 1989.
4. Interview, *Présent National*, March 31, 1989.
5. "Bouchard faked probe papers, Liberals charge", *Toronto Star*, April 1, 1989.

6. Letter from Willard Estey to Dr. and Mrs. J.D. Phillips, April 21, 1989.

7. Letter from Willard Estey to Roger Lacroix, April 12, 1989.

8. CP, April 11, 1989.

9. CP, May 10, 1989.

10. "CASB member wrote crash victims' families", *G&M*, April 21, 1989, p. A4.

11. "Critics blast air safety board bill", *Ottawa Citizen*, April 18, 1989, p. A5.

12. CP May 19, 1989.

13. Memo from Thorneycroft to CASB staff, June 30, 1989.

14. William Broad and Nicholas Wade, *Betrayers of the Truth: Fraud and Deceit in the Halls of Science* (Simon & Schuster, 1982).

15. "Lax security preceded Gander crash," *St. Petersburg Times*, April 12, 1989, p. A1.

16. Biographical Sketch, Vol. X, p. 30.

17. "Young urges Gander inquiry be reopened," *St. Petersburg Times*, March 18, 1989, p. A1.

18. "Congressional letter seeks new probe of Gander Crash" *St. Petersburg Times*, Oct. 24, 1989, p. 1.

19. "Gander crash no bombing, Canadian officials say", *Ottawa Citizen*, Oct. 14, 1989.

20. "Cause of Gander crash still in question", *Washington Post*, Nov. 9, 1989.

21. "Gander crash probe likely to shock," *Toronto Star*, Jan. 28, 1990.

22. "Gander: After Five Years the Odor Lingers," *Counter-Terrorism & Security Intelligence*, Dec. 17, 1990.

23. See column by A.M. Rosental, *NYT* Nov. 1, 1988, p. 31:1

24. "U.S. TV station seeks Gander files," *Ottawa Citizen*, July 3, 1990.
25. "New Congressional group to probe Gander crash", *Ottawa Citizen*, Oct. 30, 1990, p. A3.
26. "Controversy over Gander crash revived by U.S. study," *Ottawa Citizen*, Nov. 1990, p. 1.
27. "The Gander Crash", *Ottawa Citizen*, Dec. 3 1990.
28. *NYT*, July 27, 1987.
29. Cited in William S. Cohen and George J. Mitchell, *Men of Zeal* (Viking 1988), p. 233.

INDEX

"n" beside the page number indicates a footnote